Evolving Thinking Around the Systemic Structure of Knowledge

Evolving Thinking Around the Systemic Structure of Knowledge

Steven E. Wallis

3810 N 188th Ave
Litchfield Park, AZ 85340

Evolving Thinking Around the Systemic Structure of Knowledge
Steven E. Wallis

My Works Series: Volume 1

ISBN: 978-1-938158-19-3

Library of Congress Control Number: 2021932022

3810 N 188th Ave, Litchfield Park, AZ 85340, USA

Printed in the United States of America

This book is dedicated to the next generation of scholars; those who are striving to advance scientific thinking and its practical application as they reach for new horizons.

Steven E. Wallis received his Ph.D. from Fielding Graduate University in 2006. His academic work focuses on "theory of theory" where he is pioneering insights and tools to support scholars as they create increasingly efficacious theory. Steve's interdisciplinary interests span the social sciences. He is engaged in studies of organizational theory, collaborative human systems, knowledge management, and others. Dr. Wallis has ten years of experience as an independent consultant in Northern California. There, in a variety of industries, he supports consultants, trainers, and leaders on issues related to collaboration, communication, succession planning, creativity, organizational change, and knowledge management.

Contents

Introduction: Evolving thinking around the systemic structure of knowledge

The chapters (articles) presented in this book are meant to represent my thinking around the systemic structure of knowledge—as represented by theories, models, policy models, strategic plans, and so on (we will use these terms interchangeably). The most important idea here is that knowledge is more useful (for understanding, making effective decisions, communication, collaboration, planning, and reaching goals) when it is more systemic—more structured.

When I began in this field that idea of systemic structure existed; however, it was not very well developed. It was difficult or impossible to look at two theories and decide which one had the best structure (and so would be the most useful). My efforts were focused therefore on understanding the underlying structure—what is it that makes any (and all) theoretical knowledge useful. After 14 years of effort (and the writing of this book) I have reached a good stopping point. Rather than a set of loose axioms, the field now has Integrative Propositional Analysis (IPA); a clear and simple method for evaluating the structure of knowledge, along with a few other methods that are a bit more complex but also effective for advancing useful knowledge.

In the following section, I reflect briefly on some key articles, how they represent steps forward in my thinking. My main hope here is to provide some stepping stones for those who might want to understand and/or follow this path. And, from a broader perspective, support those who want to study how thinking evolves.

The Articles and the progress

Chapter 1

Wallis, S. E. (2006). A Study of Complex Adaptive Systems as Defined by Organizational Scholar-Practitioners. Fielding Graduate University, Santa Barbara.

The first step of the path began with a dissertation (a relevant portion of which is included with the present book). There, dimensional dialectics were used to understand how parts of a theory might co-define one another. This may be thought of as a kind of reflexive approach; but it is about how each part reflects on the others (how each defines or emerges from others), not how the reader might reflect upon each part.

The method of analysis was called reflexive dimensional analysis. At this stage, the term "robustness" was used to describe the theory as a whole. But it was not well defined. Concepts were labeled as dimensions or aspects. And, transformative change due to concatenated structures was described as emergence. This idea of emergence was to be dropped for many years, until picked up again in a very different form in the 2020 paper on levels and emergence.

Chapter 2

Wallis, S. E. (2008). From reductive to robust: Seeking the core of complex adaptive systems theory. In A. Yang & Y. Shan (Eds.), Intelligent Complex Adaptive Systems (pp. 1-25). Hershey, PA: IGI Publishing.

A call for chapters on the topic of Complex Adaptive Systems theory led to the condensing of that dissertation into a single chapter. An anonymous reviewer said something like, "It's very nice but can you make the process simpler?" A great debt is owed to that reviewer because the comment started a process of re-thinking everything. A key realization emerged that the concepts and connections of a theory could be counted (very simple). And, because some sub-structures of theory support more understanding than others, those could also be counted (reasonably simple). Thus, the methodology took a big step forward because now it became possible to measure the level of structure for each theory with some level of rigor and objectivity—a formal measure of robustness.

In this chapter it became normal that arrows should represent only causal relationships. Also (although weakly described) that theories with more structure were more systemic and so more useful in application. In contrast, poorly structured theories were subject to more interpretation and would change more rapidly (though the changes would not be very useful). A large enough theory (or synthesis of theories) was likely to have a more systemic/interconnected core and a less systemic belt of disconnected concepts.

Chapter 3

Wallis, S. E. (2008). Validation of theory: Exploring and reframing Popper's worlds. Integral Review, 4(2), 71-91.

A deepening exploration of knowledge led to this study of Popper's three worlds, with an eye toward understanding how theories may be tested (through a process of falsification) and so have their weaknesses identified and so could be improved. Those worlds helped separate the world of theoretical concepts from the world of data and the world of (we might say) human interest and emotion. All three worlds are important, of course, but more useful when we are able differentiate between them.

The ability to separate, to differentiate, of course is key to science. Another key is the ability to integrate, to synthesize.

Chapter 4

Wallis, S. E. (2009). The complexity of complexity theory: An innovative analysis. Emergence: Complexity and Organization, 11(4), 26-38.

The ideas of falsification were tested in this next paper that used the new methods to explore Complexity Theory. This successful application to a new body of theory confirmed that the project was on the right track—that this suite of methods, techniques, and perspectives could be used with some level of objectivity and reliability to evaluate bodies of theory- to show where we stand and how to advance a body of theory.

Chapter 5

Wallis, S. E. (2010). Toward a science of metatheory. Integral Review, 6(3 - Special Issue: "Emerging Perspectives of Metatheory and Theory"), 73-120.

Another part of this exploration in values asks what it meant to have a metatheory—a theory about theories—and how that might inform my further investigations. The results are summed up in this paper. Importantly, this was not a simple survey of perspectives. Instead, it was done with a focus on how make the use of metatheory to improve theory a more scientific endeavor. This would help scholars to avoid wallowing in speculative philosophy and make some progress in a world that desperately needs our help.

Chapter 6

Wallis, S. E. (2010). The structure of theory and the structure of scientific revolutions: What constitutes an advance in theory? In S. E. Wallis (Ed.), Cybernetics and systems theory in management: Views, tools, and advancements (pp. 151-174). Hershey, PA: IGI Global.

The next big challenge was to show the evolution of theory. This was best done by evaluating a theory from physics. In what might be called a seminal work, this chapter on the structure of theory and the structure of scientific revolutions showed how the systemic structure of a theory changed and evolved over the centuries from relatively useless to useful knowledge. The ability to show progress is placed in contrast with the high rate of failure for social theories—and so hints at the opportunity to accelerate the development of the social sciences.

Here the methodology is approaching some level of maturity. The name of the methodology is now Propositional Analysis. The use of causal arrows and the focus on concatenated structures continues but some older terms are still in use such as robustness.

Chapter 7

Wallis, S. E. (2013). How to choose between policy proposals: A simple tool based on systems thinking and complexity theory. Emergence: Complexity & Organization, 15(3), 94-120.

With growing confidence in the methodology, its use was expanded and applied to policy models. For a method that had been developed for theories of the social sciences, and applied equally well to theories of physics, this was another step forward in broadening the scope of use.

Additionally, and importantly, this article introduces and compares five structures of causal logic (atomistic, linear, branching, circular and concatenated) and uses a case comparative study process to evaluate two policy proposals and show which one is more likely to be effective in practical application. This paper clarifies the predictive use of the methodology. That is to say, given two or more conceptual systems (be they theories or policies or plans) we can make a good prediction as to which one we should choose in order to inform our decisions and reach stated goals.

Chapter 8

Wallis, S. E. (2014). Existing and emerging methods for integrating theories within and between disciplines. Organizational Transformation and Social Change, 11(1), 3-24.

With increasing confidence in the methodology, and having analyzed theories from multiple sciences, this paper was written to explore and demonstrate how existing approaches to integrating (synthesizing) theories between disciplines did not lead to progress—but rather to fragmentation and confusion. And, that using IPA could serve as a more rigorous tool to reverse that process to increase our collective capacity for integrating multiple perspectives and improving our collective knowledge. Here, a strong argument is also made against "parsimony"—the mistaken idea that the simpler theory is the better theory.

This conversation on interdisciplinary synthesis would reach a peak (at least a temporary one) with the 2020 publication, "The missing piece of the integrative studies puzzle" in the Interdisciplinary Science Reviews, 44(3-4).

Chapter 9

Wallis, S. E., & Wright, B. (2015, March 4-6). Strategic Knowledge Mapping: The Co-creation of Useful Knowledge. Paper presented at the Association for Business Simulation and Experiential Learning (ABSEL) 42nd annual conference, Las Vegas, CA.

Another paper broadening the scope of use was presented (and won a "best paper" award) at the ABSEL (Association for Business Simulation and Experiential Learning) conference. This paper presented a "gamified" approach to the co-generation of useful knowledge. Briefly, by building the analytical process into the game process means that the more the game is played, the more that the resulting knowledge map will become a representation of useful knowledge. This push towards gamification it reaching new heights with an online version at: https://cauzality.com/.

This is the first paper to introduce the term Integrative Propositional Analysis (IPA)—which is used to this day—with the emphasis on integration found in the collaborative activity of the game participants. This paper also solidified the importance of looking at knowledge as maps. Something that has proved quite useful for strategic planning and implementation.

Chapter 10

Wallis, S. E. (2016). The science of conceptual systems: A progress report. Foundations of Science, 21(4), 579-602.

Returning to the theme of studying theories, a good touchstone paper is in this progress report. Here, IPA is the method, causality is the best way to represent connections, Systemicity is used to represent "how systemic" a theory is, and complexity is used to represent the "simple complexity" of the number of concepts within a theory. This paper summarizes a few other papers to show the evolution of theories over time—from both the natural and social sciences. Importantly, this paper talk about using the IPA method to *accelerate* the development of the social sciences. Speculating that this kind of science multiplier could be used to double to usefulness/efficacy of our theories in a very few years.

Chapter 11

Wallis, S. E. (2014). Abstraction and insight: Building better conceptual systems to support more effective social change. Foundations of Science, 19(4), 353-362.

Backing up in time, and twisting the dial on our focus, this paper broke new ground by exploring individual concepts (and their interrelationships) in a whole new way. Here, the argument is made that we can develop more effective theories by building them of concepts that are at or about the same level of abstraction.

Chapter 12

Wallis, S. E. (2020). Orthogonality: Developing a structural/ perspectival approach for improving theoretical models. Systems Research and Behavioral Science, 37(2), 345-359.

A more sophisticated version of that "abstraction approach" was finally developed in the paper on Orthgonality. From the orthogonal perspective, each concept of a theory must be orthogonal to all other concepts in the theory. That is to say represent something that is completely different from all other things represented by concepts in the theory. This is, perhaps, the most challenging of ideas presented yet well worth further exploration for developing better theories.

Because it has not been proven in multiple publications, orthogonality is considered somewhat experimental. Hence, its designation as an "IPAx" method.

Chapter 13

Wallis, S. E. (2021). Understanding and improving the usefulness of conceptual systems: An Integrative Propositional Analysis-based perspective on levels of structure and emergence Systems Research and Behavioral Science, in press.

Another member of the IPAx family presents the challenging idea is that theories might have differing "levels." And, importantly, that those levels might represent levels of emergence. This new approach holds great promise for developing more sophisticated theories to more effectively address highly complex multilevel problems. It may also be seen as bringing closer together the world of theory and the world of our lived experience.

Chapter 14

Wallis, S. E. (2020). Evaluating and improving theory using conceptual loops: A science of conceptual systems (SOCS) approach. Cybernetics and Human Knowing, 27(3).

The third member of the IPAx family starts with the idea that theories should contain feedback loops. While, generally, that idea has broad acceptance, few grasp the idea that the structure of those loops may be evaluated with objectivity and rigor. This paper provides breakthrough techniques for the evaluation of loops in theories.

The more useful theory will have more loops; and, a higher percentage of its concepts will be included in those loops. Importantly, it is also necessary to consider the "signs" (positive or negative) of those loops; an interesting balance is required.

Chapter 15

Wallis, S. E. (2020). Commentary on Roth: Adding a conceptual systems perspective. Systems Research and Behavioral Science, 37(1), 178-181.

The final paper on this tour is the shortest. And, in some ways, the most humble. Yet, it serves to clarify an important perspective of "what it is all about." This paper shows how ideas of "conceptual systems" (theories, models, plans, etc.) within the world of concepts are related to "physical systems" (social systems, communication systems, environmental systems, etc.). And, in doing so, helps to clarify our collective understanding of the similarities and differences of these two worlds.

In doing so, and in concert with the other parts of this book, the stage is set for each reader to extend this exploration; to accelerate the evolution of useful knowledge for understanding and resolving the seemingly impossible problems of the world.

Chapter 1
A Study of Complex Adaptive Systems Theory as Defined by Organizational Scholar-Practitioners

Wallis, S. E. (2006). A Study of Complex Adaptive Systems as Defined by Organizational Scholar-Practitioners. Fielding Graduate University, Santa Barbara.

Abstract: *This dissertation is concerned with Complex Adaptive Systems (CAS) theory, as developed by scholar-practitioners, and how it might be better understood as a body of theory. In an important sense, this dissertation reflects a "turn" away from forms of analysis that might be considered atomistic, or mechanistic, and toward a form of analysis that might be considered "reflexive."*

In this dissertation there are two studies. The first essentially deconstructs the theories and recombines their conceptual components to gain new insights into the theory. The second study builds the concept of "dimensional dialectics" from the ideas of Hegel, Marx, and Nietzsche. Using the notion that such dimensions might be "co-emerging" or "co-defining," this dissertation develops the notion of a "robust" theory and develops "Reflexive Dimensional Analysis" (RDA) as a method to find new insights. This method is used to code of a body of theory, categorize those components, and identify those categories as dimensions. Finally, dimensional analysis is used to investigate the co-emergent relationship between the core dimensions.

This study finds that most dimensions of CAS theory may be seen as co-emergent. Within the theory, however, there are contradictory notions that might be resolved. Further, the study suggests that the notion of "time" must be effectively co-defined before CAS theory might be understood as robust. Some important implications of this study include the opportunity to analyze and "crystallize" a body of theory to gain insights into its robustness. Insights such as these may suggest opportunities to clarify relationships between theories and/or between bodies of theory.

Dimensions

This step may be seen as a short, yet important conceptual shift. Rather than looking at the categories as collections of codes, this step is to begin thinking of them as root dimensions—each one having a direction (indicated by the category title) and a relative level (indicated by higher and lower).

Identifying Dimensional Dialectics

In this section, I identify the dimensional dialectics that may be found in the original definitions. For most authors, I first note key aspects of their definition, then provide a close rephrasing of their definition in what might be thought of as the terminology suggested by the root dimensions (with original terminology in parentheses adjacent to the terminology used for the root dimensions). Following each reframing, I generally present a diagram representing the dialectic. In this dialectic analysis, each dimension noted above (Change, System, Interaction, Levels, and Time) may be shown as emerging from some combination of the others. I did not, as was done in the exemplar, seek to identify the co-emergent properties of each dialectic found in every definition. That robust relationship emerges over the course of this analysis, as each of the five root dimensions are shown to emerge from the others.

In the definition provided by Ashmos *et al.* (1998), the focus is on the connections between parts of the system. The core of their definition may be stated as, "Systems have the capacity for self-organization and this capacity is enhanced by the quality of connections among organized sub-systems." If that statement were to be rephrased in terms of the emergence above root dimensions, it might be stated as, "More System (agents), with more Interactions (connections) causes the emergence of more Change (self-organization). This relationship might be graphically depicted as:

In the definition provided by Axelrod & Cohen (2000), the authors noted, "Agents ... use their strategies, in patterned interaction, ... thus changing the frequencies of types within the system (p. 154)."

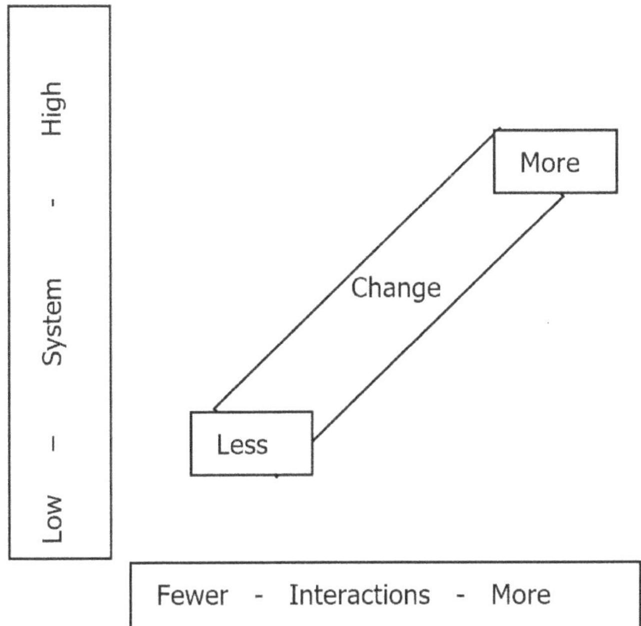

Figure 1 *Dialectic emergence of change from system and interactions.*

In the terminology of the root dimensions, this might be rephrased as, "Systems (agents) Interact (patterned interaction) with Systems (their strategies) and so increase the Levels (frequencies of types)." This relationship might be depicted as shown in Figure 2.

The fact that the authors specifically note the interactions of various levels of systems (agents and strategies) suggests that the "System" dimension might not be a linear one. If it were represented mathematically, it might be emphasized, for example as, "Systems (squared) times Interactions will equal Levels. However, this model is not yet to the level of sophistication that would allow such a precise representation. Another way to look at their definition, and represent it graphically (see Figure 3), might be to say that the System dimension is, in essence, a dimension that emerges from two other dimensions—one of "quantity" of systems" and another of "quality" of systems.

Such an investigation is an interesting opportunity for future exploration and, I speculate, may suggest a link between the "observ-

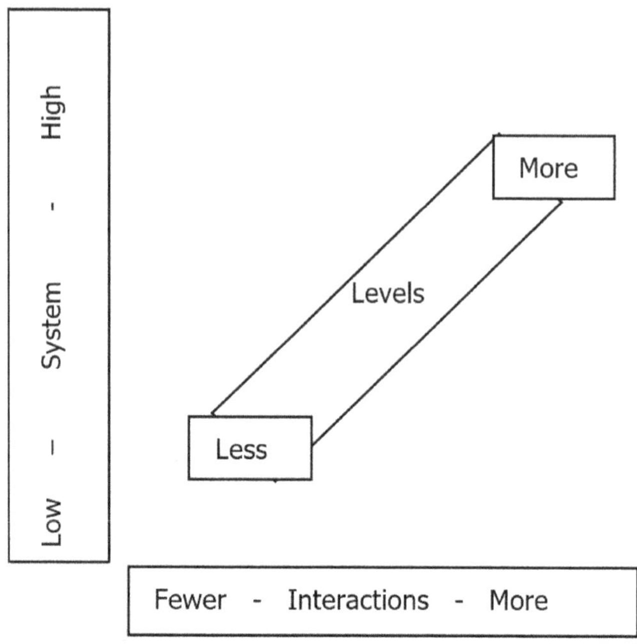

Figure 2 *Levels emerge from system and interactions.*

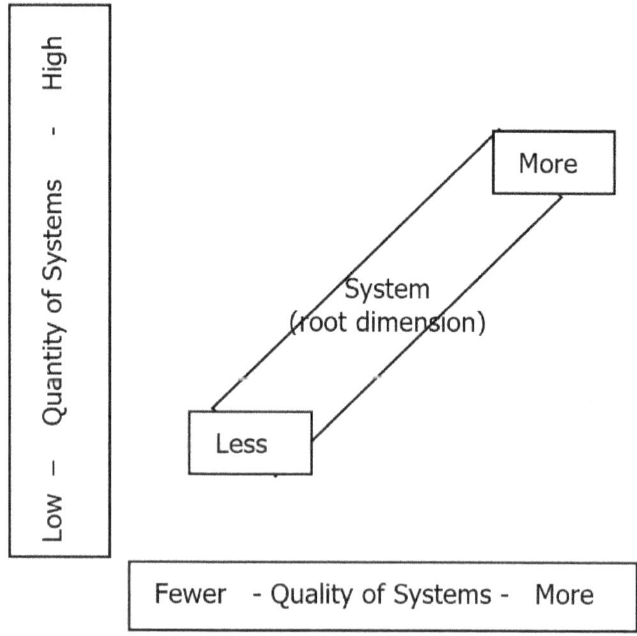

Figure 3 *Root dimension of system emerging from quantity and quality dimensions.*

able" systems each of us discerns, and the theoretical sense of "Systemness" suggested by this analytical method.

Hunt & Ropo (2003) stated, "the component parts interact with sufficient intricacy that standard linear equations do not predict their behavior (p. 317)." To restate this in terms for the root dimensions it might be said that, "Systems (components) Interact (interact) across Levels (intricacy) to cause Change (unpredictable behavior)." Such a relationship does not contain the same level of "quantitative" relationship that many have. A more relational definition might say, for example that, "More systems and more interactions across more levels will lead to more unpredictability. However, for the purpose of this analysis, it may be possible to assume that such a relationship was the intent of the authors. In such a case, the relationship might be depicted as:

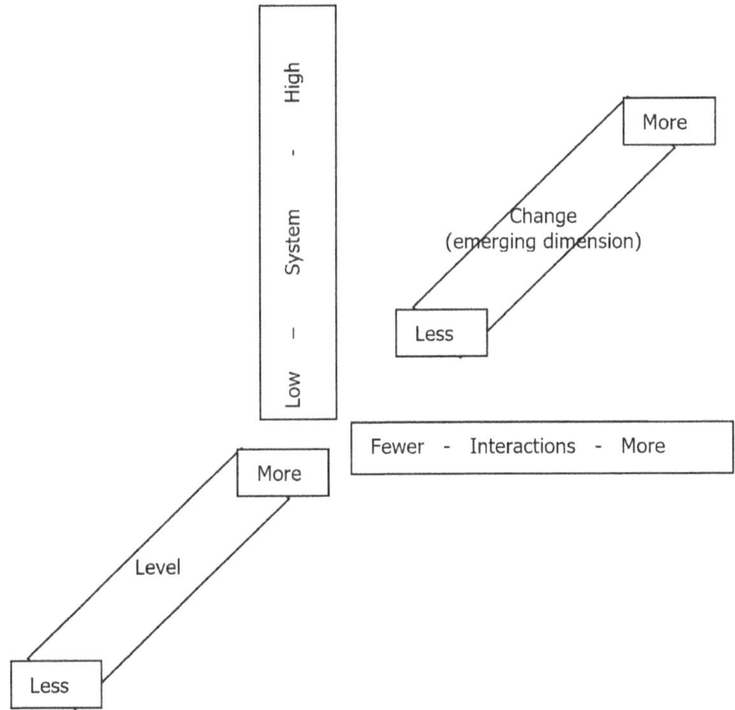

Figure 4 *Emergence of change from system, levels, and interaction.*

Although this form of dimensional dialectic is not developed by Nietzsche (2005, original 1886/1886), there does not seem to be any overt reason to reject it. Indeed, as this relationship is based on the authors' understanding, the diagram may be accepted as a contribution to the body of thought being developed here. However, it should be noted that this relationship suggests an area for additional study.

Stacey (1996) states,

> *A complex adaptive system consists of a number of components, or agents that interact with each other according to sets of rules [schemas] that require them to examine and respond to each other's behavior in order to improve their behavior (p. 10)*

According to the conventions suggested by this analysis, The Stacey definition may be rephrased as, "Systems (agents) Interact (interacting, investigating behavior) across Levels (agents and schemas) causing Change (their own schemas/behaviors)." This description does not seem to provide a sense of relationship (e.g., More Systems cause more change). However, by inferring such a relationship, a more complex dialectic may be suggested where the Change may be seen as emerging from the dialectic interaction of Systems, Interaction, and Levels.

McDaniel *et al.* (2003) highlighted a different aspect of the emerging dimensions with their definition, "Complex adaptive systems (CAS) are composed of a number of agents interacting over time... are able to self-organize (p. 269)" In the terminology of the root dimensions, this might be re-described as, "Levels (number) of Systems (agents) that are Interacting (interacting) as Time leads to Interaction (self-organize)." This stated relationship suggests something of a loop, where interactions, in a sense, lead to interactions. This does not seem to be a tautology, however, as the direct relationship is complicated by the influence of the dimensions of Time, Levels, and Systems. As this investigation moves forward, it may become increasingly clear that there may be some sort of circular line of reasoning here. Indeed, this seems to reflect the inherent "connectedness" of the universe where one might say that everything is connected to everything else. And, while that may be true, the question seems to arise in the description of exactly how those things are connected. For example, Senge *et al.* (1994) discussed the notion of "reinforcing loops" where one action leads to another in a reoccurring, or recursive, process. In that sense, these dialectic relationships might be seen as reinforcing loops where more systems will tend to lead to more systems; however, there will be the influence of time, interactions, and the levels of similarity between the systems. Therefore, in order to gain some insight from their definition, it may be best to reduce the redundancy in the relationship and suggest that Systems over Time lead to Interaction. Given the complexity of the definition, however, that statement does not seem to bear as much strength as others.

Harder *et al.* (2004: 84) stated in their definition that,

> *Furthermore, the distinct and diverse parts of a living system engage in patterns of interaction that serve to maintain the homeostasis of the system, a dynamic equilibrium in which there is constant change and adaptation in the context of holding a steady state within a certain band of parameters (Boulding, 1985).*

It may be that the authors are suggesting that, "System (parts) Interact (in patterns) to reduce Change (maintain homeostasis). This notion would seem to be in opposition to those above which seem to suggest that interactions and Systems serve to increase Change. However, it may also be recognized that these authors are noting certain changes at certain Levels. It may be understood that the "patterns of action" to which they refer and the "homeostasis of the system" that exists "within a certain band of parameters" might be understood as the "context" or (broadly stated) the "meta-system" (under which the agents operate). That is to say that Change and Interaction on one level (for example, the level of parts) may support stability on another level (for example the level of the context in which the agents operate). This is an area that seems to call for additional investigation along the lines of "complexity shifting" suggested by Hutchins (1995) in his description of a group that accomplishes a task by "shifting" the complexity of the task to the appropriate specialist, or even the appropriate tool.

Olson & Eoyang (2001) noted, "Agents are semi-autonomous units that seek to maximize some measure of goodness or fitness by evolving over time (Dooley, 1996: 7)." Where these authors begin with agents, and move to define them, the reverse might also be possible. It may be said, in the spirit of a process definition, that a "System (agent) emerges when there are Interactions (evolving) over Time (time)." That relationship is suggested in Figure 5.

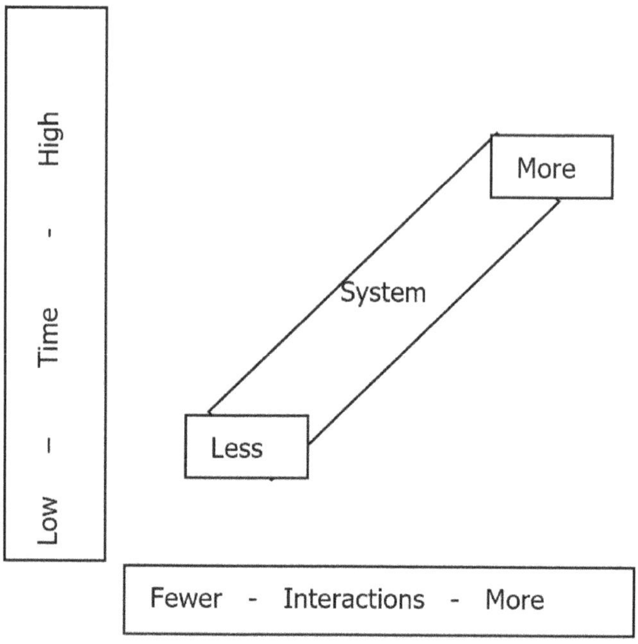

Figure 5 *Systems emerging from interactions and time.*

It should be noted, however, that their definition also includes multiple Levels of Systems (i.e., seeking—which may be seen as a schema from above conversations). Such a relationship suggests that Interactions and Time cause the emergence of System and Levels.

Pascale (1999) noted that a CAS, "First, it must be comprised of many agents acting in parallel (p. 84)." "Second, it continually shuffles these building blocks and generates multiple levels of organization and structure (p. 84)." Which, in the terminology of the root dimensions might be understood as, "Systems (agents) Interacting (acting) causes the emergence of Levels (of organization and structure)." Such a relationship might be portrayed as shown in Figure 6.

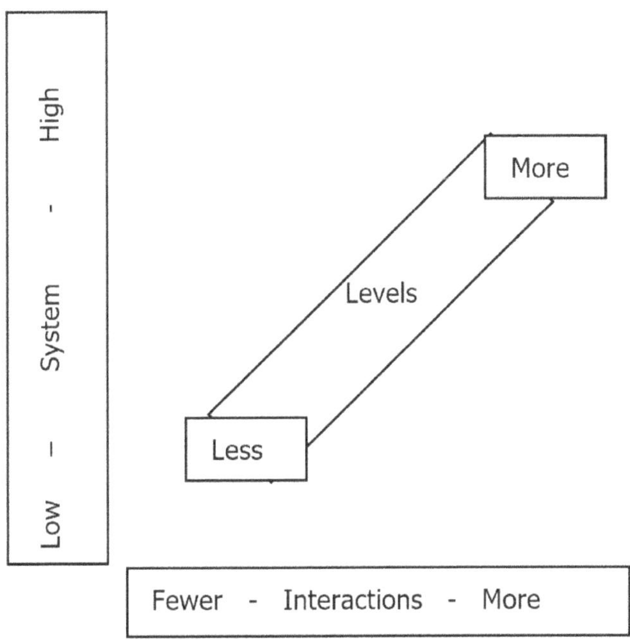

Figure 6 *Levels emerging from system interactions.*

Dent (2003) drew from Brown & Eisenhardt (1998) to note, "Each agent is constantly acting and reacting to what the other agents are doing. They are coadaptive, taking 'mutual advantage of each other in order to change more effectively' (p. 83)" which, in the parlance of the root dimensions might be rephrased as, "Systems (agents) that are Interacting (coadaptive) cause the emergence of Change (change more effectively). If the assumption is made that more System and More interaction create more change, the relationship might be graphically depicted as shown in Figure 7.

This suggests the same relationship as this analysis found with Ashmos *et al.* (1998). Similarly, in the definition by Bennet & Bennet (2004), the authors noted, "The term complex system means a system that consists of many interrelated components with nonlinear relationships that make them very difficult to understand and predict (p. 26)." In accordance with the root dimensions, this might be rephrased as, "Systems (elements) Interact (interrelated) to cause Change (difficult to predict). This relationship might be depicted in the same way as Hunt & Ropo (2003).

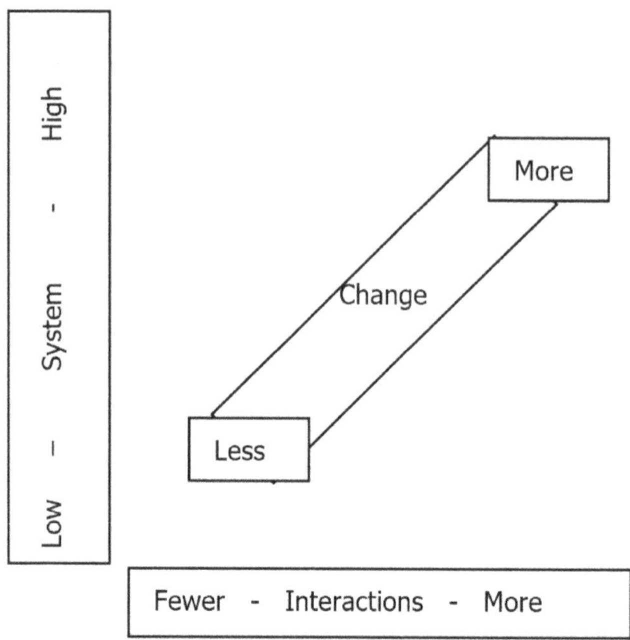

Figure 7 *Change emerges from system and interactions.*

Moss (2001) noted, "members may nonetheless organize themselves into stable patterns of activity and organization that provide them with a common frame of reference (p. 217)" which, In the terms of the root dimensions, might be characterized as, "System (agents) Interact (organize) enabling the emergence of System (stable patterns of activity) on what might be termed a higher Level (organization / common frame of reference). This might be depicted as shown in Figure 8.

This figure seems to suggest another relationship that does not fit "neatly" into the Nietzschean dimensional dialectic. Again, it may be accepted as a useful understanding of a relationship by the author; and, an area for future study.

Chiva-Gomez (2003) noted that, "these systems self-organize themselves at the 'edge of chaos' (p. 101)" he goes on, drawing from Stacey, to describe the edge of chaos (EOC) as, "...a form of bounded instability found in the transition phase between the order and disorder zones of operation for a CAS (p. 101)." This explanation does not seem to add much to the understanding; it might be rephrased by

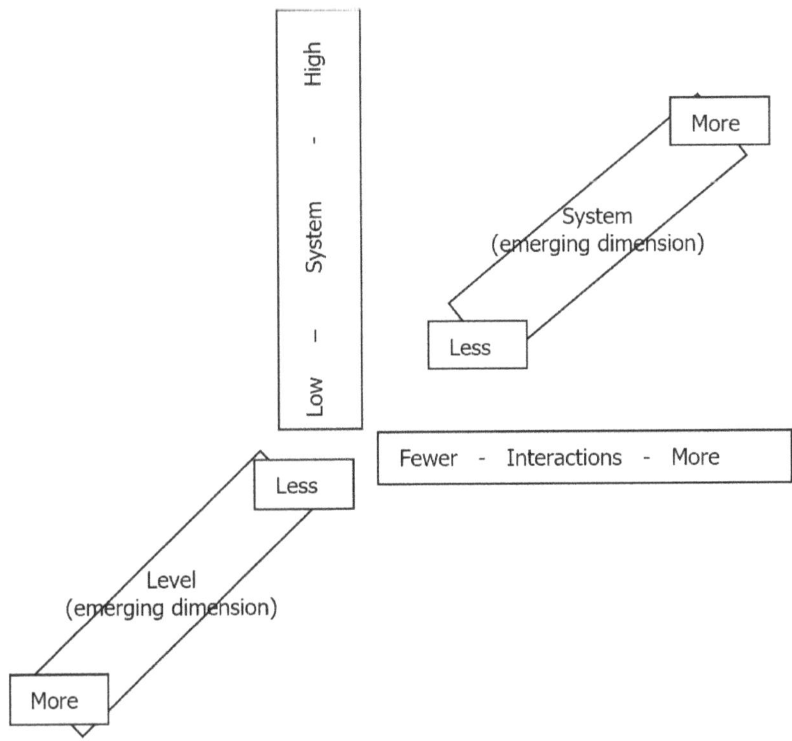

Figure 8 *System and level emerging from system and interaction.*

saying, "A CAS self-organizes when it is in its own EOC." In this dissertation, the EOC is essentially recognized as a form of interaction. Certainly, the opportunity exists for more exploration in this area. For example, I speculate that (in the terms of the root dimensions) that the EOC phenomena might be explained in terms of the ratio of System to Interaction. In such a relationship, many connections between few systems might be seen as a stable environment, and few connections between many systems might be seen as a chaotic environment.

Shakun (2001) noted, "(1) players, agents, negotiators or decision makers in a group (coalition); (2) values or broadly stated desires; (3) operational goals, or specific expressions of those values; (4) decisions, actions, or controls taken to achieve these goals (p. 98)" Which, broadly stated, might be reframed as, "Levels of System (agents, coalition, values) cause the emergence of Interaction (action)." That might be represented as shown in Figure 9.

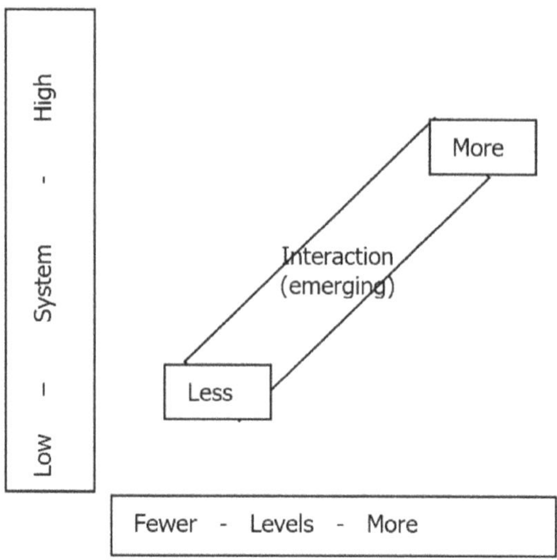

Figure 9 *Interactions emerging from system and levels.*

The definitions provided by some authors, do not lend themselves to this form of analysis. For example, McKelvey (2004) states, "The Arthur, *et al.* (1997) characterization of complex adaptive systems are what are missing [from mature sciences, orthodoxy, and much of evolutionary economics]: agent, nonlinear, level, coevolutionary, far-from-equilibrium, and self-organization effects (p. 69)." This is a more atomistic description of what a CAS is, versus how the components interact. Similarly, Lichtenstein (2000) did not suggest relationships, only stated four "basic assumptions. Also, Daneke (1999) described a CAS as, "A simulation approach that studies the coevolution of a set of interacting agents that have alternative sets of behavioral rules (p. 223)." That description does not seem to provide useful descriptions of relationships. Yellowthunder & Ward (2003) also describe a CAS in terms of "three key principles" (rather than the relationship between those principles).

McGrath (1997) made clear distinctions of levels in his description of CAS. He described people, intentions, and resources, complex systems, smaller systems (members), larger systems (organizations), and boundaries. In the terminology of the root dimensions, all of these are thought of as systems. McGrath also noted that these levels of systems are "embedded" within one another, which may suggest "Interaction." Although, with this definition, there does not seem to be a causal relationship, so it is unclear whether it might be said that interaction causes levels, or systems, or the other way around.

The following diagram is presented as a summary of the above conversation. While the display of this process is (for reasons of available space and dimensions of representation) problematic, I have depicted some of the relationships suggested by the authors. To enhance the clarity of this presentation, I have used the terms of the root dimensions, rather than the varied descriptors used by the authors. For example, where Hunt & Ropo (2003) might have stated, "the component parts interact with sufficient intricacy that standard linear equations do not predict their behavior (p. 317)." the diagram might state, "With more Systems and more Interactions across more Levels, there is increasing Change."

The following figure summarizes (and simplifies) the dialectic relationships described by the above authors. The authors are noted on the outside of the diagram, their causal dimensions just in from their citations, and the emergent dimensions listed in the center.

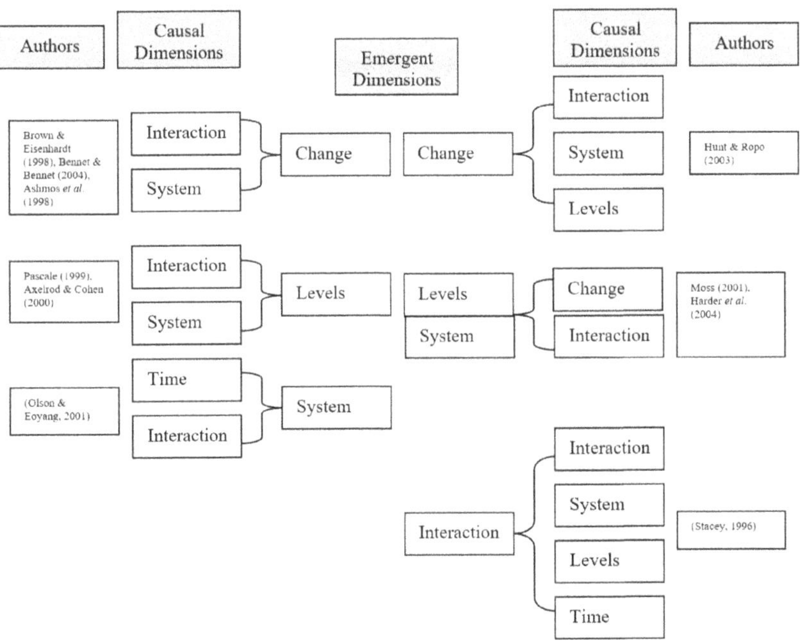

Figure 10 *Combined and simplified summary of dialectic relationships.*

This figure seems to indicate some similarities and some differences between the insights of the authors. While such differences may be understandable, given the relative "youth" of CAS theory, it seems as through there is significant opportunity to continue the conversation—both to reconcile understanding, and to obtain new insights.

Despite what might be thought of as dissimilarities between authors, this diagram seems to suggest the opportunity to identify co-causal dimensions within CAS theory. This relationship may be inferred from the above diagram wherever an "emergent" dimension can also be seen as a "causal" dimension. In other words, the self-referent, or reflexive nature of these dimensions suggests a robust relationship for these dimensions. While the temptation may be to reconcile these disparities by delving deeper into the writings of each author, Study Number One suggested that such an atomistic approach may be no more successful. Instead, I speculate that future studies (perhaps in computer modeling, or studies within human organizational systems) might be used to show which authors' versions are closer to the "truth."

Conclusion

In this chapter, two studies have been presented. Both studies used, as their data, 20 concise definitions of CAS theory that had been developed by scholar-practitioners. In Study #1, these data were categorized in an essentially *ad hoc* process that identified 26 categories of CAS components. Scientometric and bibiliometric techniques were applied to these components to surface alternative formulations of CAS theory (e.g., "expert CAS theory," "practitioner CAS theory," etc.). This process seemed sufficient to answer Q1: What aspects, components, or dimensions of CAS theory are described by organizational scholar practitioners?

Several insights were generated, including multiple versions of CAS theory, areas of publication, which authors used the most citations, which were the most cited, and so on. However, the forms of CAS theory developed in Study #1 contained fewer components than were contained in the larger body of CAS theory—as a whole. Those gaps were seen as suggesting that the new CAS theory was less complex and therefore capable of less explanatory capacity than the body as a whole.

In Study #2, the techniques of dimensional analysis were linked with the concepts of dimensional dialectic and the concept of reflexive analysis to suggest the Reflexive Dimensional Analysis (RDA) method. Using the RDA method, the original data were coded, and a critical-reflexive analysis was performed to combine the coded data into five, coherent, root dimensions of CAS theory. The dimensions were examined through the literature to identify the robust, reflexive, co-emergent nature of the dimensions. In that process, the dimension of Time could be seen as influencing other dimensions, but not being influenced itself. This may indicate one weakness of CAS theory and/or the RDA methodology. Some of the strengths and limitations of R-CAS and the RDA process will be discussed in chapter 4.

References

Ashmos, D. P., Huonker, J. W., & Reuben R. McDaniel, J. (1998). Participation as a complicating mechanism: The effect of clinical Professional and middle manager participation on hospital performance. *Health Care Management Review, 23*(4), 7-20.

Axelrod, R., & Cohen, M. D. (2000). *Harnessing Complexity: Organizational Implications of a Scientific Frontier*. New York: Basic Books.

Bennet, A., & Bennet, D. (2004). *Organizational Survival in the New World: The Intelligent Complex Adaptive System*. Burlington, Massachusetts: Elsevier.

Brown, S. L., & Eisenhardt, K. M. (1998). *Competing on the Edge: Strategy as Structured Chaos*. Boston: Harvard Business School Press.

Chiva-Gomez, R. (2003). The Facilitating Factors for Organizational Learning: Bringing Ideas from Complex Adaptive Systems. *Knowledge and Process Management, 10*(2), 99-114.

Daneke, G., A. (1999). *Systemic Choices: Nonlinear Dynamics and Practical Management*. Ann Arbor: The University of Michigan Press.

Dent, E. B. (2003). The complexity science organizational development practitioner. *Organization Development Journal, 21*(2), 82.

Harder, J., Robertson, P. J., & Woodward, H. (2004). The spirit of the new workplace: Breathing life into organizations. *Organization Development Journal, 22*(2), 79-103.

Hunt, J. G. G., & Ropo, A. (2003). Longitudinal organizational research and the third scientific discipline. *Group & Organizational Management, 28*(3), 315-340.

Hutchins, E. (1995). *Cognition in the Wild*: MIT Press.

McDaniel, R. R., Jordan, M. E., & Fleeman, B. F. (2003). Surprise, surprise, surprise! A complexity science view of the unexpected. *Health Care Management Review, 28*(3), 266-277.

McGrath, J. E. (1997). Small group research, that once and future field: An interpretation of the past with an eye to the future. *Group Dynamics: Theory, Research, and Practice, 1*(1), 7-27.

McKelvey, B. (2004). Toward a 0th law of thermodynamics: Order-creation complexity dynamics from physics and biology to bioeconomics. *Journal of Bioeconomics, 6*(1), 65-96.

Moss, M. (2001). Sensemaking, Complexity and Organizational Knowledge. *Knowledge and Process Management,* 217.

Nietzsche, F. (2005, original 1886, 12/12/2005). Beyond Good and Evil. Etext. Retrieved from Olson, E. E., & Eoyang, G. H. (2001). *Facilitating Organizational Change: Lessons From Complexity Science.* San Francisco: Jossey-Bass/Pfeiffer.

Pascale, R. T. (1999). Surfing the edge of chaos. *Sloan Management Review, 40*(3), 83.

Senge, P., Kleiner, K., Roberts, S., Ross, R. B., & Smith, B. J. (1994). *The Fifth Discipline Fieldbook: Strategies and Tools for Building a Learning Organization.* New York: Currency Doubleday.

Shakun, M. F. (2001). Unbounded Rationality. *Group Decision and Negotiation, 10*(2), 97-118. Stacey, R. D. (1996). *Complexity and Creativity in Organizations.* San Francisco: Berrett-Koehler Publishers, Inc.

Yellowthunder, L., & Ward, V. (2003). Designing and Implementing Organizational Change in a Complex Adaptive System. In G. H. Eoyang (Ed.), *Voices From the Field: An Introduction to Human Systems Dynamics* (pp. 125-144). Circle Pines, Minnesota: Human Systems Dynamics Institute.

Chapter 2
From Reductive to Robust—Seeking the Core of Complex Adaptive Systems Theory

Wallis, S. E. (2008). "From reductive to robust: Seeking the core of complex adaptive systems theory." In A. Yang & Y. Shan (Eds.), Intelligent Complex Adaptive Systems (pp. 1-25). Hershey, PA: IGI Publishing.

Abstract: *This chapter seeks to identify the core of Complex Adaptive Systems (CAS) theory. To achieve this end, this chapter introduces innovative methods for measuring and advancing the validity of a theory by understanding the structure of theory. Two studies of CAS theory are presented that show how the outer belt of atomistic and loosely connected concepts support the evolution of a theory; while, in contrast, the robust core of theory, consisting of co-causal propositions, supports the validity and testability of a theory. Each may be seen as being derived from differing epistemologies. It is hoped that the tools presented in this chapter will be used to support the purposeful evolution of theory by improving the validity of Intelligent Complex Adaptive Systems (ICAS) theory.*

What is the Core of CAS Theory?

Where other chapters in this book may use Intelligent Complex Adaptive Systems (ICAS) theory as a framework to understand our world, we strive in this chapter to understand theory, itself. Through this process, the reader will gain a new perspective on the theory that is applied elsewhere in this book. To gain some perspective on ICAS, we will study the literature of Complex Adaptive Systems (CAS) as developed in the field of organizational theory. As such, this chapter may be of interest to those discussing organizational theory and organizational change, multi-agent systems, learning methods, simulation models, and evolutionary games.

CAS theory originated in the natural sciences as a tool for understanding non-linear dynamics (Kauffman, 1995) and has gained popularity in organizational studies through the efforts of many authors (including: Axelrod & Cohen, 2000; Brown & Eisenhardt, 1998; Gleick, 1987; Stacey, 1996; Wheatley, 1992). As CAS expanded into this discipline, every author seems to have placed a personal mark by revising CAS for interpretation and publication. Indeed, in researching the literature, 20 concise, yet different, definition/descriptions of CAS theory were found.

Within these 20 definitions, "component concepts" were identified. For example, Bennet & Bennet (2004) note (in part) that a CAS is composed of a large number of self-organizing components. The concepts of "self-organization" and "large number of components" may be seen as conceptual components of CAS theory as described by those authors. These conceptual components might also be thought of as the authors' "propositions." It is important to note that among the 20 definitions, no two contained the same combination of component concepts. This raises a serious question: When we talk about CAS theory, are we really talking about the "same thing?" After all, if one author states that a CAS may be understood through concepts "a, b, and c" while another author states that the relevant concepts are "c, e, and f," there may be some conceptual overlap but there are also inherent contradictions.

In the social sciences, this issue has been of concern for decades. In one attempt to make sense of the issue, theory has been described as consisting of a "hard core" of unchanging assumptions, surrounded by a more changeable "protective belt" (Lakatos, 1970). When a theory is challenged, a theorist may rise to defend it with a new proposition that changes the belt, but presumably leaves the core intact. Phelan (2001) suggests that complexity theory has its "hard-core assumptions;" however, among the 20 definitions discussed here, there is no one concept that is held in common by all of the authors. If there is no concept, or set of concepts, held in common, where then is the core of CAS theory? Motivated by this apparent lack of commonality, we seek to identify the core of CAS theory.

A Chinese proverb states, "The beginning of wisdom is to call things by their right names." (Unknown, 2006: 1). The difficulty of engaging in conversations with imprecise definitions was famously illustrated when Plato called man a "featherless biped." He was forced to add, "with broad flat nails" in response to Digenes, who arrived with a plucked bird, proclaiming "Here is Plato's man." (Bartlett, 1992: 77). We may speculatively ask if it was the atomistic nature of Plato's definition that left it so open to misinterpretation. In short, we must wonder how we can know what we are talking about, if the name keeps changing? In contrast to Plato's rapidly evolving definition, Newton's laws (e.g., F=ma) have proved effective, and unchanged, for centuries.

As scholars, of course, we are continually engaged in the discovery (or social construction, depending on your view) of understandings and definitions. And yet, we need some level of shared understanding of existing concepts, so we may communicate effectively as we work to understand new concepts. In short, as scholars, we might see the increasing clarity and stability of our definitions as an indicator of "progress" in a given field. We dig the clay, form it into paving stones, place them in front of us and walk on them to find more clay. In this chapter, we will suggest some tools for identifying the milestones along our shared road.

Central to the exploration presented in this chapter, we must ask, "Is it possible to ascertain the legitimacy of a theory through its structure?" Dubin (1978) suggests that there are four levels of efficacy in theory; and these levels do reflect the structure of the theory. They are: 1) Presence /Absence (what concepts are contained within a theory). 2) Directionality (what are the causal concepts and what are the emergent concepts within the theory). 3) Co-variation (how several concepts might impel change in one another). 4) Rate of change (to what quantity do each of the elements within the theory effect one another). Parson and Shills note four similar levels of systemization of theory—moving toward increasing "levels of systemization" (Friedman, 2003: 518). Reflecting the validity of these assertions, Newton's formulae might be seen as residing at the highest level because it is possible to identify quantitative changes in one aspect (e.g., force) from changes in other aspects (e.g., mass & acceleration). Such a high level of understanding has been long sought in the social sciences but has, as yet, remained elusive. One goal of this chapter is to advance CAS theory along this scale—and identify how further advances might be enabled for similar forms of theory.

To find the core of a theory, two studies are presented in this chapter. One study is based on content analysis (essentially, looking at the words used by the authors as reasonable representations of the concepts that they are conveying). The second study uses a more traditional narrative analysis. The first study is a reductive look at CAS theory—focusing on the axiomatic propositions of the authors. This method will be seen as adding to the outer belt. The second study focuses on the relational propositions of CAS theory. This method suggests that there is a core to CAS theory. However, it also shows that the core (based on the current state of CAS theory) has only a limited internal integrity. A path for developing a more robust CAS theory is then suggested. The process of developing a robust theory is expected to provide great benefits to scholars (based on the successful use of Newton's robust formulae).

Due to limitations of space, the studies in this chapter will be focused on the level of "concept" (with concepts presented as they are named by the authors and as they may be generally understood by most readers) and theory (as a collection of concepts). These studies will generally avoid the sub-concept level of interpretation and what might be called a post-theory level of application and testing.

The next section includes a relatively linear and reductive analysis of the concepts of CAS theory. This process might be seen as a thought experiment—a cognitive construction that represents the creation of new definitions in an ad-hoc manner.

A Reductive Study of CAS Theory

In this section, we engage in the development of new theory where theory might be seen as a collection of concepts. This process identifies the range of concepts in CAS theory and develops new versions of that theory. The new versions of CAS theory created here may be seen as newly evolved definitions. Although such definitions may be tentatively used to identify various perspectives of CAS theory, they may also be seen as adding to the outer belt of theory rather than clarifying the core.

We begin with a review of literature. Searches of the ProQuest database yielded nearly 100 articles in academic journals where CAS theory was discussed in the context of a human organization. Within those articles, 13 were found to contain concise (less than one page) definitions of CASs. Additionally, those journals (and other sources) suggested other scholarly publications. Promising books were reviewed and seven additional concise definitions were found. In all, this study (although not exhaustive) found 20 relatively concise definitions of CASs. Concise definitions were used so that the study could cover as much ground as possible. It is also assumed that a concise definition includes the most important aspects of each author's version of the theory. It is also expected that a sample of this size will provide a sufficient representation of the body of theory.

Although the authors' definitions are not listed for reasons of space, this study uses concise definitions from Ashmos, Huonker, & McDaniel (1998), Axelrod & Cohen (2000), Bennet, & Bennet (2004), Brown & Eisenhardt (1998), Chiva-Gomez (2003), Daneke (1999), Dent (2003), Harder, Robertson, & Woodward (2004), Hunt & Ropo (2003), Lichtenstein (2000), McDaniel, Jordan, & Fleeman, (2003), McGrath (1997), McKelvey (2004), Moss (2001), Olson & Eoyang (2001), Pascale (1999), Shakun (2001), Stacey (1996), Tower (2002), Yellowthunder & Ward (2003).

In the process described above, the study deconstructed each definition into the authors' propositions, or component concepts. For example, Daneke describes a CAS as, "A simulation approach that studies the coevolution of a set of interacting agents that have alternative sets of behavioral rules" (Daneke, 1999: 223). The concepts describing the CAS here would be: coevolution, interaction, agents, and rules. While another reader might develop a different list, it is expected that such lists would be substantively similar to the one developed here where a total of 26 concepts were identified, consisting of:

Agent, Non-Linear / unpredictable, Levels, Co-evolutionary, Adaptive, Agents evolve, Far From Equilibrium / Edge of Chaos, Self-Organizing, Many agents, Interrelated / interacting, Goal seeking, Decision-making, Emergence / surprise happens, Act in rules/context of other agents and environment, Simple rules, Permeable boundaries, Evolves toward fitness, Boundary testing, Iterative process, Agents are semi-autonomous, Evaluate effectiveness of decisions/results, Self-defining, Identity, Morality, Irreversible, Time.

This list might be seen as representing the whole of CAS theory from an atomistic perspective. It should be noted that at this "survey" stage, no component appears to be more "important" than another, and no component seems to be closer to the core than any other.

Of the 20 publications, three could clearly be seen as the "most cited" (each having been cited by hundreds instead of tens, or fewer). Between them, there are six concepts used by at least two of the three, including:

Co-evolutionary, Many agents, Interrelated / interacting, Goal seeking, Emergence / surprise happens, Simple rules.

This focus on what might be considered the "authoritative" versions essentially creates a new definition of CAS theory built on the shared conceptual components of the authors. However, it should be noted that this new definition has lost some conceptual breadth when compared to the whole body of CAS theory. Moving from one form of popularity to another, the following is a list of those six concepts that seemed most popular among the 20 definitions:

Non-Linear / unpredictable, Co-evolutionary, Many agents, Interrelated / interacting, Goal seeking, Emergence / surprise happens.

Again, a new definition of CAS theory has been created with a new focus. Again, the conceptual components have shifted—both in comparison to the whole body of CAS concepts and in comparison to the authoritative version.

Additionally, while most authors identified themselves as scholars, others identified themselves as scholar-practitioners. Those whose affiliations were uncertain were left out of this ad-hoc, demonstrative study. The five concepts most commonly described by those authors who identified themselves as scholars were:

Non-Linear / unpredictable, Self-Organizing, Many agents, Interrelated / interacting, Emergence / surprise happens.

Among those who identified themselves as scholar-practitioners, the four most popular concepts used were:

Non-Linear / unpredictable, Many agents, Interrelated / interacting, Goal seeking.

There are obvious limitations to this ad-hoc study. However, a number of insights and benefits become apparent here. First, that this study creates a comprehensive view of the concepts within a body of theory. Of course, in this study, that view is limited to the level of the concepts, rather than delving deeper which is another possible level of exploration. Second, that each group of concepts suggests those specific concepts that might be most appropriate for a given application or area of research. In a sense, each group of concepts might be viewed as a "school of thought" for CAS theory within its specific venue. Importantly, this brief study essentially began with 20 definitions and generated four more. The number of theories in the outer belt was easily increased, yet we do not seem to have increased our understanding of the core. Our lack of core insight may be related to the form of analysis, or the type of data used. Importantly, we approached the data as lists of concepts and rearranged them into new lists. Each attempt to identify a new perspective resulted in a new list. As with the broader survey (the first list), each subsequent list has no discernible core.

The analysis presented in this section has served to demonstrate some strengths and weaknesses of a reductive form of study. In the next section, we look at alternative approaches to the ordering of conceptual components and, in the section following, we apply those ordering ideas to clarify the structure of CAS theory.

Looking at the Structure of Theory

Drawing on Southerland, Weick (1989) discusses theory as, "an ordered set of assertions" (p. 517). If a theory is defined (in part) as consisting of ordered assertions, it begs the question of just how well ordered those assertions might be. By "ordered," we might understand those assertions to be arranged alphabetically, by apparent importance, or any number of possible methods. This "disposition of things following one after another" (Webster's, 1989: 1013) do not seem to add much to our understanding of theory, however. It is not clear, for example, that a theory might be considered more valid if the assertions are in alphabetical order instead of ordered (for example) by the year each concept was added to the literature. However, based on that simple interpretation of order, there would seem to be no epistemological preference between ordering the assertions by their historical appearance in the literature, by the first letter in the concept, or ordering the assertions by the apparent importance ascribed to them by an author.

By ordered, therefore, it may be that Weick was implying something more significant than a list. A more useful (or at least an alternative) epistemological validity, therefore, might be developed by looking at the assertions or propositions of a theory as being "interrelated," where the propositions might be seen as, "reciprocally or mutually related" (Webster's, 1989: 744). With such a view, a body of theory might be seen as a kind of system and, "... any part of the system can only be fully understood in terms of its relationships with the other parts of the whole system." (Harder *et al.*, 2004: 83, drawing on Freeman). It seems, therefore, that every concept within a theory would best be understood through other concepts within that body of theory. Significantly, this perspective seems to fit with Dubin's (above) assertion that theories of higher efficacy have explanations and concepts that are co-causal.

To briefly compare and contrast levels of interrelatedness, we might say that the lowest level of relationship may be found in some jumble of random concepts. A higher level of interrelatedness might be seen in a book where an author describes concepts (thus causing each to exist in closer relationship with others). Other authors have used a wide variety of methods for increasing relatedness such as placing them in a list (as above), a flowchart showing a cycle (e.g., Nonaka, 2005, for social construction), a matrix (e.g., Pepper, 1961, for metaphors), or a combination of lists and flows to create a meta-model (e.g., Slawski, 1990). With each increasing level of relatedness, a given reader might understand a concept in relationship with other concepts, and so find new insights based on the relatedness between concepts. In short, an increasingly systematic relationship might be viewed as having increasing relatedness. One example of a systemic theory may be seen in Wilber's integral theory of human development (e.g., Wilber, 2001). In his theory, Wilber describes four quadrants that represent categorizations of insights from numerous disciplines. Wilber claims that each of these quadrants is co-defined by the others—in essence, that no quadrant can be fully understood except in relation to the other three. This claim suggests a high degree of relatedness.

Another way to look at interrelatedness might be seen in the concept of "reflexive." Hall (1999) suggests that some forms of inquiry represent a "third path" of inquiry that is primarily neither objective, nor subjective; rather it is essentially reflexive, where meaning is created in a socially constructed sense. In contrast to reflexive forms used in the sense of the interaction between individuals, however, the second study of this chapter looks at reflexive analysis in the sense that suggests a relationship within, or between, the concepts of CAS theory.

Combining the idea of relatedness with the idea of theory having a tight core and a loose belt, we might see the concepts in the core as being more closely interrelated than the belt. For example, the above reductive study of CAS produced definitions with low levels of interrelatedness, as might be found in the loosely defined belt of a theory, because the new theories are presented essentially as lists of concepts.

In addition to the concept of relatedness, another important concept for this chapter is that of "robustness." Wallis (2006a) explored a number of interpretations of this term. Following insights developed from Hegel and Nietzsche, Wallis settled on an understanding of robustness that might be familiar to those working in the natural sciences, where a robust theory is one where its dimensions are "co-defined." An example of this would be Newton's law of motion (F=ma) where each aspect (e.g., mass) may be calculated, or understood, in terms of the other two (e.g., force and acceleration).

It is very important to differentiate between theories whose structure might be seen as robust, and theories whose structure might be understood as an ordered list. For example, a list of assertions might be understood as an atomistic form of theory and might be represented abstractly as: "A" is true. "B" is true. "C" is true. In contrast, the propositions of a robust theory might be seen as: Changes in "A" and "B" will cause predictable changes in "C." Changes in "B" and "C" will cause predictable changes in "A." And, Changes in "C" and "A" will cause predictable changes in "B." The interrelatedness of concepts in a robust theory suggests that the theory may be validated from "within" the theory.

In this book and elsewhere (e.g., Richardson, 2006), the concept of robustness is used to describe the stability of a network experiencing external perturbations. A system that is completely unstable would have a robustness of zero, while a perfectly stable system would be assigned a robustness of one. While it could be legitimately argued that no system can have its measure of robustness at the extreme ends of the scale (zero, or one), this chapter will use zero and one as approximations to facilitate discussion.

An understanding of perturbations might be used to determine what might be called the "dynamic robustness" of a system of theory by identifying the ratio of stable concepts to changing concepts. In this two-step proves, one first identifies the concepts contained in each form of the theory and assigns each a numerical value based on the component concepts. Next, the ratio between the two (earlier and later) versions of the theory is taken. If the two theories are identical, the robustness will be equal to "unity" (or one). If the two theories have no concepts in common, they will have a robustness of zero. For example, if theory "A" has four distinct concepts (a, b, c, d) and subsequently evolves into theory "B" with four concepts (c, d, e, f,), we may see theory A and B together as having a total of six concepts (a, b, c, d, e, f) with only two concepts held in common (c, d). This relationship suggests that in the process of evolving from theory A to theory B, the theory exhibited a robustness of 0.33 (two divided by six). Of course, such measures might only be considered valid when the concepts themselves are unambiguous. This method may be seen as responding to Hull's (1988) deep discussion on the evolution of theory—and providing a tool to aid in the mapping of that evolution.

If we look at each author's influence on CAS theory as a perturbation, CAS theory may be seen as having a low level of robustness. For example, Yellowthunder & Ward (2003) describe a CAS using four concepts drawn from Olsen & Eoyang (2001) who used those four in addition to three additional concepts. This change suggests a robustness of 0.57 (the result of four divided by seven). Other times, CAS does not fare even that well. Dent (2003) states that he drew his conceptualization of CAS from Brown and Eisenhardt (1998). However, between the six concepts listed by Dent and the eight concepts listed by Brown & Eisenhardt, there are only two concepts that clearly overlap. This suggests a low robustness of 0.14 (the result of two divided by fourteen). In contrast, Newton's formula of force (F=ma) may be seen as having a robustness of one as the formula is unchanged in any non-relativistic application. The widespread use of Newton's formula may suggest that theories of greater robustness are more useful (and may have more predictive power) than theories of less robustness.

While low robustness may be seen as enabling "flexibility" (where a theory changes and evolves with rapidity), it may also be seen as an indicator of confusion or uncertainty. It recalls our original question as to the core of CAS theory and suggests that axiomatic, atomistic, or reductive definitions (e.g., theories with concepts that are structured as lists) have shown too much flexibility to provide an adequate representation of the core. In the following section, we analyze the body of CAS theory to identify more robust relationships between the concepts.

Investigating Relational Propositions

In this section, we investigate the relational propositions described by the authors of the above 20 concise definitions to identify the core of CAS theory, where that core may be seen as shifting CAS theory toward Dubin's (1978) second level of theory efficacy (where concepts are directionally causal).

In this process, as with the reductive study, we deconstructed each of the concise definitions into propositions. Rather than use all of the available conceptual components, however, those statements that were essentially axiomatic are left out. For example, Bennet and Bennet (2004) state, "There are some basic properties common to many complex adaptive systems. Examples are some level of self-organization, nonlinearity, aggregation, diversity, and flow" (p. 26). Those concepts would be considered axiomatic or atomistic. In contrast, their statement that, "The term complex system means a system that consists of many interrelated elements with nonlinear relationships that make them very difficult to understand and predict" (p. 26) may be seen as a relational proposition because nonlinear relationships are seen to cause unpredictability.

Of the few relational propositions, the first is Stacey (1996) who states that agents follow rules (or schemas) in their interactions to improve on their behavior. This proposition shows that there is some relationship between the agents, their schemas, behaviors, and the subsequent improvement in behavior. Many authors echo this same general idea.

Axelrod & Cohen (2000) note that varied schemas (situational decision-making rules) differentiate, or provide variety, among agents. Agents are also differentiated by geography (differences in physical and conceptual space). These agents interact with one another (and with tools) in an essentially evolutionary process that might be seen as being based on the agent's fit with the environment. In that process, the agents are changed through changes in their schemas. Changes might be seen as increasing or decreasing the similarity of those agents. Similarly, Shakun (2001) states that agents take actions to reach goals. In this conversation, goals may be seen as generally similar to schemas as both seem to have some influence over the actions of the agents.

McDaniel *et al.* (2003) also suggest that agents interacting over time leads to self-organization. Time may be seen as important, although most authors include it only implicitly. Moss (2001) notes that members (agents) self-organize toward more stable patterns of activity. This may be seen as generally similar to the process of agents interacting to cause self-organizations—with the added idea that the process of self-organizing causes more stable patterns of activity. These stable patterns of activity are a result of common frames of reference, an idea that seems generally similar to rules or schemas as noted above.

Other authors (e.g., Bennet & Bennet, 2004; Hunt & Ropo, 2003) start from generally the same position—that the components (which might be seen as agents) of a CAS interact. However, according to these authors, the interactions lead to uncertainty. Dent

(2003) also agrees and adds that the interaction is to find fit. Dent then notes the results of the agentic interactions may be seen as causing change (that may be thought of as a form of uncertainty).

Somewhere between, or combining, these camps, Harder *et al.* (2004) state that varied agents interact to maintain a system (homeostasis, in their words) rather than create a new system. However, their description of homeostasis is described by terms such as "dynamic equilibrium," "constant change," and "adaptation." Thus, it seems that the authors are suggesting that agents do not change—so much as they enable their CASs to interact, change, and evolve.

Pascale (1999) notes that the process of agents engaged in interaction will lead to more "levels" of organization. This idea of levels might be seen as conceptually similar to Axelrod & Cohen's description of the geographic differentiation of agents (physically and/or conceptually). Additionally, it does not seem as though the creation of a new level of organization should be significantly different (within the context of this conversation) than the creation of a "new" system. It simply seems that this particular new organization is one that is already nested in an existing one. A larger difference between Pascale's version and that of other authors is that Pascale states that the agents are "shuffled" by the larger system.

Drawing on Dooley, Olsen & Eoyang (2001) note that agents evolve over time to reach fitness. Again, we may note the explicit surfacing of the temporal aspect of CASs that many authors leave tacit. Also, where most relational statements discuss agents interacting to achieve fitness, these authors might be seen as leaving the step of interaction as tacit. In both versions, however, agents do tend toward fitness.

Finally, Chiva-Gomez (2003) draws on Stacey to note that CASs that are closer to the edge of chaos (EOC) will experience more self-organization. This concept of EOC might be considered synonymous with "bounded instability," which Stacey describes as a balance between formal and informal systems. More broadly, EOC might also be described as the boundary between stability and ambiguity. In developing the OEC concept, Stacey (1996) notes Kauffman, Wolfram, Gell-Mann, and Langton among his influences. Seen from the perspective of the present conversation, Kauffman's (1995) description of bounded instability may be understood as occurring where the number of agents approximates the number of interactions. If there are too many agents, and insufficient interactions, chaos reigns. On the other hand, if there are few agents and many interactions, stability prevails. Therefore, it seems reasonable that we may integrate the EOC concept with the other co-causal statements above because the EOC may be understood as a ratio between the number of agents and their interactions.

Generally speaking, there appears to be considerable overlap between many of the above causal propositions. Specifically, many authors discuss the existence of agents (including parts), schemas (that may be seen as including goals, rules of interactions), interactions (that may include communication, and also implicitly or explicitly assumes the passage of time), and fit (including evolutionary tests of success). Additionally, it may be seen that the fit test is based on the existence of an external environment (although that environment may be seen as one or more other agents). There are a variety of changes that may result from a fit test including adaptation, change in interaction, change in schemas, increasing uncertainty, increasing certainty, self-organization, and the maintenance of existing organization. Also, as change is seen as a "general" result, change may also be seen to alter agents.

Looking at the concepts in their causal relationship, we may define the core of CAS theory as: Agents, with schemas, interacting over time. The results of those interactions are maximized at the EOC and are subject to a fit test with the environment. The result may be changes in schemas, changes in interactions, the creation or maintenance of larger systems, increased stability and increased instability. Finally, the status of the EOC may be changed by the creation of new agents, schemas, or CASs. This definition is represented graphically in Figure 1.

Each arrow represents causal direction, where one aspect of CAS theory will have an effect on another. As the concept of nonlinear dynamics is an important aspect of CAS theory and complexity theory (e.g., Dent & Holt, 2001; Lichtenstein, 2000), it may be worth noting that the relationship between the causal aspects

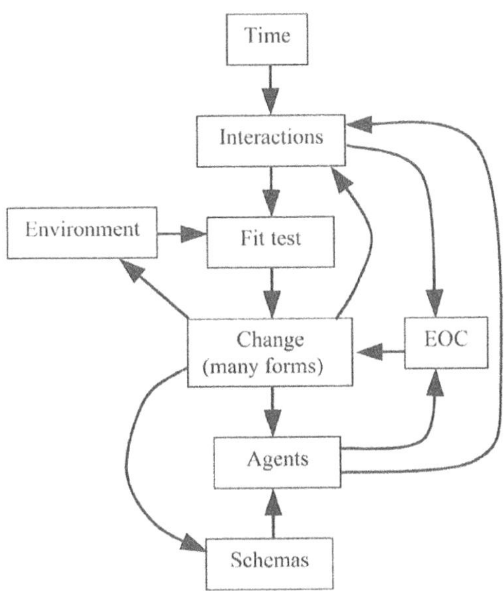

Figure 1 *Relationship between causal concepts.*

of this definition of CAS theory seems to support the idea of a non-linear or non-deterministic structure. This view is in contrast to an atomistic list of conceptual components.

The presented definition surfaces additional challenging questions. These questions might be seen as stemming from potentially contradictory statements by the authors. For example, Axelrod & Cohen note that changes occur at the level of the schema while Moss suggest that change occurs at the level of interactions. These should not, however, be seen as mutually exclusive. Instead, we might ask, "How much change will occur?" Then we might ask, "Where will that change occur?" For example, if a group of individuals self-organize into a corporation to take advantage of a new business opportunity, how much change is seen from the perspective of the business environment (with the inception of a new entity) compared to how much change is seen by the agents (as they organize themselves into new relationships and interactions), and how much change might be seen as occurring at the level of the schema of those individuals?

The structure of this model does not seem to be "perfectly" causal, however, in at least three areas. First, time (along with agents and change) enable the occurrence of interactions. However, nothing seems to "cause" (or alter) time. Therefore, time seems to be seen as an atomistic concept. To become more fully robust, this model should identify what causes (and alters) time or/and its perceived passage. Second, schemas seem to be caused by change, alone. Such a linear relationship would seem to suggest that any change in schemas would create a corresponding change in agents. This relationship would suggest that the concept of schemas might be bypassed. Alternatively, the concepts of schemas might be enhanced by adding causal influences. Finally, while the EOC seems to control the "quantity" of change, the variety of possible "forms" of change is still open to interpretation. Similarly, the environment seems to be changed only by change (another linear relationship, like schemas). However, the environment may represent numerous CASs—each with their own nested agents and schemas.

Although this model has left out many component concepts of CAS theory, the closely related nature of the included aspects suggest that it may be a good representative for the core of CAS theory. That is because that causal model more closely meets Dubin's requirement for stage two of efficacy of theory, while the reductive models only reach stage one. In the next section, we compare three models of CAS theory.

Comparisons and Insights

Drawing on Davidson and Layder, Romm (2001) suggests that researchers use "triangulation" (multiple research methods to reduce subjectivity in research). Thus far, we have presented two forms of analysis. A third form of analysis of the same data set may be found in Wallis (2006a) where Reflexive Dimensional Analysis (RDA) was used to understand CAS theory as consisting of five related conceptual dimensions. Developed by Wallis (2006b) from grounded theory and dimensional analysis, RDA is a method for the investigation of a body of theory.

Looking at the broad range of concepts in CAS theory, the above reductive study has the benefit of including all concepts of the body of CAS theory—presented as a list of atomistic concepts. The RDA version also purports to include all the concepts of CAS theory; however, it presents those concepts as "enfolded" so that concepts not directly represented by the theory might be understood indirectly by combining the dimensions of the theory. For example, the concept of "evolution" (not directly represented) might be understood as "change" over "time" (both of which are directly represented in the model). In contrast to either of the above methods, the causal model leaves out concepts that are considered to be axiomatically atomistic (although those concepts may be seen as still existing in the belt). This suggests some benefits and detriments to each method. Where the atomistic/reductive approach might be more complete, the RDA version might be understood as more abstract (and so has the opportunity to be applied to a wider variety of systems) and more parsimonious (where aspects are related by the minimum number of laws possible, per Dubin).

Comparing the flexibility of these forms of theory (setting aside their common requirement for scholarly justification), it seems that reductive forms enable the easy addition or removal of concepts, essentially adding or removing them from the list. In contrast, forms of theory that are reflexively structured (RDA and causal) are not so easily altered. For example, referring to Figure #1, if we were to remove the concept of interactions, our understanding of other concepts would be imperiled. Time would become "disconnected" from the model, agents would become linearly causal to change (thus eliminating the concept of EOC), and agents would go straight to fit test without interactions (a wholly unsatisfactory description).

In short, it seems that the reductive (and less integrated) forms of theory are more easily changed and therefore may be seen as more easily manipulated or evolved by theorists. That, in turn, suggests that the closer we get to a fully robust form of theory, the more difficult it will become to make further progress. The next few decades, therefore, may see the evolution of competing very-nearly robust theories. Those theories may then be tested (in real-world and computer modeling venues). Then too, each version of theory may find its own specific niche (e.g., one model may be applied to organizations, another to individuals, and a third to schemas).

Additionally, the reductive study suggests a technique for identifying which concepts within a general body of theory might be more closely connected with a given focus (e.g., scholarly version of CAS). The RDA and causal versions, in contrast, are both highly integrated so that the researcher is encouraged (almost required) to utilize all of the concepts for any given analysis. In the causal model depicted in Figure #1, for example, a researcher focusing on interactions would be impelled to describe how those interactions are changed by changes in agents, change, and time. Additionally, the causal relationships would suggest that the researcher describe how changes in the interactions aspect altered the EOC and the fit test. In a sense, this creates a road map that might be of benefit to researchers and students alike.

Shifting our focus to the core, we should note that the core concepts of the RDA version of CAS are (atomistically): Interactions, Agentness, Change, Levels of difference, and Time. The causal version of CAS may be represented as, schemas, agents, interactions, time, fit test, change, EOC, and environment. The two models hold in common the concepts of Change, Interactions, and Time. In the RDA version, the schemas, agents, and environment of the causal version might be explained as Agentness seen at different Levels. Both of these models suffer from understanding Time as an atomistic concept. Where Time enables change (for example), nothing enables Time. This could seem to be a relatively innocuous concept; however, the concept of "flow" (Csikszentmihalyi, 1991) suggests that the idea of time, especially as a subjective representation of productivity, may be an important area for investigation. The RDA version leaves the idea of agents (and other forms of systems) loosely defined based on the idea that it is the observer who determines what the system "is." The causal version leaves the concept of Levels relatively tacit, seemingly accepting that there are three levels of systems (which may be broadly understood as schemas, agents, and environment). Similarly, there are differences within each causal concept (there are different forms of change, a variety of agents, and schemas include a variety of rules).

In contrast to what might be called the "dynamic" measure of robustness discussed previously in this chapter (where we quantified the level of robustness of a theory as it evolved between authors), we might apply here a more "static" measure of robustness. Static robustness might be applied to quantify the internal integrity of the structure of a body of theory and so provide a point of comparison between the two views of the core. Alternatively, it may be used to differentiate between core and belt concepts within a body of theory. A measure of robustness may be achieved by comparing the total number of aspects within the theory to the number of aspects that are both causal and emergent. On one end of that scale, a theory that simply lists its component concepts, without identifying how the concepts were related to one another would have a robustness of zero. On the other end of the scale, a fully robust theory (e.g., Newton's F=ma) might have three dimensions, each of which is both causal and emergent, and therefore would have a robustness of one (3 / 3 = 1). By "emergent," it is important to note that a given concept must be understood in relation to two or more other concepts, as in Newton's model. If changes in one concept were to be determined directly by only one other concept, that change would be seen as linear, and so adding little to the model. For example, if we were to say that changes in the environment cause changes in the agent that cause changes in the schema, we might as well say that changes in the environment cause changes in the schema. If, on the other hand, there is something about the agent that ameliorates, filters, or accentuates, the changes from the environment, we may then say that the changes in the environment and the effect of the agent result in changes to the schema.

In the above reductive study, all lists have robustness of zero (e.g., 0 / 29 = 0). In the RDA CAS model, there are a total of five dimensions (Agentness, Levels, Change, Time, and Interactions). However, of those five, only three are both emergent and causal (Agentness, Interactions, and Levels). Therefore, we might understand this form of the theory as having a static robustness of 0.6 (3 / 5 = 0.6). In contrast, the causal version presented in this chapter contains eight concepts—only five of which may be seen as both causal and emergent providing a robustness of 0.625 (5 / 8 = 0.625). In short, the causal version may be seen as an improvement over the RDA version and this chapter serves as an example of theory-advancement towards a robust core.

Shifting to the relatively non-robust aspects, we see that time appears in the causal model to be atomistic. Schemas and environment are seen as linear/determinant. In the causal model, it seems that the more important area for investigation are the concepts of time, environment, and schema. Understanding how those concepts may be understood as emerging from two or more other concepts within the core should indicate how the model might be rearranged to make sense from a robust perspective. For example, the multiple levels of this model (including schema, agents, and environment) may each be seen, in some sense, as a CAS. That may indicate the opportunity to create a model where each level is represented by that same (simpler) model. Such repetition of simpler models might, in turn, allow the elimination of some redundancies from the model. Indeed, until such an investigation is undertaken, those linear components of an otherwise robust model might be seen as existing somewhere between the belt and the core in an intermediate, or connecting zone.

Finally, based on Dubin's list (where increasing relatedness suggests higher efficacy of theory) and Weick's inference (that propositions should be effectively ordered, or interrelated), it may be suggested that increasing robustness suggests a higher level of epistemological validity of a theory. In general, however, it is important that the causal core may be seen as being derived from a different epistemological validity than the loosely related list of concepts that comprises the belt. Where the belt may find validation from any one of a wide range of research methodologies and points of view, the core finds validation only in the relationship between its own concepts. This is a significant epistemological shift that suggests a rich opportunity for additional study.

Will CAS and/or ICAS Theory Survive?

In this chapter, 20 concise versions of CAS theory were found in the discipline of organizational theory. The conceptual components were identified for each theory and subjected to two forms of analysis. The first, a reductive study, was beneficial in the identification of concepts representing the range of CAS theory, for linking specific concepts within the body to specific uses of the theory, and the creation of additional versions of CAS theory. Each additional definition was seen as adding to the outer belt of the theory rather than clarifying the core. The second study focused on the causal statements found within the body of CAS theory and identified the core of CAS theory by identifying relationships between those concepts. It is suggested that by focusing on causal relationships, we may be able to accelerate the evolution of CAS theory.

This chapter suggests that the flexibility of loosely connected forms of theory may support the spread of that theory through the social sciences. However, the very flexibility that allows CAS theory to grow may also obscure the core. The lack of core, in turn, may limit the effectiveness of that theory in application. Another significant contribution of this chapter is the creation of specific measures of robustness as tools for examining theory. These include measures of dynamic robustness (as an indicator of the evolution and stability of a theory as it passes between authors), and static robustness (a numerical indicator of how well co-defined a theory may be). The more static robustness a theory has, the more it may be considered to be part of the core.

Will CAS theory survive? Or, will it join the 90% of social theory that rises rapidly only to disappear just as quickly (Oberschall, 2000)? This chapter represents a significant step toward a new understanding of theory (in general) and CAS theory (in particular). If CAS theory is to retain its validity and even gain credibility in the face of the next wave of theory (that will, undoubtedly, arrive), it seems that it must develop a robust core. In an important sense, a robust theory might be seen as possessing epistemological justification, not from external testing, but internally, as the understanding of each aspect of the core is tested against the other aspects of the model. Conversations on the structure and construction of theory are likely to continue, and even increase, as our understanding of theory-creation increases. Measures of theory-robustness would seem to provide useful tools for advancing that conversation.

Looking to achieve a more optimal future of CAS theory, investigations should first clarify the causal relationships suggested in this chapter. With a fully robust version of CAS, the next step should be to test that model in the field and through computer modeling to clarify (or deny) relationships suggested by the co-defined aspects. Additionally, the "inter-testability" of the core aspects hints that such a model might be disprovable in the Popperian sense. For example, if the model pictured in Figure #1 were to experience a change in interactions that did not result in a change in the fit test, the model would be disproved. This would, of course, open the door to improving the model (Popper, 2002). Then too, if everything observable may be explained in terms of the robust aspects, then each application in the field becomes a test of the theory, and we have another opportunity to accelerate the evolution of CAS theory through practice.

This chapter has focused on CAS theory, but what about its close cousin ICAS? While this book may, or may not, contain all concepts related to ICAS theory, it certainly provides a cross-section of that theory. As such, the reader may apply the insights and techniques presented in this chapter as he or she reads other chapters. That way, the reader may identify the full breadth of concepts for ICAS theory. Similarly, within each chapter, the reader may find an emphasis on those concepts specifically related to that particular area of study and so suggest a "school of thought" within ICAS. Finally, the reader may range between the chapters seeking causal relationships between concepts, and so develop a robust model of ICAS theory. Each of these opportunities suggests how the reader might develop an alternative point of view that may be useful for further study and developing new insights.

In conclusion, we have presented three major insights that may support theory development and the progress of ICAS theory. First, a robust form of theory provides the best description of a solid core of a theory and so avoids the growing belt of loose concepts that obscures it. Second, it is possible and desirable to measure a theory's level of robustness and by so doing, to measure the progress of that theory. A corollary here is that measuring the progress of a theory opens the door to advancing that theory in a more rapid and purposeful way. Finally (although less deeply explored), robust theories may provide a path to more effective analysis and application. As the data for developing the core came from the belt, it should be noted that no core is possible without a belt. This suggests that both belt and core, with their separate epistemological justifications, are necessary to the advancement of theory.

Shifting to an evolutionary perspective, CAS might be seen as a recently evolved (and rapidly evolving) "species" of theory—derived from its fecund progenitors in systems theory and complexity theory. As a relatively recent species, it is well adapted to fit its niche as an insight-generator for theorists. Theorists, theory, and this book (as a representative of the conversation) may then be understood as three species—all engaged in a co-evolutionary process. This co-evolutionary process, in turn, suggests the opportunity to improve ourselves by accelerating the evolution of CAS. In the sense of an evolutionary landscape, this book might be seen as a path leading CAS off of the plains (inhabited by herds of "big-belt" theories) and up the slope of Mt. Kilimanjaro.

References

Ashmos, D. P., Huonker, J. W., & Reuben R. McDaniel, J. (1998). Participation as a complicating mechanism: The effect of clinical Professional and middle manager participation on hospital performance. *Health Care Management Review, 23*(4), 7-20.

Axelrod, R., & Cohen, M. D. (2000). *Harnessing Complexity: Organizational Implications of a Scientific Frontier.* New York: Basic Books.

Bartlett, J. (1992). *Familiar Quotations: A Collection of Passages, Phrases, and Proverbs Traced to their Sources in Ancient and Modern Literature* (16th Edition). Toronto: Little, Brown.

Bennet, A., & Bennet, D. (2004). *Organizational Survival in the New World: The Intelligent Complex Adaptive System.* Burlington, Massachusetts: Elsevier.

Brown, S. L., & Eisenhardt, K. M. (1998). *Competing on the Edge: Strategy as Structured Chaos.* Boston: Harvard Business School Press.

Chiva-Gomez, R. (2003). The Facilitating Factors for Organizational Learning: Bringing Ideas from Complex Adaptive Systems. *Knowledge and Process Management, 10*(2), 99-114.

Csikszentmihalyi, M. (1991). *Flow: The Psychology of Optimal Experience.* New York: Harper Perennial.

Daneke, G., A. (1999). *Systemic Choices: Nonlinear Dynamics and Practical Management.* Ann Arbor: The University of Michigan Press.

Dent, E. B. (2003). The complexity science organizational development practitioner. *Organization Development Journal, 21*(2), 82.

Dent, E. B., & Holt, C. G. (2001). CAS in war, bureaucratic machine in peace: The U.S. Air Force example. *Emergence, 3*(3), 90-107.

Dubin, R. (1978). *Theory Building* (Revised Edition). New York: The Free Press.

Friedman, K. (2003). Theory construction in design research: Criteria: approaches, and methods. *Design Studies, 24*(6), 507-522.

Gleick, J. (1987). *Chaos: Making a New Science.* New York: Penguin Books.

Hall, J. R. (1999). *Cultures of Inquiry: From Epistemology to Discourse in Sociohistorical Research.* New York: Cambridge University Press.

Harder, J., Robertson, P. J., & Woodward, H. (2004). The spirit of the new workplace: Breathing life into organizations. *Organization Development Journal, 22*(2), 79-103.

Hull, D. L. (1988). *Science as a Process: an Evolutionary Account of the Social and Conceptual Development of Science.* Chicago: University of Chicago Press.

Hunt, J. G. G., & Ropo, A. (2003). Longitudinal organizational research and the third scientific discipline. *Group & Organizational Management, 28*(3), 315-340.

Kauffman, S. (1995). *At Home in the Universe: The Search for Laws of Self-Organization and Complexity* (Paperback Edition). New York: Oxford University Press.

Lakatos, I. (1970). Falsification and the Methodology of Scientific Research Programmes. In I. Lakatos & A. Musgrave (eds.), *Criticism and the Growth of Knowledge* (pp. 91-195): Cambridge University Press.

Lichtenstein, B. M. B. (2000). Emergence as a process of self-organizing: New assumptions and insights from the study of non-linear dynamic systems. *Journal of Organizational Change Management, 13*(6), 526-544.

McDaniel, R. R., Jordan, M. E., & Fleeman, B. F. (2003). Surprise, surprise, surprise! A complexity science view of the unexpected. *Health Care Management Review, 28*(3), 266-277.

McGrath, J. E. (1997). Small group research, that once and future field: An interpretation of the past with an eye to the future. *Group Dynamics: Theory, Research, and Practice, 1*(1), 7-27.

McKelvey, B. (2004). Toward a 0th law of thermodynamics: Order-creation complexity dynamics from physics and biology to bioeconomics. *Journal of Bioeconomics, 6*(1), 65-96.

Moss, M. (2001). Sensemaking, Complexity and Organizational Knowledge. *Knowledge and Process Management*, 217.

Nonaka, I. (2005). Managing organizational knowledge: Theoretical and methodological foundations. In K. G. Smith & M. A. Hitt (eds.), *Great Minds in Management: The Process of Theory Development* (pp. 373-393). New York: Oxford University Press.

Oberschall, A. (2000). Oberschall reviews "Theory and Progress in Social Science" by James B. Rule. *Social Forces, 78*(3), 1188-1191.

Olson, E. E., & Eoyang, G. H. (2001). *Facilitating Organizational Change: Lessons From Complexity Science.* San Francisco: Jossey-Bass/Pfeiffer.

Pascale, R. T. (1999). Surfing the edge of chaos. *Sloan Management Review, 40*(3), 83.

Pepper, S. C. (1961). *World Hypothesis: A Study in Evidence.* Berkeley, California: University of California Press.

Phelan, S. E. (2001). What is complexity science, really? *Emergence, 3*(1), 120-136.

Popper, K. (2002). *The Logic of Scientific Discovery* (J. F. Karl Popper, Lan Freed, Trans.). New York: Routledge Classics.

Richardson, K. A. (June, 2006). *The role of information 'barriers' in complex dynamical systems behavior.* Paper presented at the NESCI, Boston, MA, USA.

Romm, N. R. A. (2001). *Accountability in Social Research: Issues and Debates.* New York: Kluwer Academic / Plenum Publishers.

Shakun, M. F. (2001). Unbounded Rationality. *Group Decision and Negotiation, 10*(2), 97-118.

Slawski, C. (1990, July 8-13). *A small group process theory.* Paper presented at the Toward a Just Society for Future Generations - 34th Annual Meeting of the International Society for Systems Sciences, Portland, Oregon, USA.

Stacey, R. D. (1996). *Complexity and Creativity in Organizations.* San Francisco: Berrett-Koehler Publishers.

Tower, D. (2002). Creating the Complex Adaptive Organization: a Primer on Complex Adaptive Systems. *OD Practitioner, 34*(3).

Unknown. (2006, 3/6/07). *Quotation,* link.

Wallis, S. E. (2006a, July 13, 2006). *A sideways look at systems: Identifying sub-systemic dimensions as a technique for avoiding an hierarchical perspective.* Paper presented at the International Society for the Systems Sciences, Rohnert Park, California.

Wallis, S. E. (2006b). *A Study of Complex Adaptive Systems as Defined by Organizational Scholar-Practitioners.* Unpublished Theoretical Dissertation, Fielding Graduate University, Santa Barbara.

Webster's. (1989). *Encyclopedic Unabridged Dictionary of the English Language.* Avenel, New Jersey: Gramercy Books.

Weick, K. E. (1989). Theory construction as disciplined imagination. *Academy of Management Review, 14*(4), 516-531.

Wheatley, M. J. (1992). *Leadership and the New Science.* San Francisco: Barrett-Koehler.

Wilber, K. (2001). *A Theory of Everything: An Integral Vision for Business, Politics, Science, and Spirituality.* Boston, Massachusetts: Shambhala.

Yellowthunder, L., & Ward, V. (2003). Designing and Implementing Organizational Change in a Complex Adaptive System. In G. H. Eoyang (Ed.), *Voices From the Field: An Introduction to Human Systems Dynamics* (pp. 125-144). Circle Pines, Minnesota: Human Systems Dynamics Institute.

Chapter 3
Validation of Theory—
Exploring and Reframing Popper's Worlds

Wallis, S. E. (2008). Validation of theory: Exploring and reframing Popper's worlds. Integral Review, 4(2), 71-91.

Abstract: *Popper's well-known arguments describe the need for advancing social theory through a process of falsification. Despite Popper's call, there has been little change in the academic process of theory development and testing. This paper builds on Popper's lesser-known idea of "three worlds" (physical, emotional/conceptual, and theoretical) to investigate the relationship between knowledge, theory, and action. In this paper, I explore his three worlds to identify alternative routes to support the validation of theory. I suggest there are alternative methods for validation, both between, and within, the three worlds and that a combination of validation and falsification methods may be superior to any one method. Integral thinking is also put forward to support the validation process. Rather than repeating the call for full Popperian falsification, this paper recognizes that the current level of social theorizing provides little opportunity for such falsification. Rather than sidestepping the goal of Popperian falsification, the paths suggested here may be seen as providing both validation and falsification as stepping-stones toward the goal of more effective social and organizational theory.*

Introduction

Within the social sciences, and more specifically within the study of organizations, numerous papers are implicitly or explicitly creating and/or promoting some form of social theory. Each of these papers indicates by its existence how relatively easy it is for someone to create and (in some sense) validate a social theory. However, despite the importance of advancing social theory, it remains nearly impossible to test, falsify, and improve such a theory (Popper, 2002). As a result, the efficacy of social theory and the opportunity for advancing organizational science is also limited. My interests are to understand and advance social theory, thus these are material concerns to address. My approach in this paper is to provide a tool for advancing theory.

Indeed, the limitations of social and organizational theory are becoming increasingly evident (e.g., Boudon, 1986; Burrell, 1997). There are even ongoing questions as to the viability of academia (Shareef, 1997). The limited effectiveness of social theory seems to have encouraged some to adopt an essentially a-theoretical approach (e.g., Shotter, 2004) that appears as an "epistemology of practice" (Schön, 1991). However a-theoretical it may seem, "practice is never theory-free" (Morgan, 1996: 377). It is doubtful to me that such a focus on application, rather than theory, enhances the development of the social sciences. Dichotomies, unresolved, rarely enhance any endeavor over the long term because they represent a failure to integrate the dialectic.

The difference between the ease of creation and the difficulty of falsification suggests an explanation as to why 90% of social theory rises rapidly, only to fall again (Oberschall, 2000). This is a distressing situation for the continued validity of academia and also for the practitioners who use those theories to improve life on personal, interpersonal, and transpersonal levels. Shareef (2007) suggests that we use Popperian falsification to improve our management theories. Yet, he only echoes a call that is decades old and mostly unheeded.

I suggest that Popper's falsification criterion presents a too-high hurdle, and in its present state, social theory is not strong enough to make that leap. In order to more effectively advance theory, I suggest a reframing of Popper's views, not as a means to avoid falsification, but to suggest multiple paths—each with relatively easy steps leading to the lofty summit of empirical falsification. In this paper, I explore a philosophical pathway from present forms of scholarship toward a more integral view of scholarship and its results.

To begin that exploration, I will discuss the nature and structure of theory. Next, I will describe and briefly discuss Popper's views on empirical falsification as well as his lesser-known views on "three worlds" (empirical truth, emotions/concepts, and theories) that are closely related to his goal of empirical validation. In that section, I discuss some weaknesses of his three worlds.

I propose resolution of these structural weaknesses by reframing and applying integral thinking. This reframed version of Popper's three worlds suggests a new way of validating theory, one that more easily supports the advancement of theory towards empirical falsification. Finally, I then apply this framework to test a theory and suggest how theory may be advanced more effectively.

Theory and Validity

In the social sciences, there is a growing conversation around the nature, structure, and validation of theory. I provide a brief background to that conversation and suggest that the multiplicity of views may be more useful than any single criterion. As prelude, I will draw an analogy to two primal forces of our universe.

As we know, the force of gravity holds together our galaxy, solar system, and planet. Life without gravity would not be possible. Therefore, it is tempting to say that gravity is "the" best way to understand life. Yet, life as we know it would not be possible without atoms. And atoms are held together by the strong nuclear force (Considine, 1976: 1629). Understanding our universe requires an understanding of the galactic-scale force of gravity as well as the atom-scale nuclear force. Similarly, we might consider that the process of validating theory requires the understanding of multiple influences.

Sutton and Staw (1995) characterize the conversation around theory by saying, "Strong theory, in our view, delves into the underlying processes so as to understand the systemic reasons for a particular occurrence or nonoccurrence" (p. 378). Their version of theory might be summed up as possessing, "a single, or a small set of well-developed, logically linked research ideas that lend themselves readily to empirical testing" (Ofori-Dankwa & Julian, 2001: 415). Similarly, Weick (1989) works with Southerland's definition that a theory is, "an ordered set of assertions about a generic behavior of structure assumed to hold throughout a significantly broad range of specific instances" (p. 517). Wacker (1998) pushes for a more atomist than systemic approach, asserting that a theory has four basic criteria: conceptual definitions, domain limitations, relationship building, and predictions.

In their philosophical debate on epistemic justification, Bon-Jour and Sosa (2003) suggested that the validation of a theory might be centered on either an internal or an external perspective. The interlinked assertions or basic criteria noted above may be understood as an internal test of the theory. In contrast, an empirical test is one that occurs on the outside of the theory.

From another external perspective, Quine (1969) suggests in his essay on ontological relativity that we may make sense of a theory from the perspective of a larger theory—or the view from society, as a whole. For example, just as one who knows a complete language is able to make sense of any sentence in that language. The sense-making arises in and applies to the larger societal context. However, in the quest for a more effective social theory, such an approach is problematic: it invites regress and irrelevance. Sense-making contracts to apply to the "nearest" context. For example, within the largest context of our shared society, the sub-context of academia seems to make sense. Nested within academia are various disciplines and publications that make sense. Nested within each of those are many theories, and each of those may also be said to make sense. However, those theories remain ineffective in practice—out in the larger society. Thus, Quine's methods do not seem adequate to his own challenge of adjudicating among rival ontologies.

In contrast to Quine's "outside-in" approach testing that uses the inside-out approach mentioned above might be considered a test of the logical structure of the theory—what might be called the "logos." I am defining the logos of a theory as the structure of the logical arguments that support the theory. While the academic process may be seen as an exercise in the development of the logos, I suggest we *expand and deepen how theorists understand the structure of the logos.* Whetten (2002) appears to agree, noting how a theory may be criticized for its lack of logical internal consistency. This integral perspective goes beyond Popper's decades-old understanding of theories that were comprised of and therefore structured by separable axioms.

Such an advance in thinking seems to parallel the development of applying an integral perspective. For example, in Wilber's (2001) four quadrant model (where each quadrant is said to be defined or understood through the perspective of the other three), we can take the perspective of viewing the relationship among the quadrants as the internal structure of the theory. That structure must be rigorously defined, or the model will be open to criticism.

In short, for a theory to be considered effective, it must be subject to both internal and external testing. This leads us to a potential problem of regress, because in one test we are looking for empirical facts; yet, "there are no facts that are independent of our theories" (Skinner, 1985: 10). As a result, the empirical/applied test becomes problematic because the results of that test must be viewed through a worldview, that is, through the lens of another theory. Thus, the question: if theories guide perceptions and actions change theories, where then is there any certainty to conclusions based on perceptions?

To recognize the difficulty of testing social theory through practice is not new. Echoing reports above, some say social theories have not proven very useful for effecting social change (Appelbaum, 1970; Boudon, 1986). The success of economics as a theory was questioned long before the late 2008 meltdown (Rapoport, 1970, as cited in Dubin, 1978).In organizational studies, even though theory does inform practice (Weisbord, 1987) in some quarters, an increasing number of practitioners describe organizational theory as "irrelevant" to their work (Tsoukas & Knudsen, 2005). In such cases as these, theory is ignored in practice rather than tested through practice.

On the other side of the coin, we can imagine an excellent organizational change practitioner who, if required to employ "bad" theory, makes modifications "on the fly" either consciously, and/or unconsciously. Thus, a good practitioner might compensate for bad theory, further compounding the difficulty of testing theory through application. A version of this approach is implied by Weick and Sutcliffe (2001) where they advise practitioners to manage more mindfully (adopting a range of useful techniques) to move away from the trap of too heavy a reliance on plans and procedures.

Of course, "good" theory might be applied with bad results—as anyone can attest whose fingers have slipped while using a calculator, and subsequently generated an answer that they considered believable... but useless within a mathematics examination.

Thus, both internal and external approaches have their possibilities and limitations when it comes to determining the validity of a theory. In such a case, we would do well to understand how both approaches are necessary, or at least useful, to understand the validity of a theory, just as we do well to understand how both the gravitational force (large scale) and the nuclear forces (small scale) are needed to understand our physical universe.

In the next section, I provide some background on Popper's worldview and his approach for advancing theory. I also go into some detail describing Popper's three-world model, and how, with slight revisions to that model, with slight revisions, we can use it to advance our understanding of theory validation.

Popper: Background and Worlds

Broadly, Popper's view suggests that science is in continual revolution. There, in the pursuit of objective knowledge, science is advanced by the rigorous testing and falsification of theories, "designed to replace existing theories with more accurate theories based on empirical research outcomes;" (Shareef, 2007: 275), and "each step brings us closer to the truth" and, importantly, Popper's "concept of truth is based on the intuitive idea of truth corresponding to empirical facts" (p. 276). Necessarily, this would be an iterative process beginning with inaccurate theories, and moving toward theories of greater accuracy.

Popper's version of scientific progress through the conjecture and refutation of theory has influenced a wide range of disciplines including organizational theory (Shareef, 2007). Thus, I find his ideas useful in my efforts to understand and advance social theory. Despite wide understandings of Popper's call, and many decades of research, progress is clearly elusive. The tree of knowledge grows slowly and thus far is not bearing fruit. In the field of organizational studies, of many theories proposed, few have been adequately falsified. I find no well-documented advancement toward truth.

Turning attention to develop my earlier point, we have not advanced because the requirements for falsification set a standard too difficult for most scholarly investigations. Realistically, then, while falsification remains a useful goal, a series of steps toward the goal rather than a single hurdle would be more achievable for the social sciences.

Popper's own work has the insights needed to identify those steps. He developed his "three worlds" model to resolve the mind-body problem of knowledge. Although this approach is not so well known as his call for falsifiability, I explore those three worlds and the system they create, to suggest a useful key for validating theory. In this, I hope to provide a tool for advancing theory more easily towards falsifiability and thus closer to effective use by practitioners.

In brief, Popper's (1978, 1996) three worlds are as follows. World one (W1) is the physical world, consisting of physical objects, energy, and plants. World two (W2) is the mental world, consisting of feelings, thoughts, decisions, and perceptions. Popper suggests that W2 may be subdivided into the conscious and the subconscious.

World three (W3) is the world of more complex conceptual constructs including stories, mathematical constructions, artistic compositions, and, importantly for this paper, theories. The contents of each world are called "objects;" so that, for example, a rock would be a W1 object, a feeling of joy would be W2 objects, and a theory would be a W3 object.

These worlds can interrelate. For example, a painting (as a physical thing) would have some W1 existence (as do all physical objects). The aspect of the painting that makes it a great work of art would be a W3 object. When an individual views a great painting and experiences an emotional reaction, that reaction is a W2 object.

In developing the idea of a third world (of theory and complex constructions), Popper broke with many contemporaries and classical philosophy that suggested there was only one world or at most two (one of body, and one of mind). Popper called himself a pluralist for this reason, suggesting that all three worlds were needed to understand the mind-body problem and the situatedness of knowledge. In this, Popper suggests, somewhat hierarchically, that each world is founded on the one below it. W3 Theories are based on a combination of W2 ideas, and those ideas are based on perceptions of W1 objects.

Objects are subject to classification schemes. Popper discusses W1 objects (such as a block of marble), as a class of things. He also discusses W3 objects (such as a theory or symphony) as a class of things. He also notes how W3 objects, such as a plan, may be changed and altered as if they were W1 objects (one may chip away at a theory, in much the same was as one may chip away at a block of marble). While he mentions that knowledge is situated in W2, he also asserts that both W1 and W2 objects are things that might be known. Logically then, there are classes of that which might be known.

Part of Popper's discussion in these texts involves proving the "reality" that he refers to as "truth" of these worlds, essentially seeking to validate, justify, or prove the existence of the objects within each world. He does this primarily by recognizing the importance of interaction of objects within and between worlds; for example, "observation is always observation in the light of theories" (Popper, 2002: 47). Thus, no W2 observation is possible without W3 theory.

Wilber (1999), perhaps accurately though much too briefly, notes one main thread of Popper's investigation; he characterizes the three worlds as relating to objective and subjective knowledge. In lumping Popper's three worlds with a variety of other concepts, Wilber misses an opportunity to use the three world concept in a way that might have proved useful in supporting the arguments he presented in *The Marriage of Sense and Soul* thus advanced integral thinking. For the present paper, however, I will use the three-world idea—but focus on science, rather than soul.

For his AQAL model, Wilber suggests that each of the four quadrants exists in relation to the other three. That is to say, each quadrant may be understood only through the perspective of the other three. In a sense, if one quadrant were removed, the other three would lose their validity. This is the heart of integral thinking—recognizing and learning from interrelatedness. In contrast, Popper sees each of his three worlds as arising from the one below it. The W1 physical universe must have existed before humans evolved to perceive it (W2), and those perceptions must have preceded the sensemaking needed to create W3 theories.

This generation of worlds, however, is not a one-way street: "Our minds are the creators of world three; but world three in its turn not only informs our minds, but largely creates them" (Popper, 1978: 167). This view resonates with the idea that "we shape our world and our world shapes us" or "our mindset affects how we see the world."

Morgan (1996: 274) presents a similar view when he draws on theories of complexity, chaos, and autopoiesis to encourage managers to think in terms of "loops, not lines." That is to say, we should avoid thinking mechanistically (e.g., A causes B), and rather incorporate "the idea of mutual causality, which suggests that A and B may be co-defined as a consequence of belonging to the same system of circular relations." We humans frequently seek a single cause to link to a single effect. We need deeper understandings about the complete system and its multiple causes, effects, feedback loops, limits to growth and exponential growth.

In contrast to Morgan's mutual causality, the emergence of Popper's worlds one from another, seems more linear. However, Popper's discussion of those worlds goes deeper. In addition to the above description of how W3 theories may impact W2 perception and knowledge, he also described how changes in theory would enable changes in W1. This makes good sense, as a person with a plan might make more significant changes to the world than a person without one. Yet, he suggested that such changes would only occur through the intermediary of W2. In short, W3 affects W2, which affects W1. This relationship may be understood as a three-stage linear relationship. Importantly, as Stinchcombe (1987) notes in his extensive discussion on theory, such a relationship seems to render the intermediary term (W2, in this case) redundant. We might just as well say that changes in W3 cause changes in W1. However, this would leave out Popper's very reasonable suggestion that W2 objects such as emotional reactions and insights are of a very different nature than the W1 empirical facts, and W3 theories. The three worlds are intimately interrelated while qualitatively different.

Popper saw a theory as being constructed of a set of axioms, suggesting that falsification of one axiom need not result in the falsification of another axiom. His view of axiomatic severability is troubling because the most effective theories (those that allow us to most reliably make changes or predictions in W1) are the theories of physics. Those theories are robust in the sense that each axiom is co-defined by the other axioms of the theory. For example, in Ohm's law (a theory of electronics), there are three aspects (volts, ohms, and amps). Each of those is defined by the other two. If we were to falsify volts (for example), the entire structure would collapse: there would be no way to describe amps and ohms. Thus, effective theories of physics are not amenable to axiomatic severability.

Popper pushes further in this direction suggesting that there is "no hope" of developing a useful understanding of a theory based on the relationships between the component statements because those statements (and the objects represented by them) can always be dissected to smaller pieces (Popper, 2002: 112). Instead, he suggests that a theorist begin with "relatively atomic statements" as long as those statements are linked to some sort of objective instrument of measurement.

However, as is well known (then as now), even atoms may be split into components. In short, Popper's preference for atomistic structure of theory must be called into question. Further, as noted above, our philosophies have advanced since Popper's time, allowing for new insights into the structure of theory.

Popper can appear contradictory. "All of our actions in world one are influenced by our world two grasp of world three" (Popper, 1996: 142), yet, this suggests actions are a W1 object, while actions are not typically mentioned as an integral part of this theory. The act of research, as an example of an action, is mainly an implied part of the model, rather than an explicit aspect. This lack of clarity is an opportunity to improve on, or at least reframe, Popper's model.

This analysis indicates that Popper's worlds might be better understood by looking at them in a different relationship: moving from a linear view to a co-causal view of the three worlds. This approach would be a more integrated view of theory, one I expect to provide a closer coupling—a tighter integration—between worlds, and so enable a better understanding of scholarship in general, and the validity of theories, in particular. In short, I hope that by re-examining and reframing Popper's ideas, I might draw some novel conclusions that will aid in the advancement of social theory.

Reframing Relationships between Popper's Worlds

If, as noted above, the linear relationship between the three worlds is inadequate, an alternative, co-casual, model is an alternative with which to experiment. To start by iterating the observation above, each of the three worlds might be considered co-casually related to the other two. To do so would suggest that W3 theories would be determined by W2 facts and feelings and W1 reality/truth. This throws us into an immediate tailspin of regress, as we have no way of establishing that truth, outside of our feelings and theories.

In order to develop an adequate model of co-causal worlds, the content of those worlds must be reframed. An important part of Popper's equation may have been overlooked. He carefully describes how the combination of W1 truth and W3 theories will conjoin to create W2 knowledge. Yet, he seems to take for granted that such knowledge would be accepted as such. We should ask, why that truth is accepted as truth.

If the objectivity of knowledge were the only criterion, belief would not be necessary (Powell, 2001). Yet, the W2 aspect of belief does seem to be an inescapable aspect of the Popperian philosophy. The answer, which we experience on a daily basis, is that it "makes sense." That "sense of truth" would be a necessary part of the equation. Without it, how would we "sense" that one truth had any more validity than another?

We could argue that one could read two versions of theory, and simply identify that set of arguments that appears to be more valid than the other. However, that may be understood as the very definition of "sense making." After all the arguments, conversations, examples, dissections and connections, each reader must experience some feeling that the conversation makes sense in order for it to be considered correct or right.

For example, I occasionally enjoy a good trivia quiz. In those events, I have noticed in myself and observed in others that there appear to be instances where knowledge is accompanied by a sense of "rightness," even if that sense does not necessarily correspond with the version of truth held by others. While that feeling might, conceivably, be lumped into W2 with Popper's version of knowledge, such an approach interferes with our ability to differentiate between knowledge and feelings. And, for the reasons just mentioned, that distinction is an important one. In short, there are important differences between knowledge and the sense of rightness associated with that knowledge that is not related to a change in theory.

That sense of rightness seems to emerge from the intersection of theory and knowledge. When the facts fit the theory, we say that there is epistemic justification, that is, that the information seems to be true. Therefore, that sense seems to have important distinctions from what Popper would describe as knowledge even though both of them are what Popper would call W2 objects. Therefore, I recommend reframing those three worlds to separate these vital aspects.

Additionally, while there may be a world of absolute reality or truth, and our improved theories may bring us closer to that ultimate truth, I think it is safe to say that we will never fully understand that ultimate reality. Therefore, I'm not sure it makes sense to have W1 as part of the model. We would not, as an example, include the moon in a theory of continual advancement in running techniques, for no amount of running will ever reach the moon. There may be absolute truth. But if that knowledge is unobtainable, there seems to be little use to including it in a theoretical model.

For this reframing, it is more useful to understand W1 as the repository of perception and knowledge. The W2 would remain the domain of feeling and emotions, and W3 remain the domain of theory. Significantly, this distribution of worlds creates the opportunity to understand each of the three worlds in integral, co-causal relationship to one another.

For those who might object to removing Popper's version of reality, some absolute truth from this model, I would suggest designating "world-zero" as some place of absolute truth. Then too, as we are observes of this model, and we humans are also observers to our own selves (knowledge, emotion, and theories) the self or observer might be represented by a "world-infinity." Those two worlds, however, are bracketing the present conversation, and would require additional development. So, let me return to the focus of validating theory.

I propose Popperian falsification may be adequately represented by the reframed relationships between three worlds. There is a W3 theory, a W2 sense of rightness, and a W1 knowledge. I think, however, that there is room for additional insight into theory validation that may be seen within each of these three worlds.

Reframing Validation within Worlds

In the spirit of applying multiple perspectives, I will now shift our point of view. Instead of looking at the relationship between worlds, I will seek to clarify the situation as it exists within worlds.

According to the previous discussion, a theory might be falsified by the information derived from the theory, in conjunction with the feeling of rightness accompanying the theory and the information. This other, internal, perspective might look at the truths within a world as a class unto themselves. Or, as Shareef (2007) notes, Popper recognized that within each world there were levels—essentially "degrees of truth."

For example, one might have the same theory, the same feeling of rightness; yet have a greater quantity of knowledge. Therefore, we may understand at least a quantitative level within each world. Similarly, as a new version of theory emerges, each advanced degree of truth or succeeding theory might be met with an increasing sense of rightness. These may be understood as varying levels of certainty within W2. For example, I know when I am asked a question, there are times when I have an answer... and times that I am more or less certain of that answer. As discussed above, these forms of validation might be seen as internal to the theory, rather than the external tests of validation through application and the generation of data, and the accompanying sense of rightness (or lack of same).

For example, in W1 it may be said that each object of fact has validity because it is differentiable from other objects. Also, internally to W2, it may be said that the feeling of love has some validity because it is differentiable from the feeling of hate. Moving to look at W3, it is important that Popper notes, "What is most characteristic of this kind of world 3 object is that such objects can stand in *logical relationship* to each other" (Popper, 1978: 158). In short, there is a sense of validity afforded by the categorization of objects within a world. They are also validated by the ability to differentiate objects within a world, as well as how they are related to one another within that world. Focusing on the validation of theory within W3, Popper notes the validity of mathematics (including the idea of infinity) as existing relatively independent of W1 (where no one has ever counted to infinity).

Of course, relying on internal justification alone creates the opportunity for a pitfall. Powell (2001), for example, suggests that "thick illusion" by definition, is not falsifiable. Therefore, he concludes, our uncertainty never vanishes. And so--and this is quite important—no theory is ever complete: there is always the opportunity for improvement. Exactly what constitutes "improvement" in a theory is topic for additional dialog. While Popper suggests that a more advanced theory is one that is closer to the "truth," such a claim is difficult to confirm because that truth, itself, is part of the thick illusion.

Gödel's Proof hints at this limitation. Essentially, Gödel proved that nothing could ever be proved. Or, more formally, that no theory could be both complete and self-contained. However perfect and complete the theory, there is always someone reading the theory and so formulating a new insight that goes beyond the original theory (for an interesting discussion on this topic, see Hofstadter. 1980, *Gödel, Escher, Bach*).

We might create a theory that appears to have a high level of internal consistency; yet has no basis in reality. For example, Tolkien's *Lord of the Rings* describes a carefully constructed world (including mythic languages and fire-breathing dragons). There is a high level of internal consistency in this work. Despite this W3 internal strength, fire-breathing dragons remain fictional. This same theoretical limitation exists in texts of fiction, fantasy, metaphysics, and religion.

While this paper is focused on interrelationships within and between worlds, the traditional need for clarity and consistency of logical arguments is not discarded. Consistency is especially important. If a theory says that a dollar may be redeemed for three apples, we cannot change dollars for kiwi-fruit, nor apples for diamonds without also describing the changes in the rate of exchange.

Our understanding of clarity may shift, however, under the present model. For example, we might develop (ad hoc) a three dimensional theory that includes aspects of People, Places, and "Stuff." While the use of Stuff as an aspect of theory might, on the face of it, seem hopelessly vague, the vagueness is due to the lack of context. We might enrich that context by specifically describing the relationship between the aspects within the theory. For example, we might say that more People and more Places will result in more Stuff. Thus, Stuff might be seen as roughly analogous to "memories" (for the limited purpose of this ad-hoc example).

When co-validation is said to occur between the constituent propositions of a W3 theory, that co-validation may be repeated (fractal-like) in W1 and W2. In those other worlds, the relationship between the propositions are played out and falsified by actions and observations. For example, a theory that is essentially a metaphor might have little or no W3 validation. It would, if published, have greater W2 validation. If applied successfully to an organizational change effort, it would have greater W1 validation. At the other end of the spectrum, a theory with infinite propositions might be perfectly accurate (with high W3 validity); however, the application would be nearly impossible.

For W3 objects, which may be said to pose significant validity in one world, I suggest that the call for greater validity may be answered from within other worlds, and in the relationship between worlds. Indeed, as a theory or other W3 object advances in its internal validity, it may be said to generate a call for validation in and between the other worlds. *Broadly, a complex W3 fantasy calls for a reality check in W1 and a sensibility check in W2.* A profundity of W1 data that does not "make sense" calls for the creation of a W3 theory to weave them all together. Similarly, the level of sense between W3 objects may also be compared to suggest which one theory might seem more sensible than another.

Popper's conversation suggests that the best way to validate a theory can be seen in the way it allows for predictions in W1. Thus, a theory of physics (W3), which fails to predict change in W1, will have been falsified. In the social sciences, however, the process becomes a highly complex issue of recursion. For example, changes in theory (W3) will cause changes in productivity (W1) but may also alter individual sense of importance (W2), so that an individual may no longer place the same value on the items being produced. For example, individuals working on an assembly line might (based on an emerging mindset) change their minds and decide that the human producer is more important than the objects on the conveyor belt.

Driven by a modern, rather than integral mindset, Popper may have pushed too far in pursuit of his deductive agenda, a "Kantian epistemological activism" to the point where, "we attempt to impose our interpretations onto the world rather than being passively instructed by it" (Shearmur, 2005: 265). Although Popper concludes that measurable W1 results represent the highest level of validation, because W1 represents empirical outcomes, both the process and the content of his conversation on the relationships within and between worlds suggests that the door is still ajar for alternative forms of validation.

For example, by way of recursion, Popper's arguments in favor of falsifiability may themselves be understood as a form of theory.

As a set of logical arguments that are not backed up by empirical data, Popper's argument does not stand up to his own test. It does, however, retain validity in this reframing of his worlds where his theory may be understood as an advanced W3 object. By reframing his model from a linear/mechanistic one to an integral/co-casual form, I am advancing that model to a higher W3 standing. However, that higher W3 standing also heightens the imbalance between worlds, resulting in a call for W1 validation.

Thus, no one world, or relationship between worlds, should serve as the single criterion for the validity of theory. Instead, a theory should be understood as increasingly valid as it increases in validity within and between all three worlds. Because my focus here is on the validation of theory, I discuss validation within W3 in the next section. Following that section, I apply these concepts in an analysis of a theory.

Validation within World Three

If a theory, as a set of related propositions, claims validity through logical arguments, it is legitimate, even necessary, to ask whether those arguments are essentially part of the theory, or are they part of the justification of the theory. If it is the latter, the logical arguments used to justify a theory are themselves objects of W3. This means they also must be legitimized in the same way. This creates an issue of recursion that may severely undermine attempts to validate theory through logical arguments. In the past, such an issue has been resolved by resorting to axioms. For example, Steiner (1988) describes the system of a theory, at least in part, as being constructed as a chain of definitions. Frustratingly, her conclusion was that the last link of a chain remains undefined. Her assertion seems atomistic and unusable. As I have developed elsewhere (Wallis, 2006, 2008a) the idea of a theory constructed of co-defining propositions provides a robust alternative to axiomatic terms.

Legitimacy exists for two reasons: (a) there is a co-causal relationship *between* worlds, and (b) there is a co-causal relationship *within each* world. If greater validity is derived from greater agreement (e.g., the consensus of expert opinion is more valid than the opinion of a single expert), that line of reasoning can extend to positing that *more aspects* of validity should be seen as *conferring greater* validity. That is, a combination of validating relationships (including W1, W2, and W3) confers greater validity than any single one. An object in any world has greater legitimacy when it is validated from more "directions:" (a) qualitatively, within worlds and between worlds, and (b) quantitatively, more relationships are better, whether they are inter-world validations or inner-world validations.

As the issues I am addressing indicate, the conversation on the validation of theory based on its structure has been mired in Western-traditional understandings of logic and rhetoric. For example, Steiner's (1988) work Isa framework for theory justification based on the construction of language and logic. In a logical argument, the W3 object of theory is constructed of a chain of other arguments that ultimately must rest on observation. While the components of the theory may, or may not (depending on their complexity), be considered as W3 objects, the observations are certainly W2 objects. Hence, many arguments of logic are essentially validations between worlds. Because individual understanding of the world (W2 objects) is dependent on individual theories (W3 objects) the issue of regression, as demonstrated above, is inherent even in simple observation. Or, more briefly, "Truth itself is plainly useless as a criterion for the acceptance of a theory" (Kaplan, 1964: 312).

In contrast, Dubin's (1978) work went beyond the traditional calls for rational argument by suggesting that higher validity could be obtained by higher orders of structure of the theory. While this as appear to be looking within the W3 object, another perhaps more useful view is to say validity results from comparing the components of the theory, *where those components are of sufficient complexity to stand on their own as W3 objects.*

For example, a theory that is essentially a list of facts would not be much of a theory. Indeed, decomposing any body of theory to a list of bullet points would essentially remove the structure from its W3 standing. In short, to achieve the opportunity for W3 validation, a theory must be of sufficient complexity that each of its component propositions must also qualify as W3 objects. In a sense, such a reduction is like atomizing or de-integrating the construct from W3 to W1 status.

Recently, I developed Reflexive Dimensional Analysis, a method for analyzing a body of theory (Wallis, 2006). Later, I clarified the importance of "internal justification" of a theory (Wallis, 2008a). For an example of an internally validated theory, I return to Ohm, and his I=E/R .where each aspect (volts, amps, and ohms) is defined by the other two aspects (Considine, 1976: 1678). The internal justification view I introduced considers theories of greater internal validity to be more "robust" and, as illustrated by the robustness of Ohm's law, more robust theories stand much stronger chances of enabling—and eventually predicting—W1 changes.

Although some might say Ohm's law is a relic of modernist or deterministic thinking, from another perspective, Ohm's law is only deterministic if two of the three aspects are known. For example, in order to determine the volts in an electrical circuit, both the ohms and the amps must be known. If, in attempting to determine the voltage of a circuit, the investigator is aware of only one aspect (e.g., the amps), the other two become *indeterminate.* In that case, the model, even when indeterminate, may serve as a signpost suggesting directions for further investigation. Thus, the model simultaneously provides deterministic boundaries (based on the limitations of the model) and also provides indeterminate opportunities for exploration. The context and extent of determinants account for a model's simultaneous possibilities.

Where Nonaka (2005) suggests the group creates and validates knowledge, in the same way, both individuals and groups must use theories in the process of validating knowledge. Where Nonaka discusses the idea of "ba," which includes "contexts and meanings that are shared and created through interactions" (p. 380), I seek to differentiate contexts and meanings in a straightforward process with methods for theorists and change agents to evaluate theory through multiple perspectives. Those intersubjective methods of measurement may be developed across and within Popperian worlds to objectively identify the epistemological validity of theories.

If theories can be created and confirmed to have a high degree of validity, we should expect those theories to be more helpful for processing information individuals encounter in their daily lives.

In short, when theories are more useful, individuals may develop themselves more effectively.

In some sense, delving into a turn of the regression, we can conceive the inside view of one model to be the outside view of another. That is to say, an outside view of validation of theory involves a comparison of three worlds (theory, fact, and feeling of rightness). As we move into the inside view of theory, to validate a theory from the inside, we are now adopting a new perspective. What were previously seen as W3 theories, have now become W1 facts. The W3 theory place is now held by the method of metatheoretical analysis being developed here. However, that is all a matter for a future paper. In the next section, I apply the above ideas to the analysis of a theory.

An Analysis of Theory

The ideas presented in this paper have useful implications. If validity of theory may be understood within and between worlds, each scholarly study should include reflections on how the new theory fits within and between worlds. Such an effort need not be extensive, once the ideas are generally understood.

Such a review would include observations, logical arguments, and some reflection on the authors' sense of rightness. To some extent, many scholars aim toward such goals on a regular basis. In this section, I present and apply a framework for using this analysis to show how it might be done.

This framework for the analysis of theory operates as a checklist. A benefit of such a checklist, when used for any published theory, is the clear indication of validation within and/or between worlds. The results could be shown in a table using the dimensions shown in Table 1. Although not demonstrated here, the results could be combined to create a simple, numerical, "Logos Quotient," a broad indicator for the integrity of the theory (Wallis, in preparation). Such a system would enable theories to "recommend themselves" to managers and other organizational change agents in search for a strong, appropriate theory to apply. While this is more a test of existing theory, it also indicates areas to focus on for future advancement of the theory.

	Level 1	Level 2	Level 3
World One (Facts or data)	Uses objective data.	Uses objective data from multiple sources.	Future facts are predicted.
World Two (Meaning, emotions)	Makes sense to author.	Makes sense to editor, reviewers, and readers.	Consensus of expert opinion (this theory is preferred over other theoretical options).
World Three (Theory)	Includes logical arguments.	Theory is constructed of specific propositions.	Theory is constructed of co-causal propositions.

Table 1 *Dimensions of Validity*

For this demonstration analysis, I pulled (more or less at random) an article from my files, and now analyze its contents using the reframed three worlds perspective. The article is: *Institutional entrepreneurship in mature fields: The big five accounting firms* (Greenwood & Suddaby, 2006). The study combines multiple theoretical bases to identify dynamics of change in organizations. Essentially, the focus is on institutional theory, which considers organizational change (and stability) as well as how the organization affects individuals, their mental states and their interactions. Briefly, the big five accounting firms became so large and powerful that they were no longer subject to the norms of their industry. They became more open to alternative ideas, such as the creative accounting processes of Enron. The analysis shed light on the dynamics leading to the collapse of Enron and Arthur Andersen.

World one considers the facts, or the data of the study. In this case, the authors interviewed senior partners at top accounting firms. They also drew on archival data such as annual reports, audit statistics, training records, press releases, and transcripts from hearings. As such, there was a large quantity of what might be considered facts, in the Popperian sense. The figures showed that the big five did indeed hold the lion's share of the market—nearly 80%. Based on their analysis, the authors went so far as to predict that the evolution of the field's elite will outstrip regulatory efforts." Based on these results, their study has a high level of W1 data. Their data are objective, from multiple sources, and includes their prediction of future events (another form of data).

In contrast to the world of facts, world two is focused on meaning and emotions. In this world we may assume that the paper makes sense to the authors, or they would not have written it. Additionally, in their "appreciations," the authors note the contributions of research funding, numerous collaborators, and anonymous prepublication reviewers. This explication raises the level of W2 validity because it indicates that more scholars believe the paper has merit. Finally, the paper earned the authors the Academy of Management Journal "best paper" award for 2006. The external professional recognition indicates the paper has a still higher level of meaning for experts in that field. It is too early to determine if the paper will have much effect outside of academia; if so, that would indicate another level of meaning.

While we may assume that the authors' writing process included a great deal of reflection (an important part of making meaning), one defect here is that the authors do not explicitly surface their own reflective process in the paper itself. Explicating their making meaning process transparently would add to the W2 validity of the paper. In an interesting negative example, the authors credit a specific individual with cutting a "Gordian knot." This analysis would have more data if the authors had discussed the knot and the process that led to its removal.

In world three, the theory of the article contains many logical statements. I do not spot any "contradictory axioms" that could be used by the authors to draw any conclusion they desired (as might be found in metaphysical texts) (Popper, 2002: 71)

The authors describe their theoretical orientation, and as a result of their research, suggest a process model and a set of seven propositions to explain the phenomena under investigation. Interestingly, their process model suggests two views of the process. The first view shows a coarse-grained analysis of the interrelationships between relevant theories, the second a fine-grained analysis that shows the causal relationships between the various aspects of those theories.

Propositional analysis is a useful tool for analyzing the relationships between co-causal aspects of a theory (Wallis, 2008a). Further, robustness is the measure of the internal integrity, or integrality, of a theory and as posited here, is a useful predictor of that theory's efficacy in practice. Robustness is measured on a scale of zero to one, with zero indicating no robustness (as might be found in a shopping list of concepts), while a robustness of one may be found in a very effective theory of physics (such as Ohm's law). In the present analysis, the coarse-grained view has a robustness of 0.29 while the fine-grained view has a robustness of 0.31. These levels of robustness, while not very impressive compared to Ohm's law, are typical for social theory based on my analyses.

Moving outside of the individual worlds, and into the relationship between worlds, additional insights and important contradictions begin to emerge. For example, the authors' prediction about the field's elite evolving faster than regulatory changes. Such a relationship would be considered observable data, yet those data are not accounted for in the theoretical model the authors propose. This misalignment between the world of data and the world of theory suggests the need for additional (and more careful) work in developing their model. In short, each of the three worlds is an admirable exemplar of academic work. However, these three worlds are not necessarily part of the same solar system (Table 2).

The analysis presented in this section may also be understood as a piece of W1 evidence, relating to the W3 theory presented in this paper. Of course, as discussed above, those two worlds are insufficient by themselves. I must apply my model to my use of the model, another recursion to examine validity. To form a more integral model, then, there must be an accompanying W2 feeling of validity. The same levels of analysis pattern I used on Greenwood and Suddaby is the pattern I now use on this work I am writing that you are reading. On one level, there is some feeling of rightness on my part, although I certainly feel that I need to do continued investigation and explication of my theorizing. I assume there is an implicit sense of rightness on the part of the reviewers and the editors, else this article would not be published here. A more important level of W2 validity is the feeling of rightness that may be felt by you, the reader.

	Level 1	Level 2	Level 3
World One (Facts or data)	Uses objective data.	Uses objective data from multiple sources.	Future facts are predicted. However, data predicted do not match source data.
World Two (Meaning, emotions)	Makes sense to authors. Authors should explicitly surface reflections.	Makes sense to editor, reviewers, and readers. Authors recognized by peers.	Does not have consensus of expert opinion. Model is not compared with other models.
World Three (Theory)	Includes logical arguments.	Theory is constructed of specific propositions.	Theory has co-causal propositions to 0.3 level of robustness. Room for improvement by increasing robustness.

Table 2 *Dimensions of Validity for Greenwood & Suddaby*

Conclusion

Popper (1996: 4) is mainly interested in the growth of objective knowledge. This direction is in keeping with his call for empirical validation of social theory. However, in his discussion of the three worlds, he is highly focused on developing the W3 perspective of theory. In that focus, he may have missed some interesting insights. Significantly, Popper was a product of the modern age. That milieu seems to be reflected in the linear structure of his three worlds model.

Popper's view of theory may have opened the door for the explosion of theory in the social sciences without necessarily advancing any specific theory. This is because he saw a theory as being comprised of and constructed by a set of axioms. He thought falsification of one axiom need not result in the falsification of another axiom. This view may have provided philosophical ammunition to scholars, emboldening them to revise theories by changing their axioms, almost at random. It seems strange to me that this approach would actually be acceptable to Popper (Popper, 2002: 113).

While, on the surface, this may seem a perfectly reasonable approach to some, demands for rigor should prevent such a disconnection within the bounds of a single theory. After all, if a theory is composed of two axioms, and one is falsified, the theory would no longer be a theory: it would be a single axiom. An alternative would be to replace the falsified axiom with a new one. However, in this example, that means half of the theory has been changed. Can we reasonably call that theory by the same name? I don't think so. For a deeper discussion on the problems of changing theory, and the measurement of that change, see my discussion on "dynamic robustness" (Wallis, 2008b).

My thesis is that instead of holding ourselves to the nearly impossible standard of using empirical validation to create higher quality theory, we of the social sciences instead need to begin to build and amass a higher quantity of theories.

Drawing on and reframing Popper's idea of three worlds, I proposed a co-causal version of the three world's model. I presented a range of possibilities for the validation of social theory including methods within and between worlds. Additionally, I presented a checklist style matrix for the easy validation of theory. Essentially, this matrix indicates that theories might be validated more effectively from the inside and the outside rather than one or the other. Finally, I tested that matrix in an analysis of a theory. In this, I have identified an important way to advance an otherwise admirable paper (i.e., Greenwood & Suddaby, 2006).

Because our theories shape the way we work in the world, the opportunity to purposefully advance a theory is also the opportunity to indirectly impel social adaptation and evolution. In short, by advancing our theories, we advance humanity and more.

Despite the apparent failure of the social sciences, there is progress toward success. The shared understanding of metatheory is advancing. For example, in addition to many other valuable insights, Whetten (2002) and Van de Ven (2007) both provide useful metatheoretical exercises for students. This is a significant improvement over the previous approaches to understanding theory because the ability now exists to engage in useful classroom exercises on theory creation where previously, that ability was tacit or non-existent. Such a capacity suggests that the understanding of theory has advanced to a particular level. On the other hand, this might also indicate that the opportunity exists for theorists to step back and not be as concerned as to whether a theory is good or bad, and simply note that the mass of theories that have been developed over the past century may now be described as data for metatheoretical studies.

My approach to validation of theory does have limitations. For example, if we wanted to analyze a complex structure such as a high-rise office building containing multiple organizations, the difficulty of listing each item, to say nothing of categorizing each of those items into its sub-aspects of each of the three worlds would be intimidating, to say the least. Perhaps the most effective use of my model is in metatheoretical analysis. There, for example, the scholar may review his or her own work, or the work of another scholar, and readers could easily identify the strengths and weaknesses of articles they reflect on and make useful suggestions for improvement.

Reflecting on the above analyses, and the levels of validity noted in Table 1, indicate another area for possible improvement of this model. The W2 levels of sense-making starts with the article making sense to the author. Then, at progressively higher levels, it essentially asks if the paper makes sense to editors, reviewers, and other experts. The usefulness of this scale may be called into question because any article that is subject to this evaluation will have already passed the first and second levels. Few will ever reach the third. Some reframing of this scale may be useful. Similarly, most or all papers will easily reach the W3 level of including specific, logical propositions, while few will ever reach a level of high robustness of their co-casual propositions.

One alternative might be to revise the scale into something more quantitative. A meta-theoretician might, for example, count the total number of facts, the number of propositions, and the number of people who agree with the model. This would provide specific and objective data for comparing and contrasting theories.

The generally high levels of efficacy found within worlds for the analysis of the Greenwood and Suddaby paper may be expected for an article of their high academic caliber. This, in turn, may suggest that the form of analysis I presented represents an evaluation based on traditional academic and/or Western values. More reflection and analysis should reveal more insight into this concern. I, for one, would be interested in seeing alternative evaluation methods, and an argument might be made for non-evaluation (although such an argument would be an evaluation).

Future studies might include an investigation into the measurement of each world's objects and how they are seen and measured from the combination of other worlds. Another opportunity is to identify a "triple helix" of iterative interactions between the three worlds. Nonaka (2005) suggests a spiral relationship in the knowledge creation process, resulting in the generation of more knowledge and knowledge that is more useful. If this process is replicated for W1 and W3, we might imagine three spirals, with interesting opportunities for measurement and insight. Looking within each world, there is the possibility to "unfold" each world into additional dimensions to identify the quantitative and qualitative nature of each world object. The opportunity also exists for a similar quantification of objects that exist simultaneously in all three worlds.

Another form of validation might be seen in the process of creation. For example, the social construction perspective suggests that knowledge and social norms are both created through the general process of social interaction (Burr, 1995), that a socially constructed culture strongly determines how each individual sees the world, and finally, that knowledge and social action are linked. Social construction might be seen as analogous to the process of theory creation, where the theory is a cultural frame of reference, and that frame of reference is continually changed by its interaction with other theory, application, and conversation.

Both processes (social construction and theory construction) seem to be recursive in that they are continually repeated processes; the theory changes the way individuals see and gather information, the information changes the theory. In short, within the process of communication, theory may be understood as being created by knowledge and that the process of using theory generates knowledge. It seems that this reframed view might dovetail with a social constructionist point of view. However, more study would be required to develop this idea.

The relationship between worlds presented in this paper could be tested through a longitudinal study that tracks the knowledge, theories, and effectiveness of an organization over time. The resulting information could suggest, for example, the amount of effort that should be allocated between action, knowledge creation, and theory creation within the environment of a particular industry. Such knowledge would contribute the effectiveness of translative and transformative development.

For example, one could analyze economic theory used at the state and national level, including the robustness of those theories. Those results might be compared (over time) with level of consumer confidence and the number of people living above the poverty line. Similarly, one could study the theories and policies used by corporations and community groups. The results of those studies could be compared with organizational success and/or organizational adaptability.

Here, I have focused on W3. Future research might focus on other worlds. One big question might be an investigation into spirituality. If spirituality is associated with and evidenced through a spiritual awakening, and that awakening is reflected in a sense of wonderment, that sense might be understood as a W2 object. As such, there may be internal analyses that seek to analyze the quantity and quality of such moments of amazement. More interestingly, however, we could investigate how that W2 feeling might be understood in co-causal relationships with W1 evidence and W3 theory.

While Table 1 offers a matrix-checklist approach for investigating the validity of theory, that could be expanded by identifying a scale within each box. For example, if a theorist is seeking validation on a W1 dimension at Level 1, he or she could (and should) note how many situations have shown improvement from the application of the theory. As noted above, there is no specific progression between worlds. While W2 validation may move from individual opinion to consensus of opinion, that consensus does not guarantee increasing validity in W1 or W3.

Perhaps neither knowledge nor theory can be fully understood (or validated) except in relation to the other. If so, the primacy of that relationship suggests that the processes of theory creation, the structure of theory, and especially the relationship between action, knowledge and theory are worthy of significant exploration.

When climbing a technically challenging mountain, such as Everest, the expedition establishes a “base camp” where supplies are easily stockpiled. At a convenient location, higher up the mountain, they establish an “advanced camp” where climbers recuperate and acquire additional supplies ferried up from the base camp. Should the first attempt on the summit be unsuccessful, additional attempts launch from the advance camp. Thus, a well-organized expedition does not lose valuable time and energy returning to the foot of the mountain to re-energize. This is a more efficient way to make repeated attempts until the summit is reached. .

In that spirit, I hope the model presented in this paper indicates how we can establish a conceptual camp for advancing theories. At such a camp, an evolving theory might gain energy from validation within worlds and falsification between worlds, on its way to the summit of empirical falsification. In short, I anticipate that increasing our capacity to understand theory will lead to improvements in theory, and so inform developments in practice for the benefit of the many domains of life.

References

Appelbaum, R. P. 1970. *Theories of Social Change.* Chicago: Markham.

BonJour, L., & Sosa, E. 2003. *Epistemic Justification: Internalism vs. Externalism, Foundations vs. Virtues.* Malden, MA: Blackwell.

Boudon, R. 1986. *Theories of Social Change* (J. C. Whitehouse, Trans.). Cambridge, UK: Polity Press.

Burr, V. 1995. *An Introduction to Social Constructionism.* London: Routledge.

Burrell, G. 1997. *Pandemonium: Towards a Retro-Organizational Theory.* Thousand Oaks, California: Sage.

Considine, D. M. (ed.). 1976. *Van Nostrand's Scientific Encyclopedia* (5 ed.): Van Nostrand Reinhold.

Dubin, R. 1978. *Theory building* (Revised ed.). New York: The Free Press.

Greenwood, R., & Suddaby, R. 2006. Institutional entrepreneurship in mature fields: The big five accounting firms. *Academy of Management Journal*, 49(1): 27-48.

Kaplan, A. 1964. *The conduct of inquiry: Methodology for behavioral science*. San Francisco: Chandler Publishing Company.

Morgan, G. 1996. *Images of Organizations*: Sage.

Nonaka, I. 2005. Managing organizational knowledge: Theoretical and methodological foundations. In K. G. Smith, & M. A. Hitt (eds.), *Great Minds in Management: The Process of Theory Development*: 373-393. New York: Oxford University Press.

Oberschall, A. 2000. Oberschall reviews "Theory and Progress in Social Science" by James B. Rule. *Social Forces*, 78(3): 1188-1191.

Ofori-Dankwa, J., & Julian, S. D. 2001. Complexifying organizational theory: Illustrations using time research. *Academy of Management Review*, 26(3): 415-430.

Popper, K. 1978. *Three Worlds*. Paper presented at the Tanner Lecture on Human Values, University of Michigan.

Popper, K. 1996. *Knowledge and the Body-Mind Problem: In Defence of Interaction*. New York: Routledge.

Popper, K. 2002. *The logic of scientific discovery* (K. Popper, J. Freed, & L. Freed, Trans.). New York: Routledge Classics.

Powell, T. C. 2001. Fallibilism and organizational research: The third epistemology. *Journal of Management Research*, 4: 201-219.

Quine, W. V. O. 1969. *Ontological Relativity and Other Essays*. New York: Columbia University Press.

Schön, D. A. 1991. *The Reflective Turn: Case Studies In and On Educational Practice*. New York: Teachers Press.

Shareef, R. 1997. A Popperian view of change in innovative organizations. *Human Relations*, 50(6): 655-670.

Shareef, R. 2007. Want better business theories? Maybe Karl Popper has the answer. *Academy of Management Learning & Education*, 6(2): 272-280.

Shearmur, J. 2005. Karl Popper: The logic of scientific discovery. In J. Shand (Ed.), *The Twentieth Century: Moore to Popper*, Vol. 4: 262-286. Montreal, Canada: McGill-Queen's Press.

Shotter, J. 2004. Inside the moment of managing: Wittgenstein and the everyday dynamics of our expressive-responsive activities.

Skinner, Q. 1985. Introduction. In Q. Skinner (Ed.), *The return of grand theory in the human sciences*: 1-20. New York: Cambridge University Press.

Steiner, E. 1988. *Methodology of theory building*. Sydney: Educology Research Associates.

Stinchcombe, A. L. 1987. *Constructing social theories*. Chicago: University of Chicago Press.

Sutton, R. I., & Staw, B. M. 1995. What theory is not. *Administrative Science Quarterly*, 40(3): 371-384.

Van de Ven, A. H. 2007. *Engaged scholarship: A guide for organizational and social research*. New York: Oxford University Press.

Wacker, J. G. 1998. A definition of theory: Research guidelines for different theory-building research methods in operations management. *Journal of Operations Management*, 16(4): 361-385.

Wallis, S. E. 2006. *A Study of Complex Adaptive Systems as Defined by Organizational Scholar-Practitioners*. Fielding Graduate University, Santa Barbara.

Wallis, S. E. 2008a. From reductive to robust: Seeking the core of complex adaptive systems theory. In A. Yang, & Y. Shan (eds.), *Intelligent Complex Adaptive Systems*: 1-25. Hershey, PA: IGI Publishing.

Wallis, S. E. 2008b. From reductive to robust: Seeking the core of institutional theory. *Under submission*.

Weick, K. E. 1989. Theory construction as disciplined imagination. *Academy of Management Review*, 14(4): 516-531.

Weick, K. E., & Sutcliffe, K. M. 2001. *Managing the Unexpected: Assuring High Performance in an Age of Complexity*. San Francisco: Jossey-Bass.

Weisbord, M. R. 1987. *Productive Workplaces: Organizing and Managing for Dignity, Meaning, and Community*: Jossey-Bass.

Whetten, D. A. 2002. Modeling-as-Theorizing: A Systemic Methodology of Theory Development. In D. Partington (Ed.), *Essential Skills for Management Research*: 45-71. Thousand Oaks: Sage.

Wilber, K. 1999. *The marriage of sense and soul: Integrating science and religion*. New York: Broadway Books.

Wilber, K. 2001. *A theory of everything: An integral vision for business, politics, science, and spirituality*. Boston, MA: Shambhala.

Chapter 4
The Complexity of Complexity Theory—
An Innovative Analysis

Wallis, S. E. (2009). The complexity of complexity theory: An innovative analysis. Emergence: Complexity and Organization, 11(4), 26-38.

I am indebted to two anonymous reviewers whose suggestions helped me to create a stronger paper.

Abstract: *As more scholars join the conversation around complexity theory (CT), it seems a useful time to ask ourselves if we are talking about the "same thing?" This concern is highlighted by the present survey, which finds more conflict than agreement between definitions. In contrast to the conflict, a path toward common ground may be found by applying the idea of a "robust" theory. A robust theory is expected to be more effective in application and more reasonably falsifiable. In this paper, Reflexive Dimensional Analysis (RDA) is used to analyze existing definitions of CT. These definitions are deconstructed, redefined as scalar dimensions, combined, and investigated to identify co-causal relationships. The robustness of CT is identified as 0.56 on a scale of zero to one. Paths for advancing the theory are suggested, with important implications for complexity science.*

Introduction: Seeking the Core of Complexity Theory

Given the breadth, depth, and growth of the current conversation, it seems reasonable to ask—exactly what is this thing called "complexity theory?" For although there are many definitions of CT, it has been suggested, that there is no unified description (Axelrod & Cohen, 2000: 15; Lissack, 1999: 112), respectively. While this plurality may reflect the many voices engaged in the conversation, it also calls into question the validity of the theory because there is no common sense as to what the theory "is." Indeed, the general assumption seems to be that we are all talking about the "same thing." Like blind men discussing an elephant, such assumptions may lead to false conclusions and unnecessary conflict.

While the academic process thrives on the differences between points of view, the extent of those differences calls into question whether scholars are, indeed, talking about the same thing. After all, if one author states that CT may be understood through concepts "A, B, and C" while another author states that the relevant concepts are "C, E, and F," there is some conceptual overlap, but there are also inherent contradictions. Although according to their authors, these descriptions fit under the general rubric of CT, these differences may be seen as representing a conflict in the common understanding of CT, and so reflect differences in our understanding of systems from atoms to institutions.

The issue of understanding of a body of theory has been of concern for decades. In one attempt to make sense of the issue, theories are described as having of a "hard core" of unchanging assumptions, surrounded by a more changeable "protective belt" (Lakatos, 1970). When a theory is challenged, a theorist may rise to defend it with a new concept that changes the belt, but presumably leaves the core intact. In the present paper, I seek to identify the core of CT. This effort will provide general and specific support for the continued development of CT.

If the core is defined as "that which is generally accepted," it might be easy to define the core of CT. Unfortunately; no such commonality seems to exist (as will be explored in greater depth below). Some other indicator is then needed for the core.

Where the social sciences might be generally said to have highly variable protective belts of theory, it should be noted that Ohm's I=E/R is a robust theory. I use the term robust in the same way that it is used in physics and mathematics, to describe a theory where each dimension of the theory may be determined by the other dimensions (this will be discussed in greater detail below). In the present article, I will identify how an understanding of CT might be shifted from the shifting obfuscation of Lakatos' outer belt, toward an enduring and useful law. When our theories attain this level of advancement, we may anticipate meaningful changes in the way we study institutions.

Leaving that lack of effective theory unquestioned is like ignoring our fundamental assumptions. And, as Lichtenstein (2000a: 539) suggests, "...since these assumptions are rarely discussed, many of the potential insights from complexity science have not been fully developed."

Although a complete study is beyond the scope of the present paper, it may be suggested that a more robust version of CT may prove more effective in application. This claim is based on the idea that robust theories of physics are more effective than the ever-flexible theories commonly found in the social sciences.

"A proposition is a declarative sentence expressing a relationship among some terms." Van de Ven (2007: 117). Further, a proposition may be of three types. Atomistic propositions are very simple, essentially claiming, "A is valid." Linear propositions are more complex noting a causal linkage between two concepts such as, "Change in A cause changes in B." More complex relationships are found in concatenated proposition. In this, I use the sense expressed by Van de Ven (2007) where two aspects of a theory are shown to influence the third aspect. For an abstract example, a concatenated proposition might state that changes in A and B will cause changes in C. Similarly, a theory may be generally understood as a collection of interrelated concepts—an idea that will be explored in depth below.

In this process, I will adopt a Knowledge Management (KM) approach. KM includes the study of creating, transferring, and sharing knowledge (Kakabadse, Kouzim & Kakabadse, 2001). Codification is seen as a critical step in the process of social interaction (Leydesdorff, 2002). Codified knowledge is considered more easily transmittable, although it may seem less related to a given context. The following analysis might also be seen as a form of scientometrics and bibliometrics. The concise descriptions of CT as found in OT will be taken as codified knowledge and this paper may be seen as another step in the process of codification.

Although the present studies will avoid a historical approach to the analysis of theory, it may be useful to note here that Van Dijkum (1997) suggests that CT is closely related to cybernetics as both include questions of subjectivity, self-organization, and self-steering founded primarily on the work of Prigogine, Haken, and Casti. In the present paper, the focus will be on eight publications.

Although each author studied in the present paper is writing within the organizational field, they draw their influences from a variety of sources. Pascale (1999) traces his influences to the interdisciplinary work at the Santa Fe Institute. Frederick (1998) draws on mathematics, chemistry, biology, and Darwinian evolution. Axelrod & Cohen (2000) pull from Darwin, Smith, Simon, and "many fields of study," while Brown & Eisenhardt (1998) mention physics, biology, economics, and strategy. Stacey, Griffin & Shaw (2000) note their preference for the works of Prigogine, Kaufman, and Goodwin, and against the works of Gell-Mann and Holland. Kernick (2006) suggests that the study of complexity began with computer-based models and spread to other disciplines, while Hurtado (2006) draws primarily on more recent works from organizational theory. Finally, Dagnino (2004) suggests the origins were in biology and physics and that these origins represent the Santa Fe approach and the European approach.

It should be noted here that the diversity of sources, including fields of study and specific authors, does not seem to suggest a shared sense of what CT is.

Drawing on Davidson and Layder, Romm (2001) suggests that researchers use "triangulation" (where multiple research methods are used to reduce subjectivity in research). In the present paper, I provide two studies of CT, where both studies use the same source of data found in those eight publications.

In the first study, I draw on techniques of content analysis to identify the range of concepts found in the theory. Content analysis (e.g., Hjørland, 2002; Hood & Wilson, 2002) essentially involves looking at the words used by the authors as reasonable representations of the concepts that they convey. Grasping the range of concepts helps to identify the similarities and differences between the various versions of theory. The first study identifies a high level of disagreement between authors.

The second study is informed by narrative analysis (e.g., Pentland, 1999) to focus on the propositions found in CT. In this study, I use an innovative, though easily replicable, method of analysis (inspired by insights from CT) to identify relationships between propositions and calculate an objective measure of the robustness of CT. Importantly, an objective path for developing a more robust CT is suggested. A robust version of CT may be expected to provide meaningful benefits to academics and practitioners alike.

Study #1—The Atomistic Concepts of Complexity Theory

This study identifies the range of concepts of CT within organizational theory (OT) and analyzes that collection of concepts from two perspectives. As developed by Wallis (2008), these diverse views, in some sense, may be seen as reflecting the existing diversity in the field of CT. With each perspective, a different view of CT is provided. The new versions of CT created here may be seen as schools of thought, or as newly evolved versions of CT. In addition to identifying the range of concepts present in CT, this study also suggests that more differences than similarities exist between versions of the theory. And, these differences represent an inherent conflict.

Based on a word-search of the ProQuest™ database, I found 683 matches for "complexity theory" with "definitions." By focusing on those papers within the subject area of organizational theory, this number was reduced to 85 matches. Additionally, I followed promising leads and reviewed books from my own shelves. Next, the versions of CT were tested for conciseness to see if they were of a reasonable level of complexity for this kind of study. They should not too long (extending across many pages because such extensive descriptions of CT might suggest a study that would not be of suitable length for this publication.

From this search, I found eight definitions of CT in the field of OT where those definitions were relatively concise (less than one page). For this paper, I will analyze CT based on relatively concise definitions developed by scholars working in the area of OT. The authors drawn on for this study are: (Axelrod *et al.*, 2000; Brown *et al.*, 1998; Dagnino, 2004; Frederick, 1998; Hurtado, 2006; Kernick, 2006; Pascale, 1999; Stacey *et al.*, 2000).

From that collection of writings, I deconstructed each description into the authors' component concepts. For example, Dagnino (2004: 61) suggests (in part), "These subsystems are therefore subject to evolutionary pressures." From such a statement, the concepts of nested systems, and evolutionary pressures may be drawn. While another reader might develop a different list, it is expected that such lists would be substantively similar to the one developed in this study.

From the above eight sources, I found 47 easily differentiable concepts—from the idea that Agents act, to the existence of Unexpected change. The complete set of concepts will not be listed, due to limitations of space. That list of concepts might be seen as representing the whole of CT as found in OT from a conceptual perspective. It should be noted that at this "survey" stage, no concept appears to be closer to the core of CT than any other. The next task is to search for concepts held in common by the authors. It may be assumed that greater commonality represents some shared acceptance of some core concepts, and less conflict, around the question of what CT is. The following is a list of the concepts that seemed most popular among the eight descriptions, each concept being noted by three or more authors:

> *Meta systems have midi agents, Connections are of varied types, Firm behavior changes, Evolutionary pressures, Agents act, Unexpected change occurs, and Agents are of varied types.*

These seven concepts stand in stark contrast to the 47 concepts that comprise the complete list of concepts discussed as CT. Also, it is worth noting that a new version of CT has been created based on this focus.

Moving from one form of popularity to another, of the eight publications in this study, two could clearly be seen as the "most cited" according to Google Scholar™. The version of CT developed by Axelrod and Cohen (2000), as well as the version by Brown and Eisenhardt (1998) were each cited well over 500 times suggesting a certain level of authority in the field. The next less frequently cited source received about 280 citations while others were much lower.

While this division may be somewhat arbitrary, it serves to provide a useful observation.

Focusing on the two "most cited" versions of CT, there are significant differences and no similarities. The Axelrod and Cohen version takes note of eleven concepts focusing on agents, their variations, interactions, strategies, and success. In contrast, the Brown and Eisenhardt version notes only three concepts, focused tightly on the idea of the "edge of chaos." While it is possible to infer a connection between the two definitions, such an inference would require the interpretation of the authors' works—and so beyond the scope of the present article.

Rather than creating what might be considered a single "authoritative" version of CT, this comparison essentially creates two new description of CT, where each publication may be seen as representing its own school of thought. The lack of overlap (finding zero concepts in common) stands in stark contrast to the 14 concepts that the two do not share. Rather than creating a useful consensus, this study suggests bifurcation and implied conflict between versions. Either version, it should be noted, has less conceptual breadth than the whole body of CT.

There are obvious limitations to this study such as sample size and the amount of interpretation applied to understand exactly what the authors meant to convey with each specific word. However, it should also be noted, that any discussion around investigating some deeper level of understanding must consist of still more concepts. Therefore, a deeper exploration might be expected to produce more concepts and so more levels of conflict.

The present study may stand as an example of an expansion of Lakatos' outer belt. The number of theories was easily increased, yet this expansion does not seem to have increased the understanding of the core.

Of course, many such studies are possible. For example, a new study of this sort might be conducted based on the authors' influences, field of origin, date of publication, and so on. The point to be taken here is that each focus seems to create a new version of CT; and, that new version will have fewer conceptual components than the body of CT as a whole. Further, each new theory will similarly lack a discernible core. In short, it may be concluded that CT is a highly contested field of study, where each new theory adds to the conflict.

This form of study may be understood as useful to the extent that it identifies the range of concepts within a body of theory. This form of study is also useful because it highlights the difficulty of working with disconnected (atomistic) concepts. In the following study, I will use an innovative methodology to investigate the same body of data. Rather than focusing on atomistic concepts, however the focus will be on connected concepts found in the causal propositions of the theories. Before that study, the importance of such a focus is explained in the following section in a discussion on the structure of theory.

The Structure of Theory

Understanding the structure of theory is foundational to understanding the second study, and to identifying the core of CT. Given the diverse versions of CT, the question presents itself: is it possible to ascertain the legitimacy of a theory through its structure? The answer would seem to be in the affirmative because Popper's arguments for falsification include the idea that the structure and composition of a theory will add to the testability of that theory (Popper, 2002: 111-114). The concept is of such importance, however, that some explication seems suggested.

In the validation of theory, Kaplan (1964) describes three norms (correspondence, coherence, and pragmatism). A pragmatic test (one which is tested through application) is beyond the scope of this paper. A norm of coherence suggests that we question whether a theory fits within the existing body of theory. As the present analysis and theory development draws entirely on existing theory, there should be little difficulty in maintaining that norm. Then, there is the norm of correspondence. Kaplan takes pains to note how, for this norm, the truth is plainly useless because every appeal to the facts rests on presuppositions. Therefore, he suggests, what counts in the validity of a theory is the "concatenation" of the evidence. As noted above, concatenation is seen where two aspects of a theory are shown to influence the third aspect In this, I use the sense expressed by (Van de Ven, 2007). For an abstract example, a concatenated proposition might state that changes in concept A and concept B will cause changes in concept C.

In the present study, however, I found few propositions that were concatenated. The second study, therefore, will use Reflexive Dimensional Analysis (RDA) as an analytical tool to "shift" propositions toward increasing levels of concatenation Wallis (Wallis, 2006a, b).

Generally, RDA begins with a set of linear propositions. A linear proposition might be seen abstractly in a causal relationship between the aspects of the theory such as, "Changes in A cause changes in B." The propositions found in a body of theory are investigated to determine how they may be combined to create concatenated propositions. For example, one linear proposition (e.g., Changes in A cause changes in B) may be conceptually related with another linear proposition (e.g., Changes in C cause changes in B) to suggest a concatenated proposition (Changes in A and C cause changes in B).

The increasing concatenation may be understood as representing increasing levels of interrelationship, or increasing levels of structure. Dubin (1978) suggests that there are four levels of efficacy in theory; and, these levels seem to reflect the structure of the theory, they are: 1) Presence /Absence—what concepts are contained within a theory; 2) Directionality—what are the causal concepts and what are the emergent concepts within the theory; 3) Co-variation—how several concepts might impel change in one another; and 4) Rate of change—to what quantity does each of the elements within the theory effect one another.

The first study (above) tested for the presence/absence of concepts. In the second study, I will use RDA to combine causal propositions as a means of advancing the structure of CT toward the third level, that of co-variation. Achieving the fourth level is beyond the scope of this paper.

In a view that seems similar to Dubin's, Weick (1989) suggests that a theory may be understood as an, "ordered set of assertions." Viewing Weick's statement as a concatenated proposition seems to suggest that more assertions with more order between them, will tend to result in better theory—where better theory may be generally understood as having some preferable aspects of quantity and quality. If a theory is held to be useful in practice, we may infer that a better theory might be more useful in practice. Therefore, the structure of a theory may be said to have some bearing on the efficacy of that theory when it is applied in practice.

If we are looking at the assertions or propositions of a theory as being "interrelated," the propositions of that theory might be seen as, "reciprocally or mutually related" (Dictionary, 1993: 998). With such a view, a body of theory might be seen as a kind of system and, "...any part of the system can only be fully understood in terms of its relationships with the other parts of the whole system." (Harder, Robertson & Woodward, 2004: 83, drawing on Freeman). It seems, therefore, that every concept within a theory would best be understood through other concepts within that body of theory. This view seems to fit some assumptions of CT with regards to importance of the interrelated nature of varied components.

An objective level of relationship is easily calculated by identifying the total number of concepts within a theory as well as the concatenated aspects of the theory, then dividing the number of concatenated aspects by the total number of aspects. This method provides a measure of "robustness" between zero and one Wallis (2008). Briefly, those aspects of a theory that are more closely related to one another may be understood as being closer to the core of the theory.

As an example, Ohm's law (I=E/R) contains three aspects. Each of those aspects is concatenated from the other two. Therefore, Ohm's law has a robustness of one (the result of three divided by three. In contrast, a theory that simply lists its component concepts, without identifying how the concepts are related to one another, would have a robustness of zero. For example, in the above study, the new version of theory is a list of concepts and, as such, has a static robustness of zero because no conceptual aspect is said to have any effect on any other.

Comparing the great usefulness of Ohm's law (fully robust), with the limited usefulness of many social theories (low robustness) suggests important implications for practice.

In the following section, study #2 and use RDA to develop the concatenated set of propositions and to measure the level of interrelationship between propositions in CT.

Study #2—Investigating Relational Propositions

In this section, I analyze CT using RDA. The process of RDA consists of six steps:

1. Define a body of theory.
2. Investigate the literature to identify sources that define it.
3. Code the sources to identify relevant components.
4. Clump the components into mutually exclusive categories.
5. Define each category as a dimension.
6. Investigate those dimensions—looking for a robust relationship.

For the first step, the body of theory chosen for this analysis is defined as those concise versions of CT as found in the study of OT. The second step (investigation of definitional literature) was accomplished by a survey of the ProQuest™ database and additional sources as described above. The third step (coding propositions) proved more difficult than expected. The sheer variety among the propositions meant that few were found to be identical. Also, the complexity of some (long and convoluted) propositions inhibited the coding process.

To facilitate the process, I bypassed coding and went to the fourth step where related propositions were grouped into mutually exclusive categories. By mutually exclusive, for example, we might see that agents cannot be understood as interactions (and visa versa). Many of the categories developed were of that sort of straightforward variety. Others were more challenging. For example, there seems to be considerable congruence between concepts of adaptation, evolution, emergence, and change. Frederick (1998) links concepts of self-organization, emergence, and complexity.

The goal of RDA is to identify the fewest possible dimensions that may be used to relate the greatest range of understanding that may be attributed to a body of theory. I was able to combine a number of terms without loss of conceptual sense-making capacity by first noting the essential similarity between the concepts of evolution and adaptation as both are said to be maximized at the EOC (Brown *et al.*, 1998; Pascale, 1999). Then too, each may be understood as emergence because both are described as causing and resulting from change to a system in the context of its environment. Therefore, the concepts of evolution and adaptation may be seen as dimensions that have been "renamed" or perhaps "derived," rather than as dimensions that are essential to the model. There also appears to be general congruence between the concepts of complexity and emergence, because both may be explained as causing unpredictability at some level.

In this way, I combined the wide variety of concepts. Through the clumping process, at least one concept did not fit neatly with the others. That is the negative result of a fitness test. The propensity of a firm to exit a market is not well understood in CT, perhaps because of a more appreciative approach to firm creation and survival. We seem limited to the idea that firms may disintegrate in chaos (Dagnino, 2004) or fail through excessive stability (Pascale, 1999). Both are possible, however a greater understanding would be beneficial. Again, in the interest of parsimony, it may be said that the destruction of a firm may be described as a lack of emergence based on the idea that a firm that is created, changed and sustained through emergence will fail if the level of emergence drops below a critical level. Or, from another view, if emergence is maximized at the EOC (Brown *et al.*, 1998), it may be suggested that emergence is minimized at the extremes of stability and chaos. If firms fail at those extremes, we may conclude that extinction is the opposite of Emergence. Or, that a certain level of emergence is required to maintain a firm (nourished by a flow of people, ideas, capital, etc.).

Within the causal propositions of these eight definitions, nine mutually exclusive categories are suggested. These may be seen as mutually exclusive in that (for example) systems might engage in action, but no action seems possible without systems; therefore, the two (as concepts) are mutually exclusive. Goals, Strategies, and Information. Because of that ambiguity, it may be possible to characterize those as some form of system. This suggests an opportunity for deeper study.

Although space restrictions preclude a full investigation, this step of the RDA process may be summarized as identifying the mutually exclusive dimensions of CT as:

1. Predictability;
2. Emergence;
3. Fitness;
4. Systems;
5. Action;
6. Information;
7. Goals;
8. Strategies, and;
9. Time.

The fifth step of RDA is to define each categorical cluster as a dimension. That means the category is understood as a dimensional representation of the relevant data and/or observations. So, for example, in observing a firm, a consultant might characterize that firm as having greater or lesser Fitness. This may create perceptual difficulties in the understanding of systems. If we have a dimensional representation of systems, do we say the dimension is said to represent "more" and "fewer" systems or does the dimension represent a single system that is "larger" and "smaller?" The most general answer may prove, in the long run, to be the most useful. That is to say for the present model, what is really being represented is not so much any existing conceptualization of systems, rather a broader sense of systems. That, as yet undefined, sense of "system-ness" is the unexpected result of combining the perspectives of multiple theorists. For the purpose of the present analysis, we may hold such a dimension to represent some sense of systems that is, as yet, undetermined. The explication must be left to a future study.

In the final step of the RDA process, the interrelationships between all dimensions are investigated to identify the relationships between them—what we have been calling "concenated." In this step, the concepts and propositions within each dimensional category may be "translated" to better match the shared terminology of the dimensions. The intent of this process is not to alter the meaning of the original authors, rather it is to identify how dimensions might be related to one another as they exist within a shared body of theory and so suggest paths toward greater testability and falsifiability.

The statements of the eight authors suggest the interrelationships diagrammed in Figure 1. In this figure, each arrow represents a causal direction and each box represents a scalar dimension. For example, more Information results in more Fitness and more Predictability. This creates a richer, and we hope, more accurate view of the aspects of complexity theory than could have been obtained by any analysis of the atomistic concepts, or the linear propositions of any one author.

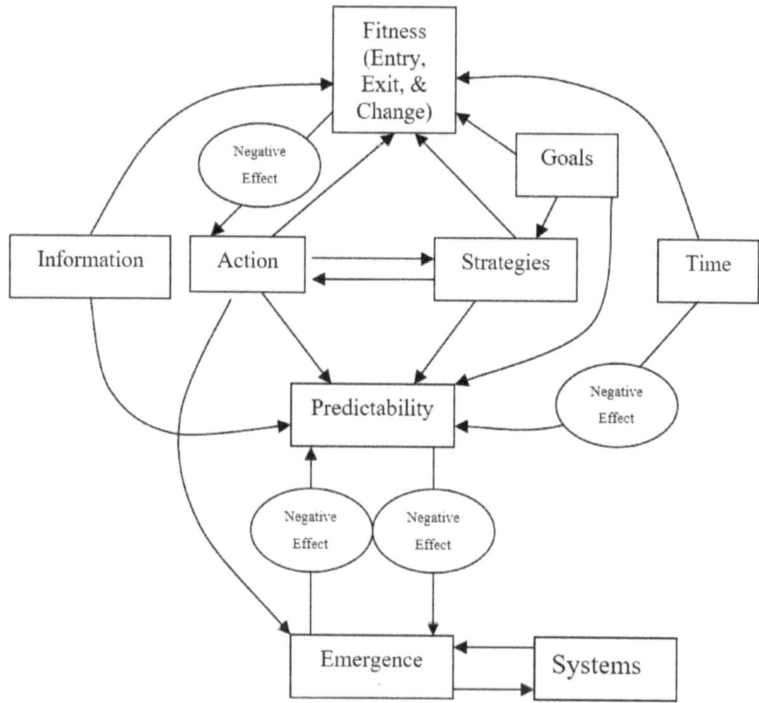

Figure 1 *Causal relationships between aspects of complexity theory*

This model may (or may not) be an accurate representation of the "real world." However, such is not the goal of the present analysis. This model is only intended to be a representation of CT as it is found in the realm of OT. This clarification of the core of CT may serve as a solid foundation for scholarly conversation around the nature of that theory.

From Figure 1, it may be seen that there are nine aspects within CT. Of these nine, there are five that may be understood as concatenated. That is to say, changes in each of those five aspects may be understood as resulting from changes in two or more other aspects. Therefore, the robustness of this model is 0.56—the result of five (concatenated aspects) divided by nine (total aspects). This measure of robustness may serve as a benchmark for the advancement of CT.

As the atomistic version of CT from the present study #1 has a robustness of zero (because it is a list), and the RDA version of CT has a higher robustness than the atomistic versions of CT, the RDA version may be seen as an improvement and a significant step toward robustness, and the present paper serves as an example of advancing theory towards a robust form. Further, the closely related nature of the included aspects suggests that this model may be a good representative for the core of complexity theory.

Conversation

With Figure 1, two challenges become immediately apparent. First, there are three aspects that appear to be strictly causal (Time, Information, and Goals). As such, these aspects are essentially undefined from the view of the model. Where more Strategies may be enabled by increased in Goals and Actions, nothing causes an increase in Goals. They seem to simply exist.

In order to improve the robustness of the CT model, Time, Information, and Goals should be described as concatenated in relation to two or more causal aspects. If those aspects may be legitimately found among the existing aspects of the present model, so much the better. We may retain some level of parsimony.

The second clear challenge to this model is that Systems seems to be both caused by, and be causal to Emergence, and nothing else. This relationship seems to suggest that the concept of Emergence and Systems may be synonymous. If they do indeed represent the same thing, than one aspect may be removed, which will increase the overall robustness of the model. If they are different, that difference must be described in terns of the causal relationships between those aspects and other aspects.

When those two challenges are met, the CT model may achieve full robustness. And, as such, may be more readily falsified, and so more rapidly improved.

Practice

CT is often applied for organizational change efforts as a vague metaphor that is useful for stimulating thought and conversation (Fuller & Moran, 2000). As such, it may be seen that every aspect of the metaphor is indeterminate. The usefulness depends on what aspects the client finds interesting. As a result, the client understands only a portion of the metaphor and does not get a picture of the whole system.

As the concept of non-linear dynamics is an important aspect of CT (e.g., Dent & Holt, 2001; Lichtenstein, 2000b), it may be worth noting that the relationship between the co-causal aspects of this model. For example, the level of Predictability might be completely deterministic if one possesses knowledge of the causal aspects (Information, Action, Strategies, Goals, Time, Emergence). If one of these is unknown, Predictability becomes less determinate. Having a model where a consultant understands which aspects of the model are more determinate and which might be less so, suggests that this kind of model may be developed into a useful tool for practitioners.

Consultants might provide a robust model, and help the client organization to "fill in the blanks." Such a process would provide a roadmap for the client that would suggest different areas for exploration. This way, unknown aspects may be inferred from known aspects. As every client has limited resources, it behooves the consultant to help the client apply those resources in the most effective way possible. If the client wants to gain a CT perspective of its situation using the present model, we might imagine that the client has more knowledge of some aspects than others.

Let us say that the client wants to gain a CT based understanding of their Fitness in an uncertain market. Let us also say, that they are well aware of their Actions, Strategies, and Goals. According to the present model, they might, instead of investigating Fitness directly, choose in investigate Information. This is because Information might be more easily investigated than Fitness, and because by knowing those five aspects, Fitness may be inferred.

Wholeness of analyses

A theory of low robustness seems to suggest the opportunity for authors to choose which concepts they use as from a menu. This may be thought of as cognitive prejudice, attentional bias, or simply a case of the author not explicating the reasons for choosing those particular concepts. In contrast, a robust model suggests the need to include all aspects of the model in each investigation because each aspect is clearly connected to the other. In other words, a robust theory may be effective in removing the blind spots from analysis. Essentially, when using a theory as a whole model, researchers would be impelled to include descriptions of each part of the model.

The dimensional relationships presented in this model suggest new requirements for the study of CT. For example, if organizational Fitness may be seen as caused by five other dimensions (Goals, Information, Action, Strategy, and Time), a study of Fitness should be reducible to those five dimensions and no others. If it is determined that this is not the case, the theory will be successfully falsified and should be modified appropriately.

Although this form of study would be necessarily more difficult, it is certainly within the realm of expertise of most contributing scholars. Importantly, by the efforts of authors, reviewers and editors, such an effort has the potential to rapidly accelerate the advancement of theory for the benefit of scholars, practitioners, and society at large.

Summary and Conclusion

There are two primary contributions of this paper, one theoretical, the other metatheoretical. The metatheoretical contribution of this paper includes arguments that link the structure of theory with the potential efficacy of the theory. The theoretical contribution of this paper is seen in the advancement of CT from a nonrobust model towards a level of greater robustness.

Where previously a reader might have believed that authors conversing around the topic of CT are describing the same thing, the present paper has suggested that is not the case.

In the present paper, I've presented two analyses each drawing on the same eight versions of CT. The first study focused on the 47 conceptual components found in those eight versions. This study found more conflict than commonality between authors. Further, this study found that approaching the whole body of CT from various perspectives resulted in more versions of the theory, but did not seem useful in identifying the core of CT.

In the second study, the focus shifted to the relationship between concepts based on the idea that more complex relationships suggest a higher level of structure—and therefore the opportunity for greater efficacy in theory. The second study used RDA to shift linear propositions of CT into concatenated propositions with higher internal integrity. In this study, I found that the robustness of CT is 0.56 on a scale of zero to one.

In a sense, it may be said that the core must be found from "within" the theory, rather than from any particular point of view from "outside" the theory. The belt, then, may be understood to consist of the 47 concepts that are relatively linear or atomistic, and so less concatenated than those at the core. It is suggested in the second study that the core of CT consists of those concepts that are concatenated. The present study suggests that the core of CT include five aspects (Action, Fitness, Strategies, Predictability, and Emergence), and, importantly, the specific causal relationships between them.

It may be seen by this analysis that a dimensional representation is a more useful tool for analysts than atomistic representations because an atomistic report might say simply whether a firm is fit or not, while a dimensional report could relate a level of fitness, or reflect how one firm were may be more fit than another. Thus, in a sense, a dimension may be said to "contain" (or support the communication of) more information than its atomistic counterpart. Thus, it may be said, that the process of "dimensionalizing" the information supports further investigation in to CT because it allows and encourages theorists to quantify their observations and insights.

In this same way, the dimensional perspective supports observations and analysis for practitioners. For example, a consultant facilitating a conversation among executives might encourage an organizational analysis from a CT perspective by asking atomistically, "Do you see emergence occurring in this firm?" The answer, according to the definitions explored here, should be yes (otherwise, the firm would not exist). In comparison, a more generative conversation would be result if the facilitator were to ask dimensionally, "How *much* emergence do you see in this firm now compared to last year (or compared to our competitors or our clients)?" The atomistic question might lead to a check mark on the flipchart paper, while the dimensional question will more likely lead to conversation, comparative analysis, and action.

While the present study suggests lines of investigation for advancing CT toward robustness, there are also limitations. Most notably, the sample size of eight concise versions of theory might not be considered statistically significant. The validity and usefulness of this line of research may be improved by repeating the study with a larger sample.

Second, it is assumed by the present author that this method of analysis is repeatable—that if performed by others, it would lead to substantially similar results. That assumption should be tested.

Finally, in this paper, I deepen the conversation on structure of theory to suggest that there may be a correlation between the robustness of a theory, and the efficacy of that theory in application. However, using robustness to measure of the potential of a theory for testability and applicability is itself untested. Further investigation will need to occur to test the a priori assertions developed in this paper—perhaps through innovative modeling techniques.

Within academia, it has been generally understood that theory is advancing due to some natural evolution of ideas emerging from the back-and-forth between scholars. Applying a measure of robustness to these theories suggests an objective method for testing that assumption.

Conversations on the structure and construction of theory seem likely to continue and increase as the general understanding of theory increases. It remains an open question, for example, whether the same level of conflict exists in other branches of CT. Measures of robustness, inspired by insights from CT, will provide useful tools for advancing that conversation.

Bibliography

Axelrod, R. and Cohen, M. D. (2000). *Harnessing Complexity: Organizational Implications of a Scientific Frontier*, ISBN 978-0465005505.

Brown, S. L. and Eisenhardt, K. M. (1998). *Competing on the Edge: Strategy as Structured Chaos*, ISBN 978-0875847542.

Dagnino, G. B. (2004). Complex systems as key drivers for the emergence of a resource - and capability - based interorganizational network. *Emergence: Complexity and Organization*, ISSN 1521-3250, 6(1-2): 61-69.

Dent, E. B. and Holt, C. G. (2001). CAS in war, bureaucratic machine in peace: The U.S. Air Force example. *Emergence*, ISSN 1521-3250, 3(3): 90-107.

Dictionary. (1993). *Random House Unabridged Dictionary* (2 ed.), ISBN 0-679-42917-4.

Dubin, R. (1978). *Theory Building* (Revised Edition ed.), ISBN 0-02-907620-X.

Frederick, W. C. (1998). Creatures, corporations, communities, chaos, complexity. *Business and Society*, ISSN 0045-3609, 37(4): 358-390.

Fuller, T. and Moran, P. (2000). Moving beyond metaphor. *Emergence*, ISSN 1521-3250, 2(1): 50-71.

Harder, J., Robertson, P. J. and Woodward, H. (2004). The spirit of the new workplace: Breathing life into organizations. *Organization Development Journal*, ISSN 0889-6402, 22(2): 79-103.

Hjørland, B. (2002). Domain analysis in information science: Eleven approaches - traditional as well as innovative. *Journal of Documentation*, ISSN 0022-0418, 58(4): 422-462.

Hood, W. W. and Wilson, C. S. (2002). Analysis of the fuzzy set literature using phrases. *Scientometrics*, ISSN 0138-9130, 54(1): 103-118.

Hurtado, P. (2006). Will the real complexity in strategic management please stand up? *Competition Forum*, ISSN 1545-2581, 4(1): 175-182.

Kakabadse, N. K., Kouzim, A. and Kakabadse, A. (2001). From tacit knowledge to knowledge management: Leveraging invisible assets. *Knowledge and Process Management*, ISSN 1099-1441, 8(3): 137-154.

Kaplan, A. (1964). *The Conduct of Inquiry: Methodology for Behavioral Science*, ISBN 0765804484.

Kernick, D. (2006). Wanted--new methodologies for health service research. Is complexity theory the answer? *Family Practice*, ISSN 23: 385-390.

Lakatos, I. (1970). Falsification and the Methodology of Scientific Research Programmes. In I. Lakatos and A. Musgrave (eds.), *Criticism and the Growth of Knowledge*: 91-195. ISBN

Leydesdorff, L. (2002). The communication turn in the theory of social systems. *Systems Research and Behavioral Science*, ISSN, 19(2): 129-136.

Lichtenstein, B. B. (2000a). The matrix of complexity: A multi-disciplinary approach for studying emergence in coevolution. In A. Lewin and H. Voldberda (eds.), *Mobilizing the Self-Renewing Organization: The Coevolution Advantage*. ISBN 9058920488.

Lichtenstein, B. M. B. (2000b). Emergence as a process of self-organizing: New assumptions and insights from the study of non-linear dynamic systems. *Journal of Organizational Change Management*, ISSN, 13(6): 526-544.

Lissack, M. R. (1999). Complexity: the Science, its Vocabulary, and its Relation to Organizations. *Emergence*, ISSN 1521-3250, 1(1).

Pascale, R. T. (1999). Surfing the edge of chaos. *Sloan Management Review*, ISSN 1532-9194, 40(3): 83.

Pentland, B. T. (1999). Building process theory with narrative: From description to explanation. *Academy of Management Review*, ISSN 03637425, 24(4): 711-724.

Popper, K. (2002). *The Logic of Scientific Discovery* (J. F. Karl Popper, Lan Freed, Trans.), ISBN 0-415-27844-9.

Romm, N. R. A. (2001). *Accountability in Social Research: Issues and Debates*, ISBN 9780306465642.

Stacey, R. D., Griffin, D. and Shaw, P. (2000). *Complexity and Management: Fad or Radical Challenge to Systems Thinking*, ISBN 0415247616.

Van de Ven, A. H. (2007). *Engaged Scholarship: A Guide for Organizational and Social Research*, ISBN 978-0-19-922630-6.

van Dijkum, C. (1997). From cybernetics to the science of complexity. *Kybernetes*, ISSN 0368-492X, 26(6/7): 725.

Wallis, S. E. (2006a). *A sideways look at systems: Identifying sub-systemic dimensions as a technique for avoiding an hierarchical perspective*. Paper presented at the International Society for the Systems Sciences, Rohnert Park, California. 0-9740735-7-1.

Wallis, S. E. (2006b). *A Study of Complex Adaptive Systems as Defined by Organizational Scholar-Practitioners*. Unpublished Theoretical Dissertation, Fielding Graduate University, Santa Barbara.

Wallis, S. E. (2008). From Reductive to Robust: Seeking the Core of Complex Adaptive Systems Theory. In A. Yang and Y. Shan (Eds.), *Intelligent Complex Adaptive Systems*. ISBN 978-1-59904-717-1.

Weick, K. E. (1989). Theory construction as disciplined imagination. *Academy of Management Review*, ISSN 03637425, 14(4): 516-531.

Glossary

- ◊ **Aspect** A broad term relating to the part of theory that represents a concept, idea, or other representation of something that is used in a model. May be represented by a scalar, atomistic concept.
- ◊ **Atomistic** Refers to any reductive proposition representation such as "A is valid."
- ◊ **Concatenated** More complex relationships are found in concatenated proposition. In this, I use the sense expressed by Van de Ven (2007) where two aspects of a theory are shown to influence the third aspect. For an abstract example, a concatenated proposition might state that changes in aspect A and aspect B will cause changes in aspect C.
- ◊ **Dimension or scalar dimension** An aspect of theory that represents quantitative or qualitative variations. For example, a dimension of "size" might be used to represent whether a system is smaller or larger.
- ◊ **Linear** Refers to propositions that describe causal relationship between two concepts. Such as, "Changes in A cause changes in B."
- ◊ **Proposition** "A proposition is a declarative sentence expressing a relationship among some terms." (Van de Ven, 2007: 117).
- ◊ **Reflexive Dimensional Analysis (RDA)** Six step process for analyzing a body of theory and shifting the component propositions of that theory from linear to concatenated.
- ◊ **Static Robustness** Ratio describing the interrelatedness between aspects of a theory on a scale of zero to one. Robustness is calculated by dividing the number of concatenated aspects by the total number of aspects in a theory.
- ◊ **Theory** An ordered set of assertions. Weick (1989: 517. Drawing on Southerland).

Chapter 5
Toward a Science of Metatheory

Wallis, S. E. (2010). Toward a science of metatheory. Integral Review, 6(3 - Special Issue: "Emerging Perspectives of Metatheory and Theory"), 73-120.

My greatest appreciation goes out to Mark Edwards for his many suggestions and support in the development of this article.

Abstract: *In this article, I explore the field of metatheory with two goals. My first goal is to present a clear understanding of what metatheory "is" based on a collection of over twenty definitions of the term. My second goal is to present a preliminary investigation into how metatheory might be understood as a science. From that perspective, I present some strengths and weaknesses of our field and suggest steps to make metatheory more rigorous, more scientific, and so make more of a contribution to the larger community of the social sciences.*

Social Science—Flapping Around in Circles?

In the integral community, as with other fields of the social sciences, many authors have worked diligently to create effective metatheories. Wilber and Laszlo, for example, have produced metatheories that are both interesting and popular. However, a storm of controversy surrounds such theories because the existing paradigm of the social sciences does not include effective conceptual tools for evaluating metatheory (or theory, for that matter). This is not the fault of the authors. Indeed, they should be lauded as brave explorers who are pushing the boundaries of human thought. The problem, and an important reason for the criticism of theory and metatheory, is that there is no generally recognized scientific framework by which we might judge them.

What is this thing called metatheory, and how might it be useful? This is a question that has been alternatively surfaced, ignored, considered, derided, and viewed with amazement around the world of the social sciences. Despite (or, perhaps, because of) these varied reactions, there has been little effort to create a coherent and comprehensive science of metatheory. The effort has not been made, perhaps because there appears to be no need for such an effort. Perhaps those in the social sciences are satisfied with their existing views of theory. More likely, the conceptual "building blocks" were not in place to support the creation of a new science.

As to the question of "need," we now face a seemingly endless array of local and global issues from psychology to policy. It is becoming increasingly clear that the challenge of those issues cannot be met with existing theory. That is at least one reason why we continue to develop new theory. Unless, of course, the bulk of academicians are insulated from these issues—which I think and hope is not the case! Because the development of more effective theory is so important, a critical question for the coming decade is, "How might we advance research and practice more rapidly to more effectively address the important issues of our time?"

The traditional approach of the social sciences is to apply expensive resources to the problem (time, money, researchers, facilities). This approach has generated thousands of theories and untold quantities of data. We have so many theories and so many data that it has become difficult to organize knowledge or resolve basic contentions within a field. That befuddling excess of information can be traced to the origin of the social sciences. There, in an attempt to mimic the success of the physical sciences, social scientists were encouraged to seek objective facts through empirical research. A goal that has continued to the present day with calls for still more empirical research.

What we seem to have overlooked is that a science, much like a bird, must have two strong wings if it is to fly. Empirical facts are understood in the light of a specific theory. And, theory can is built from a careful consideration of the facts. Indeed, theory and research are so intertwined that neither can exist without the other—and both are needed for a strong science. Theory and research are differentiable, but they are also inseparable—existing in a generative yin-yang relationship. While this is a generally accepted truism of science, the actual practice of science is rather different. Under the existing paradigm, a scholar presenting a research proposal is required to describe a specific and rigorous research methodology. However, that same scholar need present little or no justification for the theory. Indeed, some respected scholars have suggested that theory should be developed intuitively.

There is the dichotomy that has hobbled the development of the social sciences. In contrast to the well-developed wing of research, the wing of theory is withered. We have made a habit of building and testing theory without a specific methodology, essentially relying on subjective intuition. As a result, we lack a solid metatheoretical perspective, we lack a rigorous and repeatable methodology for comparing research proposals on the basis of their theories. We are left to consider the so-called facts of the situation without the benefit of a carefully developed and commonly understood context. This points to a vast gap in the literature; and, indeed, the science. Because theory and practice are (or should be) closely interrelated, the advancement as a science may be limited by the limitations of either. The science cannot advance without both.

Defining a science of metatheory is one important step. When that is done, scholars will be to seek and create theories that are more advanced—theories that can be tested according to the most rigorous and repeatable standards.

While much has been written on theory the results do not seem to have enabled the success of the social sciences. Few of those methods require any degree of rigor. For example, the most commonly used claim of validity for a presented theory is parsimony, but even that simple indicator is typically misunderstood and misapplied. Further, by way of a thought experiment, if more parsimony led to better theory, we might conclude that the smallest theory would be the best of all—a difficult conclusion indeed. Based on this understanding of the necessary interrelationship between theory and research, it is reasonable to expect that the development of rigorous methodologies for developing and testing theory and metatheory will bring balance to the social sciences. This, in turn will accelerate the advancement of that science and aid in resolving the issues that face our world.

In short, there exists in the social science, a silent prejudice. Unrecognized and unabated, it eats at the heart of our science, destroying its legitimacy and impeding its advance. To create a more effective social science, we must first develop more mature, and more scientific approaches to the study of metatheory. In this, I refer to science as a process of investigation, replication, communication and the emergence of new and useful meanings. Not, as some think, of science as a complete set of natural laws where meanings are fixed.

A parallel concern is the ongoing struggle between modernism and postmodernism. Both are forms of metatheory that shape the way we see the world. Briefly, and partially, traditional modernity emphasizes differentiation, simple-location, classification, and representation. While, in contrast, postmodernism is more about process, movement, interpretation, and change. The difference makes for interesting dialog, both inside each camp, as well as between them. Yet, drawing lines between the two is a modern approach and dialog is a postmodern approach. Both have something to offer the present paper that will help us to understand and advance scientific metatheory.

The potential benefits of advancing the science of metatheory are profound. First, by developing new metatheory, we gain the ability to become more effective in the application of theory for the alleviation of social ills and the optimization of the human condition. Second, and closely related, by working from a metatheoretical perspective, we gain the opportunity to understand and integrate theories across disciplinary boundaries. The insights and capabilities to be gained, for example, by linking human development and public policy are quite intriguing.

In first part of this article, I will review existing definitions of metatheory in hopes of finding a more complete, more useful, and less pejorative definition. In the second part of this article, I will present an outline for a scientific study of metatheory based on an authoritative, modern definition as a step toward legitimizing the study of metatheory, and the subsequent development of more effective theory. Third, I will present a postmodern view of science, combining and reflecting on multiple views of what it means to be a science. Fourth, I will combine the two in an integral approach that avoids the main concerns of each camp and combines the best of both. And, importantly, what seems to offer the most workable directions for a science of metatheory.

Many Understandings of Metatheory

The most basic problem is the lack of a clear definition of metatheorizing. Because of it, we have what appears a series of isolated works, isolated sub areas within types of metatheorizing, and isolated types of metatheorizing. Metatheorists often feel defensive about what they are doing because they lack a clearly defined intellectual base from which to respond to the critics. Thus what stand out are the criticisms that often go unanswered.

In a search of the literature, Mark Edwards (with some support from the present author) found over 20 definitions of metatheory. While this search was not comprehensive, we believe that we have identified a good cross section of the field. One of the first things I noticed in reviewing these definitions is that the authors seemed to be describing different things. While some were describing the broader field of metatheory, others were describing related, yet subordinate, areas of metatheory such as the evaluation or categorization of metatheory. In this section, I will provide an overview of the field, based on the discovered definitions.

Broadly, metatheorizing is the process of performing metatheoretical research. That process includes many different kinds of activities including sorting theories and/or sorting their components into categories, the use of reflexivity, deconstruction and reconstruction, statistical analyses of the literature, and nearly 20 other suggestions. Just as theorizing results in the creation of theory, metatheorizing results in a "metatheorum" which is a statement about theory in general or a statement about a specific theory.

Abrams and Hogg take an alternative, primarily metaphorical, approach to describing metatheory. They suggest, for example, that a metatheory makes a good travel guide. Sklair describes a metatheorum as reflecting the coherence between epistemology and objects of knowledge. This coherence extends to include a set of assumptions about the constituent parts of the world, and our possibility of knowing them. Overton, in contrast, says that metatheorems define the context in which the theories are made, and refer to the theories, themselves. He also goes on to say that a metatheory is a set of interlocking principles that describe what is acceptable and unacceptable for theory.

Broadly, the field of metatheory includes the study of the "sources and assumptions; and contexts", including the study of theorists and communities of theorists, the process of theorizing, and the analysis of the methods, findings, and conclusions of the research the use of those theories as well as their implications. Finally, the prescription that metatheorizing should produce theories that are open to empirical testing.

Although they all address the same general theme, these definitions of metatheory contain more differences than similarities. Such disparity makes it difficult to draw useful conclusions about any shared definition of metatheorizing or metatheorems. Also, these descriptions are most often very sketchy and fragmentary; indeed, they are almost entirely "atomistic." That is to say, most authors simply created a list of concepts that they considered important to the study of metatheory (e.g., metatheory is A, B, C, & D). Such an approach may be expected given the relative youth of the field. In contrast, more mature fields tend to contain concepts that are more closely integrated (e.g., A causes B and C, or A, B, & C are co-causal).

As the reader may notice, this broad range of definitions of metatheory overlaps with other areas of study. That overlap may cause confusion and concern (as well as the opportunity for additional investigation and clarification). For the present article, however, I will focus on what might be considered the core of metatheory—without which, none of the other parts would be possible. This analysis begins by identifying the area of greatest agreement between the many definitions of metatheory.

In Table 1 I list the authors of 21 definitions of metatheory and their key ideas. The more complete definitions are listed in the appendix. From this arrangement of the data, a few important conclusions may be drawn. There are also contradictions, which may prove troubling or prove to be a source of new insight.

First, and most simply, metatheory is focused on the analysis of theories. Ten of our 21 authors specifically agree on this point. In

Author/s	Date	Analysis of theories	Is (or creates) a theory of theory	Integrates multiple theories	Analysis of assumptions	Makes implicit assumptions explicit	Analysis of underlying structure	Analysis of structure of theory	Deconstructive
Finfgeld	2003	x			x				
Bondas & Hall	2007	x							
Dervin	1999					x			x
Anchin	2008			x					
Sklair	1988								
Faust	2005	x							
Clarke	2008				x				
Fuchs	1991						x		
Weinstein & Weinstein	1991	x							
Zhao	2004	x							
Craig	2009		x						
Takla & Pape	1985				x				
Thorne, et al.	2004								
Paterson, et al.	2001	x							
Ritzer	1988	x		x					x
Gadomski	2001		x						
Wikipedia	2009		x						
Turner	1990							x	
Faust & Meehl	2002	x							
Bonsu	1998	x							
Colomy	1991	x							

Table 1 *Aspects of metatheory as described by various authors*

contrast to this general agreement around the core of metatheory, few authors describe metatheory as the analysis of context. Therefore, we may reasonably conclude that studies of theory are at the core of metatheory, while other studies (e.g., contexts, groups of theorists, methods, applications, etc.) are less so. Second, it may be noted that six of our authors describe metatheory as making implicit assumptions explicit, analysis of assumptions, analysis of underlying structure, and the analysis of structure. These are essentially deconstructive approaches.

In contrast to this deconstructive approach, metatheory may also be understood to integrate multiple theories. The two approaches may be inseparable as one cannot combine integrate two theories without also integrating the assumptions, structures, and concepts of those theories. In short, metatheory (as the study of theory) may be conducted in at least two ways. It may be integrative (where multiple theories are combined). It may be deconstructive (where theories are parsed into their constituent components for analysis and/or recombination). Either way, the process leads to the creation of a meta-theory, meta-theorem, or "a theory of theory".

To summarize, the field of metatheory is richer and more complex than is expressed by any single definition to date. The following definition strives for conciseness, rather than completeness as the main focus of this article is the development of metatheory as a science. So, this definition (a simple, though useful, prelude) suggests that:

> *Metatheory is primarily the study of theory, including the development of overarching combinations of theory, as well as the development and application of theorems for analysis that reveal underlying assumptions about theory and theorizing.*

This definition, it should be emphasized, reflects the state of understanding of metatheory to date. And, as noted above, this level of understanding has not proved sufficient for advancing theory or metatheory—perhaps because the terms are too fuzzy. In the process of creating this definition, however, a potentially clarifying co-causal relationship begins to emerge. Here, the analysis, evaluation, and integration of theory leads to the development of new methods of evaluation, which may then be used to evaluate the results. Far from a useless tautology, this suggests a path for the development of some conceptual constructs through the development of other conceptual constructs. By explicating this developmental process we may accelerate the development of such constructs. This occurs in much the same way that one may use a map to more easily locate a store for buying maps. Or, the way one may use telecommunication to collaborate on the creation of more effective telecommunication systems. These are positive feedback loops, rather than useless tautologies.

In the above section, I have briefly summarized the field of metatheory. However, a larger question lurks. It is generally understood that the reason for metatheory is the development of better theory. Indeed, many or all of the same "rules" that apply to the development of theory also apply to the development of metatheory. Because the two areas share a common heritage, they may also share a common failing. Who is to say, for example, that our failure to make effective theory will not be replicated—leading to our failure to make effective metatheory? To avoid this trap, we must take metatheory to another level.

Where the creation of theory is understood to occur within a specific field of study, we might take a different view of metatheory and investigate metatheory as a field unto itself. However if the field of metatheory is to have any some degree of respect and usefulness, it must be understood as a legitimate science. As a science, we can expect metatheory to make advances and develop insights that may be applied to theories across the sciences. In the next section, I will investigate the idea of metatheory as a science from a modernist perspective.

Towards a Scientific Metatheory

Many voices have called for the advancement of social theory. To date, however, the methods for the scientific study and advancement of theory have been ineffective. For example, Meehl reports on 11 known methods for advancing theory. However, he finds them to be primarily intuitive—and so unsuited for scientific investigation. Indeed, a recent study by HERA concluded that there was no method of evaluation that could be counted on to work effectively. To achieve greater effectiveness, we must reduce the epistemological indeterminacy around metatheory—we must learn to represent more clearly our knowledge about the similarities and differences between metatheories. In this, the ideas presented in this paper support the many excellent suggestions of Tom Murray.

In this section, and the following sections, we will investigate how we might study theory and metatheory as a science unto itself. An important expectation is that following scientific methods will enable the creation of effective metatheory, effective theory, and more effective methods and practices. There are many understandings of what science is. So, in order to develop a deep and nuanced view, I will investigate the science of metatheory from multiple perspectives. In this section, I will investigate scientific metatheory from the modernist perspective. In the following sections, I will investigate scientific metatheory from a postmodern perspective. Then, I will discuss a combined perspective of a scientific metatheory.

To frame a modernist perspective of science, I begin with a recent and concise definition where, "Science is the pursuit of knowledge and understanding of the natural and social world following a systematic methodology based on evidence" ; which includes:

1. Objective observation: Measurement and data (possibly although not necessarily using mathematics as a tool);
2. Evidence;
3. Experiment and/or observation as benchmarks for testing hypotheses;
4. Induction: reasoning to establish general rules or conclusions drawn from facts or examples;
5. Repetition;
6. Critical analysis, and;
7. Verification and testing: critical exposure to scrutiny, peer review and assessment.

In this section, I will briefly review what is to have a modernist science, and explain how those standards may be applied to metatheory.

A—Objective observation

Observation includes the description (and often measurement) of subjective, objective, and relational experience. Yet, in the social sciences, where two people might observe the same phenomena and arrive at dissimilar conclusions, the idea of objective observations becomes problematic. Because "There are no facts independent of our theories" and "no observation may be totally theory-free" and, "no two scientists are ever in total agreement with each another". This situation has made advancements in the social sciences incredibly difficult and led to the fragmentation of many (perhaps all) fields of study.

In a post-positivist approach, with its attendant assumptions of uncertainty, there are great benefits associated with triangulation—combining multiple views to create a more effective understanding of our lived world. Within a post-positivist world, there is little room for so-called "objectivity." Yet, the word still caries weight. And, in the present paper, it may be understood as representing an approximately repeatable experience within a context of metatheoretical investigation. That is to say, if two researchers of similar education undertake an investigation of the same theories using the same research methods, they should reach the same result. The extent to which their results are similar is the extent to which the approach may be considered objective.

In contrast to the more general range of social sciences, the field of metatheory has an advantage in this aspect because metatheory is a primarily hermeneutic approach where, "Hermeneutics is the interpretation of texts". Thus, we meta-theoreticians have access to books and journals that are (at least for a time) stable. That means, where many social events come and go in the blink of an eye, forcing researchers and their human subjects to rely on memory, we have the opportunity to revisit the texts with all the care and patience needed to fully understand the authors. In short, we are less subject to the transience of our observed subjects. Another benefit to the metatheoretical approach is completeness. That is, we have the possibility to analyze entire "worlds," where other social scientists must make do with a small fraction. For example, if we wanted to create a metatheory based on all the works of Durkheim, we could (conceivably) collect and analyze his complete writings. In contrast, a social scientist who wanted to develop a new theory of child development could never observe all the children in the world.

In the scientific process of metatheory, the subject of analysis is primarily theory. The process of investigating theory, in part, involves the process of deconstructing existing theories. Importantly, we can share those texts with other scholars to determine if we share the same viewpoints. That means we can begin to develop some degree of objectivity based on the similarity between our observations and interpretations of texts. In short, in the study of metatheory, we have the opportunity to be a more effective science than most fields of social theory because (in the area of objective observation) we suffer fewer effects of transience, easier opportunity for completeness, and greater opportunity for objectivity.

It is critical, however, to note that the opportunity to achieve these aims is not enough to make metatheory into a science. We must show through our work that our efforts have reduced transience, increased completeness, and increased objectivity. That, in turn, suggests that our metatheoretical analyses should include a large percentage of the available writings in a given area of study to exemplify completeness. Similarly our analyses should be collaborative efforts to exemplify objectivity.

B—Evidence

The collection of evidence is an important part of any scientific endeavor. Specifically, this refers to evidence that may be used to support (or contradict) one's theory (or metatheory, in our field of study). Without the anchor of recorded evidence, the metatheorists may be accused of blue-sky speculation. Because this evidence consists of theories, the collection process may be fairly simple. One needs only collect the theories and hold them against the requirement of future reference. Evidence should also be identified and saved on the "next level" of analysis. For example, if a meta-theorist has collected a number of theories and enumerated what percentage of those theories were actually tested, that too is evidence. And, the responsible scientist should keep that evidence available in case there is a challenge to the theory and a need to "show your work."

As noted above, in the pursuit of metatheory, that evidence will come from the analysis of other theories. The evidence may be theories, the propositions within theories, or other data that is directly related to the theories. While Ritzer seems to suggests that metatheory should be focused on analyzing theories of the "middle range," that seems to present an unnecessary limitation on the field of metatheory. There is no a priori reason why metatheory should not include the analysis of theories on all subjects—from subatomic particles through galaxies, including psychological, social, and economic theories along the way. Of course, in investigating such a range of theory it is important that the meta-theorist identify some form of similarity. For example, causal and co-causal propositions are found in theories of the natural and social sciences.

Whether the evidence is collected from a broad range across disciplines and sciences, or the evidence is collected within a very narrow field of study, an important question remains as to how much evidence is enough? This is an opportunity for future investigation and clarification. The tacitly accepted practice seems to be that one has enough evidence when one is able to successfully convince others of the validity of one's model. Yet, this is a weak standard as some people may be convinced with little evidence while others will not be convinced if they face all the evidence in the world. This concern leads to the need for testing.

C—Experiment and/or observation as benchmarks for testing hypotheses

Within the field of metatheory, two intertwined branches are emerging. One branch is primarily concerned with the creation of overarching theories, while the other branch is concerned with developing and applying rigorous and repeatable tests to theory. This is an important distinction to surface at the time because the hypotheses derived from each branch will be tested differently.

Metatheory testing

Van de Ven reflects on what might be the most common point of view for theory appraisal. Drawing on Weick and Thorngate, Van de Ven agrees that the creation of each theory requires a trade-off between simplicity, generality, and accuracy. For example, a theory might be made more accurate, but that would require the theory to become more complex. This approach has led many theorists to claim that their theories are valid because they are parsimonious. This is like claiming that something is good because it is tall—without a relative measure, the claim has no value.

By far, the strongest argument for testing theories comes from Popper. Popper argues with strength and conviction that the only effective way to test a theory is to falsify that theory in practice. That view is, of course, itself a form of theory. And, importantly, has not been falsified. Adopting a metatheoretical perspective, I noted how Popper was the product of a modern age. So, his search for modernistic "objective" knowledge seems to have blinded him to other forms of knowledge. By reframing Popper's own ideas, I suggested three paths for the validation and falsification of theory.

In the "world of facts," or data, a metatheory must (at the very least) be constructed of data—explicitly drawn from existing theory. A better metatheory would use data from multiple sources. The best metatheory is one where the theory may be applied to predict what future data will emerge. In the "world of meaning," the metatheory must (at a minimum) make sense to the author. A better metatheory would make sense to the editor, reviewers and readers. At the highest level, a theory is accepted in a consensus of expert opinions and that theory is preferred over other theory of the same domain. Finally, in the "world of theory," a metatheory must (at the very least) include logical arguments. A better theory is one that is constructed of specific propositions. And, the best metatheory is one that is constructed of carefully integrated, co-causal, propositions. The tests within any one of these worlds is certainly of limited efficacy. However, when combined, they create a potent standard for the evaluation of theory and metatheory.

A common argument for advancing the validity of a theory or metatheory is the claim that it "works" in practice. However, by itself, that is a weak claim because anything can be said to work. The deeper question become whether one theory or metatheory (and its derived hypotheses) works better than any other. For example, imagine an experiment where one prays for rain using a variety of techniques and direct those prayers to a variety of deities. I imagine that the results of such an experiment would not show much difference in rainfall. For another example, one might claim that the very popular management change process of Total Quality Management (TQM) is a valid theory because it is widely used, generally accepted, and seemingly successful. However, a more critical analysis find that the TQM fails more than 70% of the time. The more effective theory, or metatheory, is the one that can be shown to work better than another when applied within a specific context.

Rigorous evaluations should also be applied to those tests, and to the understanding of the context. It is not sufficient, for example, to claim that all theories are acceptable in some context and all theories exist in some context, so all theories are valid (a useless, non-generative, tautology). So, to conclude this section, it seems that we have the opportunity to identify, clarify, formalize, and apply rigorous methods for testing theory and metatheory.

Theory bounding

An additional approach related to theory building is the idea of theory "bounding." It is an essentially metatheoretical process to delineate the boundary of a theory. Generally, a theory is bounded by additional explanation of where that theory may be legitimately applied—sometimes called the "limiting values". For example, a scholar might make a theory about team development. And, because the research was based in the telecom industry, that theory might be bounded by a statement such as, "This theory only applies to teams working in the telecom industry."[1]

In bounding the science of metatheory, we can say that our realm of study is theory. Further, the theorems that apply to the realm of theory, as a whole, are the theorems that are most central to the science. It is possible, of course, to subdivide the field of study. One might, for example, develop theorems that apply only to theories of sociology, or theories of health care. Such approaches, while they may drill-in to greater depth within a specific area of study (e.g., theories of sociology), are also less central to the science of metatheory, itself.

A legitimate science must be bounded in terms of subject matter. Further, the theory developed through the study of that subject matter might only be legitimately applied to that subject matter. For example, one cannot study galaxies and expect the resultant theories to work very well when used to choose a color for bedroom curtains. This simple idea brings with it some powerful conclusions. For example, if one is attempting to develop a "theory of everything" (TOE) the implication is that one must study everything. The impossibility of this requirement suggests the impossibility of developing a legitimate TOE. It is certainly possible to develop a generalizable abstraction—where a theory developed in the study of one organization may be legitimately applied to another organization. It seems less likely that one can studying a drop of water, develop a theory, and apply that theory to understand human development. Where exactly one draws the line is a matter for continued investigation.

D—Induction

An important part of science is the building of theory. And, in the field of metatheory, the building of metatheory. As with the creation of theory, the process of metatheory creation is essentially one of induction—where a meta-theoretician identifies related propositions within two or more theories and integrates them to generate an overarching metatheory or derive a rule for the analysis of future theories. By way of context, the inductive process begins with an abductive experience, "a surprising observation or experience. This is what shatters our habit and motivates us to create a hypothesis that might resolve the anomaly". Also, the inductive process may be seen in contrast to the deductive process, which is related to the testing of theory.

Drawing on Weick, Van de Ven suggests that from the myriad hypotheses, one might find the best theory by subjecting those hypotheses to a variety of thought-trials. "The greater the number of diverse criteria applied to a conjecture, the higher the probability that those conjectures which are selected will result in good theory". While his approach may also prove useful, it also has a fatal flaw. Because, if one uses ten criteria, nine of them might be useless. In such a situation, the other nine that are not causally related to theory-improvement would overrule the one criterion that might result in a better theory. This is the same kind of problem faced by most people during an election campaign. One receives a flood of information, yet little of it has any true bearing on the qualifications of the candidates! Instead of developing a list of criteria, Van de Ven rests on the bench of "plausibility" Thus returning to a more or less intuitive form of evaluation. He extends that idea to link plausibility with conjectures that might be interesting to readers such as problems in society that need to be explained. This "appeal to the audience" is not a strong argument. Indeed, instead of leading to good theory, such an approach might lead authors to invent problems so that they may write persuasively to "explain" those problems and create supposedly "good" theories and metatheories.

This might be good literature but it is not good science—from the modernist perspective. And, as a result, the suggestion does not rise to the level of good metatheory. At this stage, however, there may be no way to avoid the rather abductive nature of the process. Instead, that moment where we think to ourselves—"That doesn't make sense!"—is the moment when we begin a sense-making process. With careful effort, we find that something makes sense that did not before. This is part of making implicit assumptions explicit; an important part of metatheory.

As the evidence of theory is collected, combined, analyzed, and insights begin to emerge, the results begin to resemble the integration of multiple theories—an important part of metatheory. Some tools of theory building include:

- Intuition;
- Observation;
- Abstraction;
- Creating propositions;
- Defining key terms;
- Describing the domain of application;
- Identifying the units of analysis;
- Sampling some appropriate and relevant set of theories;
- Analyzing the "data" using rigorous research methods, and;
- Identifying core constructs and their interrelationships.

In a detailed review of the Handbook of Sociological Theory, Treviño notes several forms of theory construction including: Jasso's suggestion to engage in voracious reading and use of mathematics, Carley's creation of computer simulations, Joas and Becket's synthesis of existing theory, Turner and Boyns' consolidation of theory within a grand analytical scheme, Bailey's reduction in detail to focus on concepts, Lopreato's call to discover general principles, and Lindenberg's method of decreasing abstraction. From a complexity theory based approach, Ostroff and Bowen encourage theorists to note the degree of stability and change that occurs at each level under analysis. Also, Morgeson and Hofmann provide guidelines for theory-creation. Their process involves delineating the structure, function, and outputs of a given system. However, this view of an organizational system seems to be more descriptive than theoretical and Dubin notes that theory is not merely describing, categorizing, or stating a hypothesis).

Sussman and Sussman claim that the golden rule of theory development is to make one's hunches explicit by writing them down. This, they contend, also includes an assumption that the reader will be able to see more easily whether the theory is likely to be effective, and show who will benefit. To facilitate this process, they suggest four criteria for theory building (that may also be used for theory testing).

1. The theory should be plausible—it should make sense.
2. There must be enough variables within the theory so that the theory is usable.
3. The theory should be testable.
4. There should be a heuristic value to the theory—it should be usable across multiple situations.

The rules that apply to theory building also apply to metatheory building because each metatheory is also a theory—albeit one whose domain is understood as the study of theory. Importantly, combining, and integrating, multiple theories is an inherently metatheoretical process. Some call the results theory, while others call the results metatheory. However, because all theories are built from previous theories (or, a combination of existing theories and new data), and no difference has been shown in application between social theory and social metatheory, the two may be considered synonymous. This indicates an important area of study for the field of metatheory. Because, if it can be proven in application that metatheories are more effective than theories the social sciences, as a whole, will have taken a giant lean forward

The existing state of metatheory building is, unfortunately, impoverished in regards to suggestions for building theory. One important source of information on how to perform metatheory building comes from the literature on the construction of middle-range theory. For example, Mintzberg encourages theory builders to use their intuition and to "be brave." This might, conceivably, lead to brave theories, but there is no way to test the bravery of a theory (or the theorist). Nor is there any suggestion that a brave theory might be more effective in practice. Therefore, the theorist should be very wary about accepting advice that leads to untestable theories. For those theories cannot be tested cannot be improved. The same advice applies to meta-theoreticians.

When building metatheory, the data used comes from the analysis of extant theory. Meta-theoreticians investigate existing theory and extract or acquire insights from that set of theories. For example, one may look at a set of theories and perhaps categorize them (e.g., complex or simple) and abductively identify a categorical rule (e.g., parsimonious theories are easier to test and complex theories cover more ground). From what might be called a "content metatheory" perspective, one may look at a set of theories—each containing a different set of concepts. One might conclude that all valid theories must include some (or all, depending on the rule) of that set of concepts. Van de Ven suggests that the process of abstraction, when correctly applied, can increase the generality of a theory (because a more abstract theory can be applied across a broader range of situations). Abstraction also makes it possible for a theory to become less complex while increasing accuracy.

While most suggestions for theory building are "one-shot" bits of advice (e.g., be creative), a more complex and rigorous approach is to transparently follow a particular methodology. One such methodology, designed specifically for metatheory is Reflexive Dimensional Analysis (RDA). RDA was derived from grounded theory, as well as dimensional analysis. Because it involves constructing new theory from the combination of existing theories, RDA is essentially a metatheoretical process. RDA proceeds according to the following steps:

1. Define a body of theory.
2. Investigate the literature to identify the concepts that define it.
3. Code the concepts to identify relevant components.
4. Clump the components into mutually exclusive categories.
5. Define each category as a dimension.

6. Investigate those dimensions through the literature, looking for robust relationships.

Examples of this form of theory construction may be found in benchmark studies of complexity theory, complex adaptive systems theory. Another approach that deserves some attention is the process of Grounded Theory (GT). GT is a rigorous process of theory creation representing a structured methodology for theory creation with good potential for use in the creation of metatheory. Importantly, GT could provide a clear link between the data (theory, in our field) and the resulting metatheory by moving the data through a set process that can be summed up as:

1. Coding the data;
2. Thematizing the data, and;
3. Finding relationships between the themes.

Grounded theory suggests the use of an "open coding" process. There, the researcher intuitively codes the data. In that sense, the first step might be seen as an a priori one. In contrast, in the process of thematizing the data, the research is to look at the coded data and identify common categories for grouping them. This process might be seen as essentially a posteriori in the sense that the researcher begins with data, and then identifies a relationship. Similarly, the third step of finding relationships between the themes suggests an a posteriori approach—albeit one that calls for more creativity and abduction.

The iterative interactions between research and data, as well as developing relationships between the themes suggests a generative approach that is reminiscent of the mélange of social construction. Also, each step of the process is documented for transparency. However, with grounded theory, the process is surfaced for clarification. One of many examples may be seen in the development of a complex (yet useful) integrated strategic planning framework for dynamic industries using Grounded Theory. It has yet to be shown if either grounded theory or RDA might be repeatable. Although, it seems that both methods seem like they would have a higher level of repeatability than (for example) intuitive methods of theory building.

To summarize the more general conversation on the construction of metatheory, it seems that metatheory has offered little that is new in terms of theory building. Our field has adopted calls for triangulation (framing this as the use of multiple theoretical lenses). Authors with a focus on metatheory have echoed calls for techniques such as imagination and discipline. And, they have (to a small extent) formalized existing approaches to theory building. This lack of rigorous methodology may account for some criticism of metatheory. Similarly, for the process of building metatheory, I have only seen scholars use the same tools as have been used for building theory. Therefore, in looking at metatheory, theorists are able to see and criticize in our work what they do not see in their own. That is, a lack of methodological rigor. This lack of rigor may be exemplified as the lack of repeatability in the process of theory creation.[2]

It may be impossible to develop repeatable experiments for theory creation because the origins of theory seem shrouded in the mists of imagination. Who is to say exactly where Newton's ideas for laws of motion arose? In this, I do not refer to some apocryphal apple, rather the formal mathematical relationships. The inspirations and ideas are lost in moments long past. In the study of the creation of metatheory, on the other hand, explicit theories and metatheories remain for study. This may indicate a more fruitful direction for the advancement of a scientific field of metatheory—which will be explored in the following sections.

E—Repetition

Repeatability has long been a mainstay of science. For a negative example, about twenty years ago, two physicists claimed that they had achieved nuclear fusion at room temperature (instead of the millions of degrees of temperature required by generally accepted theories of physics). They shared their news with the world, and many scientists eagerly repeated their experiment. However, the same results were not forthcoming. This attempt at repetition allowed the world to quickly determine what worked and what did not.

Innovation, as with theorizing, is also important to the process of metatheorizing. Used by itself, however, it may take on the appearance of blue-sky speculation. In a science, we balance the trend toward speculation by applying the more rigorous standard of repetition. Applying the process of repetition to scientific metatheory, we might expect (from a modernist perspective) that two scholars who begin with the same source data and apply the same techniques for metatheory building, while working to address the same topics, should develop the same metatheory. However, to the best of my knowledge, this kind of study has never been attempted. This lack of repeatable studies suggests a great opportunity for metatheoretical study.

For example, a teacher or researcher could conduct a classroom exercise by providing a set of theories to three (or more) groups of students—with the assignment that each group should develop a set of metatheoretical insights based on the data set. The results should prove interesting—and highly stimulating to the science of metatheory. It may be expected that each group of students might employ a different methodology—and perhaps have different aims in their exercise. One group might simply combine all the theories to create one large (and amorphous) theory. Another group might identify similarities of structure between the theories, while a third group identifies new insights based on what the existing theories did not cover. In short, without some guide, it is unlikely that there will be repeatability in the science of metatheory.

In light of the difficulty of obtaining repeatable results in the creation of theory, it may be more useful to develop formal tools for the analysis of theory. These should be rigorous so as to enable repeatability. This is not, in any way, to restrict the imagination and creativity of the meta-theoreticians (which, fortunately, is rather irrepressible). Instead, this is to channel that creativity to where it might be more usefully applied to advancing the science[3]. In short, in order to have repeatability, we must have the rigorous methodology of critical analysis.

F—Critical Analysis

Where the tree of metatheory is beginning to blossom is in the area of critical analysis. The idea that theories may be evaluated seems to instill some with a sense of dread ; as if any method for evaluation must somehow be arbitrary. Yet, it is important that we evaluate theory to ascertain if it is, indeed, a valid theory. And, those methods of analysis need not be arbitrary. Without some form of objective evaluation, we are left with a situation of complete relativism which renders everything meaningless.

The analysis of theory is an inherently metatheoretical exercise—as in Ritzer's M_u and occurs more frequently than most imagine. For example, every academic paper involves the implicit and/or explicit examination of extant theories, a consideration of which theories (or parts of theories) might be suitable for advancing a new theory or research project. What is being coalesced and clarified here is a set of evaluative methodologies that might be called meta-theorems because they are useful for conducting metatheoretical analysis. This, of course, is rather different from the kind of "metatheories" that would be understood as overarching a range of theories and/or disciplines. Both of these, of course, are different from "metatheory" as the entire field of metatheoretical endeavor.

To the extent that the metatheoretical methodologies described in this paper are fuzzier, they will be less useful to the development of metatheory as a science. To the extent that they are more rigorous and repeatable, they will bring the field of metatheory closer to the status of a science. In this, one may conduct fuzzy metatheory and be legitimately accused of conducting bad science. However, before the accusation may be considered valid, we must have reliable, repeatable, critical methods of analysis. In this section, I will present a number of methods as a way of starting a conversation that will allow us to keep criticism in its appropriate place.

Within the study of metatheory, there is broad agreement that the analysis of theories is an important part of metatheory. Indeed, given the breadth of that agreement, critical analysis be considered central to the field of metatheory.

In his recent book, Edwards applies an impressive array of tests that may be applied to metatheories. Derived from Ritzer and others, these include:

- Nesting—ensuring that the metatheory is grounded in theory.
- Linkage—identifying and describing relationships between conceptual elements of theories.

- Comparative Techniques—qualitative comparisons and calibrations between theories under analysis.
- Conservation—integrating theories without supplanting them completely.
- Uniqueness—does the metatheory provide something new?
- Parsimony—keeping to the minimum number of explanatory factors needed.
- Generalizability—what other areas of research might the metatheory be applied?
- Level of Abstraction—how many variables and theories might be integrated?
- Internal Consistency—how relevant are the lenses of the metatheory to one another?
- New Factors—can a new theory, or lens, be identified?
- Relationships Between Factors—have the theoretical lenses been carefully identified, and linked and have the internal facets of each lens also been identified?
- Credibility—is the underlying logic reasonable?

Advancing these methods to the point where they may be applied with objective, repeatable rigor will be a boon to the science of metatheory. Among the more objective methodologies developed for the critical analysis and evaluation of theories are:

- Complexity—as a measure of parsimony.
- Static Robustness—as a measure of the internal integrity of a theory.
- Dynamic Robustness—measuring the stability of a theory as it moves between scholars.

It is important to note that while these essentially metatheoretical methods have been applied to the analysis of theories, they have not been applied to the analysis of metatheories. So, again, the field of metatheory could benefit greatly from continued investigation along these lines. In the following sub sections, I will briefly describe a number of other methods that may prove useful to the meta-theorist.

Categorization

One approach for conducting metatheoretical analysis is categorization. This is a fairly simple method for conducting metatheoretical analysis, which aids in the process of generating metatheory. One way of looking at the processes of meta-theorizing is to combine the models of Ritzer and Colomy which categorize metatheoretical activity based on the aims of the researcher. This gives us four categories of meta-theorizing: (a) reviewing the theories within some domain, (b) preparing to create middle-range theory, (c) creating an overarching metatheory and (d) determining the strengths and weaknesses of other theories and metatheories.

Another method is suggested by Gregor. Her five interrelated types of theory are distinguished as: (a) theory for analyzing, (b) theory for explaining, (c) theory for predicting, (d) theory for explaining and predicting, and (e) theory for design and action.

Kaplan suggests six forms of structure for theoretical models. His forms of structure include a literary style (with an unfolding plot), academic style (exhibiting some attempt to be precise), eristic style (specific propositions and proofs), symbolic style (mathematical), postulational style (chains of logical derivation), and the formal style. This last, the formal style avoids "reference to any specific empirical content" to focus instead on, "the pattern of relationships."

It is also possible to categorize theory in a number of other ways. For example, one might create a category based on some specific context (historical, geographical, etc.). Similarly, one might also describe a history of theory, identify schools of thought, or identify geographical differences (e.g., European and African) in the origin or use of theory. Each of these methods, and many others, may provide useful to scholars. Categorization, however, does not necessarily require a deep engagement with the theories under analysis (although, of course, it might). For example, one might engage in a metatheoretical categorization of existing theories by the date those theories were published. In such an approach, one need never look at the theory at all. Therefore, while it may be useful in some context, categorization may not be central to the field of metatheory.

General norms of validation

Kaplan discusses the validation of theories in terms of three types of norms (correspondence, coherence, and pragmatics).

Norms of correspondence. "Truth itself is plainly useless as a criterion for the acceptance of a theory." Indeed, an appeal to facts rests on a bedrock of common sense; yet, those presuppositions also stand in the way of scientific advance, "and progress has required the courage to thrust them aside." "What counts in the validation of a theory, so far as fitting the facts is concerned, is the convergence of the data brought to bear upon it, the concatenation of the evidence - beautifully illustrated, to my mind, in Ernst Jone's essay on Symbolism."

Norms of coherence. This norm suggests that new theories must be fit into the existing body of theory. In this area, Kaplan discusses the simplicity of a theory, and the need to assess that theory in comparison to other theories. Essentially asking which is the best theory to use for a particular situation.

Pragmatic norms. This norm asks if the theory seems to work in practical application. Interestingly, he also notes, "From this standpoint, the value of a theory lies not only in the answers if gives but also in the new questions it raises."

For each of these three norms, Kaplan notes strengths and weakness - essentially suggesting that no one norm by itself is sufficient for judging the efficacy of a theory. These three norms may be understood as general categories for other forms of analysis.

Maturity

Another approach to critical evaluation suggests that a more effective model might be understood to be one that is more mature. develop innovative and useful insights into the structure of theoretical models. They accomplish this by drawing a parallel between the less-understood development of models and the better-understood development of individuals, organizations and cultures. In each area, the system that is understood as more evolved is the system that is more complex. In their maturation model of theory, the stages are:

1. Abstract Stage (stories become cases, events are abstracted to data).
2. Formal Stage (two or more Abstract Stage variables are related).
3. Systematic Stage (developing simple hypotheses from Formal Stage relationships).
4. Metasystematic Stage (creating models that account for all relevant relationships).

Importantly, in this maturation model, the "Systematic Stage" is one where there is a formal description of how variables interact in relation to one another. This stage requires "multiple input variables" and may suggest multiple outputs as well. This view stands in contrast to less complex (and so less mature) models that suggest simple, linear, causality. Creating a parallel between complexity and some sense of maturity seems to have validity—and is reflected across a broad section of literature.

Strong actuarial thesis

Yet another approach (or set of approaches) is suggested by the Faust-Meehl "strong actuarial thesis" which surfaces the difficulty of evaluating theories and asks, "What features of theories predict their long-term survival?" and, "To what extent are those features similar across disciplines and domains?". Usefully, Meehl examines many forms of theory evaluation and describes the weaknesses of each (empirical studies, ceteris paribus, simplest explanation, aesthetic beauty, logical possibility, unconnected postulates, reducibility, etc). And, suggests that logical-mathematical theories are more true/accurate/useful than explanatory theories.

Causal structure

Stinchcombe suggests another form of analysis. Specifically, he investigates the causal structure between the concepts within a theory. These methods provide useful guidelines for graphically mapping the structure of a theory. Among his interesting insights, he explains how to critically analyze a theory to remove useless aspects—thus rendering the theory simpler, with little or no loss of explanatory power. For an abstract example, consider a theory that describes causal relationships between three observable things. Changes in A cause changes in B that cause changes in C. This theory has three concepts connected by two linkages. If we are using this theory to understand the end result "C," Than the intermediary "B" is extraneous. The theory might just as well say that, A causes C.

Propositional analysis

A theory or metatheory may be understood as an "ordered set of assertions". Yet, this begs the question of how well ordered these assertions actually are. One answer to this problem is seen in the use of propositional analysis which appears to be the most rigorous method for analyzing a body of theory. This multi-step process begins by finding the propositions within the body of theory. Those propositions are then compared with one another to identify overlaps, and redundant concepts are dropped. Next, the propositions are investigated for conceptual relatedness. Those aspects that are causal in nature are linked with aspects of the theory that are resultant. Those concepts that are explained by or resultant from two or more other concepts are considered to be "concatenated."

Concatenated aspects are privileged because they are better explained than others—to the extent that the better understanding is transcendent. Bateson describes how two eyes are better than one because the two views combined also provide the viewer with depth perception—something neither eye alone could achieve. Similarly, it is generally accepted that when two people combine their views on a single topic, a new and more comprehensive view emerges. Or, from a metatheoretical perspective, two lenses of theory are better than one.

The formal "robustness" of the theory is then determined by dividing the number of concatenated aspects by the total number of aspects to provide a number between zero and one. A value of zero represents a theory with no robustness—as might be found in a bullet point list of concepts. A theory with a value of one suggests a tightly structured theory, such as Newton's F=ma. For an abstract example, lets say we have a theory consisting of the following propositions: A is true; B is true; A causes B; Changes in B cause changes in C; Changes in D and changes in C cause changes in E. There are five aspects (A, B, C, D, and E). Of those five, only

E is concatenated (D and C cause E). This gives a ratio of well-integrated aspects to poorly integrated aspects of 0.20 (the result of one divided by five).

This represents an important advance in the critical evaluation of theory for at least three reasons. First, because it provides an alternative logic for understanding theories. Second, because it provides a method for objectively delineating the structure of a theory. And, third, because progress in advancing theory toward a measurably higher level of structure seems to be related to advancing a theory toward revolutionary improvement in theory and practice.

Litmus test

In conversations, some scholars have reported to me that they apply a "litmus test" to theories and metatheories as a way to determine validity. For example, one might say, "Any theory that purports to promote freedom is a good theory." While this appears to be a laudable goal, it is a weak scientific standard—primarily because it must fall back on other theories and other interpretations. It begs all sorts of questions such as, "What does it mean to promote?" or, "What is freedom?" and, "Freedom for whom? Or "Freedom for how long?" Further, it leaves unanswered questions about the actual efficacy of the theory or metatheory. A problem highlighted in the phrase, "freedom is slavery". In short, simple litmus tests do not seem to be good science. They are essentially an act of faith or intuition.

When theorists and meta-theorists resort to litmus tests or intuition in place of rigorous, repeatable, analysis, it may be because the more rigorous methods are not well understood, or there may be a cultural bias against their acceptance. Whatever the reasons for past analyses, our science would benefit from the continued development, articulation, and application of these methods in analyses of theories and metatheories, alike. These methods need to be developed both individually, and in comparison with one another. One pioneer in this area is the late Paul Meehl. Meehl found that parsimony is the most used, and most problematic of the claims for the validity of a theory. More frequently used, though less specifically cited, is the test of "logic". This test is so common

Author/s	Year	Anything goes	Avoid categorization & definition	Avoid Ontological arrogance	Deconstruction (includes undercutting privileged texts)	Contextual nature of knowledge	Narrative	Hermeneutic	Paradox	Complexity	Uncertainty / unpredictability / ambiguity	Change	Critical	Reflexive	Interpretation	Introspection	Artistic	Intuition	Humanistic	Awareness of human condition	Emancipation – Who benefits - Justice	Values	Naturalistic / Feminist / NeoMarxist	Educative	Action-Research	Praxis-oriented / Experimentation (usefulness)	Phenomenological	Observe & communicate with others	Practical / Workable	Constructivist	Develop underlying rules for understanding
Rosenau	1992	X			X		X								X	X													X		
Deetz	1996		X																												
Critchley	2001							X																							
Argyris	2005																								X						
Murray	2006								X	X	X																				
Lyotard	1984							X					X	X								X									
Smith	1999				X			X					X		X						X		X	X	X	X	X			X	
Kofman	2010			X																											
Chia	2005										X	X						X		X						X					
Law	2010					X				X	X															X					
Bentz & Shapiro	1998							X							X	X										X		X			X

Table 2 *Aspects of postmodernity as defined by various authors*

that it has faded into the background. It is simply assumed that each theory has some logic to it.

Indeed, it may be said that a theory is constructed of logical arguments. And, that logic may be accepted as a routine test for theories undergoing peer review. However, that approach has proved insufficient to the task of advancing theory. Possibly because there is no clear demarcation for what Set of logical arguments is better than another. That is to say, the logical arguments need only be "good enough" (whatever is appropriate for the needs of the journal). One problem with logic is that it is often constructed in "chains" (A is true because of B, B is true because of C, etc.). Steiner concludes that the last link of a chain remains undefined. So, although one may claim validity based on acceptable logic for a theory, there often lurks the unspoken method of intuition.

Meehl notes that there are few or no metatheoretical rules of what he calls "universal form." Therefore, meta-theorists have nothing to guide them but their intuition. And, studies have shown intuition to be unreliable. Meehl also argues at some length against intuitive methods of evaluation because intuition has been shown to be a weak tool for evaluation—even among experts working within their own field.

To conclude this section on forms of critical analysis, we should note that some of these forms might be more useful than others, depending on the needs of the meta-theoretician. For example, if one is conducting a historical analysis of theory, the process of categorization might be more useful. If one is comparing theories to determine which might be best applied in practice, one may be better off choosing propositional analysis. There is a great opportunity for investigating these forms of analysis. They could, and should, be tested and compared with one another to identify the efficacy of each approach in various fields of study and for various purposes. Given the range of tools presented in this section for the metatheoretical analysis of theory, it should be apparent to the reader that she or he need not rely on weak tests of logic, litmus, or intuition for evaluating theory or metatheory.

G—Verification and testing

It is absolutely imperative for the advancement of any science that the work be exposed to scrutiny; including peer review and assessment. Without the dissemination, comparison, and testing of theories and metatheories, one may become a lone hermit, talking to one's self and digging one's self deeper into some self-satisfying view of the world. It is much better for the science if we develop more useful, shared, understandings. Some metatheoretical models may be held and even used subconsciously—used, but never detected or explicated. In order for the theory to make sense to others, the theorist must surface tacit mental models for explanation and publication.

As analysts, we escape the trap of our own assumptions by making our theory and methodology explicit. Without that clarity, we incur at least two problems. First, we may be overwhelmed by data. Second, we are less able to perceive what is new or different. By having data, we might know *what* but not *why*. We become trapped by seemingly inescapable facts. With theory, we transcend—we gain a new perspective and are able to take action. This is the importance of metatheory—that starting assumption that we may understand the world (and ourselves) in a different way—that we may become more than we are. Following this path, we may more easily gain the ability to change our world and to change ourselves.

In Cultures of Inquiry, John Hall describes a "third path" of inquiry that is primarily neither objective nor subjective, rather is essentially reflexive where meaning is created in a socially constructed sense. Similarly, Ofori-Dankwa and Julian work to identify a bridge from the traditional notions of theory building, and novel, paradoxical approaches that reflect complexity and chaos theory. Where an individual meta-theorists might build a metatheory from one of these, or another of many approaches, those approaches are still rooted in the individual—however carefully contemplative and reflexive that individual might be. In a science of metatheory, another level of reflection is gained by exposing a metatheory to social evaluation.

In the following section, I will investigate the arguments for a postmodern approach to science, review a range of postmodern approaches to scientific inquiry, and investigate how these approaches might be applied to the study of metatheory.

From Modern to Postmodern

Kitcher notes how "there is no theory-neutral body of evidence to which scientific theories must conform." This, theory-laden nature of (so-called) facts is reflected in the study of metatheory. That is to say, there is no metatheory-neutral body of theory. Each of us has a set of assumptions (a.k.a. models, mental models, schema) that shape the way we understand and engage. The same applies to the way we engage theory (and metatheory). This view appears to hold true for modern and postmodern approaches. Historically, "The postmodern period follows the supposed triumph of science and rationality, calls into question, and produces an array of diverse and divergent conceptions of knowledge".

To understand something, one approach of modernist science is to idealize it. That is to say, to assume that everything behaves in a certain way. Applied to humans, one might assume that "all humans are rational decision makers." These kinds of assumptions may be useful in that they allow the researcher to have a set assumption upon which to build additional work. One problem with this approach is that those assumptions are always illusory. There is no way to be completely objective—to understand some absolute truth, because there are always assumptions.

There is no real way to avoid this. Modernist science tried very hard by differentiating scientific inquiry form political, artistic, and religious inquiry. As a result of those reductionist assumptions, fantastic advances were made. As another result, we are starting to see some of the weaknesses of those assumptions—including the weakness of reductionist thinking, itself. Kitcher engages this conversation at great length and depth. And, he concludes, that we should not attempt to rid ourselves of modernist science. Rather, we should seek to change it—to enrich it. One way to do that is to integrate modernist science with postmodern insights into the value of religion, art, human interaction, creativity, and values.

Objective truth does not seem to exist. And, if it does, we humans do not have access to it. What we do have is the ability to access multiple ways of understanding. We can sit in the presence of a person, we can admire the beauty of a work of art, and we can revel in the novelty of surprising insights. If we attempt to engage the world with a single method, we become guilty of methodolatry. To avoid this trap, "scholars have urged researchers to put aside their theoretical silos to uncover the potential of using interdisciplinary theoretical perspectives in research." Using, "border-crossing notion of bricolage," combines multiple methods as appropriate to the research context. This brings to mind Bateson's "double description". His essential idea is that we may reasonably ignore concerns for objectivity or ultimate reality and pragmatically accept that two descriptions of something are better than one.

Modernist approaches to science tend to seek an absolute truth about the nature of reality. This totalizing view discounts alternative views, approaches, and ways of knowing. Thus, modernist science serves as a point of view—a metatheory that both enables and restricts how a user of that view approaches the world and investigations of the world.

Postmodernism, on the other hand, stands in contrast to this monolithic view. Lyotard, for example, suggests that it makes more sense to see society as a range of dualities and oppositions, rather than some undifferentiated whole. For Lyotard's postmodern view, science is about the process of conversation, not some pile of knowledge. Similarly, Shotter and Tsoukas decry the absolutist position and argue for its replacement with approaches that are intuitive and contextual. Calás and Smircich, for example, question the faulty logic of modern knowledge and suggest a range of new logics to support research that will foster change and emancipation.

Broadly, postmodernism suggests multiple paths for investigation and ways of knowing. There is not a single storehouse of well-ordered knowledge. Rather, "Knowledges are incomplete and disordered". Therefore, it is useful to recognize that knowledge is always situated in a particular context. And, recognizing that there are multiple ways of knowing, we should learn to tolerate the differences between them so that we can explore those differences, rather than attempt to privilege one over the other. Deetz seems to agree—exhorting readers to "Fight the tendency to reduce conceptions to categories or reduce sensitizing concepts to definitions." There are difficultics and dangers in categorization because that which reveals also conceals.

Specifically, Chia notes some contrasts between modern and postmodern approaches. Modernism emphasizes differentiation, simple-location, classification, and representation. And, in contrast, postmodernism is more about process, movement, interpretation, and change. The imperatives for postmodernity, he argues, are: (a) Change—that might be understood through history; (b) Workability or usefulness of a theory; (c) Attention to surprise—seeking what is not there and appreciating accidents and novelty.

From another postmodern approach, Law talks about performativity—how we can enact realities. Generally, he suggests that good postmodern research includes: (a) Looking at our practices - what is being done and how it is being done, (b) Avoiding the assumption that there is an absolute reality beyond what we experience, (c) Asking about how processes turn representations into "windows on the world" (d) There is no way to escape from practice, (e) Look for gaps between practices and realities.

However, postmodern inquiry is not entirely about practice. For example, Rosenau notes how, "For these [skeptical] postmodernists, theory is reduced in stature, but they neither reject it altogether nor call for an absolute equality of all theories". Which is a fortunate because if postmodernism did not recognize the legitimacy of theory, there would be no opportunity to write about postmodernity (as every form of writing is a form of theory as well as a form of practice). Shotter and Tsoukas similarly decry the absolutist position that science should replace our intuitive understanding of a situation. Instead, they suggest that reflective conversation within a specific context (rich in description) and everyday use of language should be applied. However, their description of the process seems remarkably like the process of induction - as such talk may lead easily to theories and practices. So, again, we find ourselves back in the realm of theory.

There are (appropriately) a wide variety of postmodern approaches to inquiry and so science. The following table provides a distillation of some approaches by a dozen authors. From this, I hope the reader will gain an appreciation of the vast range of possibilities for postmodern scientific exploration. And, I hope, be encouraged to develop more approaches.

From a postmodern perspective, as you may infer from the above table, there are more than a few ways to conduct scientific inquiry, "So, it is not at all the case that there is something like a unified "scientific method" that governs all intellectual inquiry". Inherent in this diversity is that idea that we can continue to develop new (and more diverse) approaches that may have some validity and usefulness in a postmodern context. And, to a greater or lesser extent (as with modernist approaches), these postmodern approaches might serve as useful indicators of progress for a science of metatheory. In the remainder of this section, I will investigate some potential benefits and limitations of these postmodern concepts as they relate to metatheory.

Anything goes is a difficult thing to measure. In the present context, it seems related to the idea that we should avoid ontological arrogance; and, in that process, strive toward more creative approaches to analysis. Indeed, we are not restricted to a single mode of analysis, rather we have a range of possible methods—and that range continues to expand. When applied to metatheoretical analysis, this suggests that it is valid to develop new methods of validation. Although, it should not be seen as ruling out existing approaches.

Deconstructing existing texts (especially those that appear to be privileged) is a process that is inherent to metatheory. Because those texts are the subject of analysis, it is difficult to avoid some form of deconstruction! There is also an important piece of recursion here because metatheoretical methods are themselves theories. Therefore, it is possible to use metatheory to deconstruct metatheory. We might measure the rate of deconstruction by tracking the level at which the deconstruction of a theory occurs. For example, a postmodern researcher might write in general terms about a theory (as a whole)—and that might represent an engagement on one level. If the researcher analyzes the propositions of a theory, that is another level. Then too, if the researcher investigates individual words—that is still another level. Deconstructing on all three of these levels (or any two) would be more effective than using any one.

In contrast to the modernist desire for objective knowledge, the postmodern approach recognizes the contextual nature of knowledge. For an analysis of theory, a metatheoretical approach might simply accept and appreciate all theory as providing useful material. Or, it might be said (for example) that a particular analysis will investigate theory from a specific context, perspective, or draw on theory from a specific source (e.g., theories of the 21st century). Identifying the context may be a form of categorization. For example, only theories made in the 20th century would be considered valid in a study of 20th century theory. It is also useful to recall that a process of social construction may serve to create new contexts. For example, Lyotard explains:

> *From this point of view, an institution differs from a conversation in that it always requires supplementary constraints for statements to be declared admissible within its bounds. The constraints function to filter discursive potentials, interrupting possible connections in the communication networks: there are things that should not be said. They also privilege certain classes of statements (sometimes only one) whose predominance characterizes the discourse of the particular institution: there are things that should be said, and there are ways of saving them. Thus: orders in the army, prayer in church, denotation in the schools, narration in families, questions in philosophy, performativity in businesses. Bureaucratization is the outer limit of this tendency.*

The hermeneutic nature of metatheory also lends itself to the postmodern perspective. That is because theories, the subject of investigation, are inherently textual. Here again, it is important to qualify and quantify the range and number of theories that are involved in an analysis. It would not be good science to say (for example), "I developed a metatheory based on some theories." The sources of those theories should be clearly articulated. A good practice for our nascent field of study might be to include our data—either in quotations within the text or in an appendix. That way, it is easier for other members of the metatheory community to evaluate the process and the results.

When one is exploring existing theory, the postmodern perspective offers an important, and interrelated set of relationships to look out for. Within one theory, or between two theories, one might seek to identify instances of paradox. This offers opportunities for finding new insights when the paradox is appreciated (perhaps experienced and embodied) and resolved. Galileo's experience provides one example of this approach. In ancient times, the prevailing theory of downward motion was that small objects fell slowly while large objects fell quickly. Galileo wondered what would happen if a small object were connected to a large object. And, through that thought experiment, recognized a paradox of the prevailing theory. That recognition impelled him to drop objects from tall buildings and so develop better theories of gravity. We might, enumerate the number of paradoxes found within a between theories as a simple path to evaluation.

Similarly, in studies of theory, we have the opportunity to investigate the complexity of theories and bodies of theory. Indeed, the complexity of a theory may be an indicator of that theory's maturity—suggesting a useful way to measure the potential usefulness of that theory. This approach may be easily employed by enumerating the concepts and/or connections within a theory. Theories may then be easily compared on at least one dimension. An opportunity exists here for a meta-theorist to enumerate all the forms of enumeration within theory.

Because theory may be generally understood as a tool for reducing uncertainty (by providing an explanation of events), the idea of investigating the *uncertainty* within a theory bears a certain intriguing recursion. Often, we are able (by experiment or practical application) to test how well a theory works. The extent to which one's use of the theory enables one to predict future events may be regarded as a test of that theory's effectiveness. And, as a result, it is possible to compare two theories. However, postmodernism (along with complexity theory) holds that nothing is completely predictable. Therefore, if one finds a theory works 'perfectly' one may also be sure that one has fooled one's self.

In such situations, it is important to seek and appreciate the *ambiguity* of a situation. To purposefully seek out those events that are not predicted by the theory—those experimental anomalies that we know must exist, even though they may not be evident at the moment. This leads to a most important, difficult, and subtle postmodern technique—looking for things that are not there. In the analysis of theory, the discovery of a "missing link" can be an exciting find. As an example, from my personal experience, I am usually unimpressed by computer modeling experiments where the experiment merely confirms the researcher's expectations. Instead, I appreciate results that the researcher finds surprising.

Change is also a useful "thing" to look for in the study of theory. One might note, for example, that certain theories of physics have remained unchanged for hundreds of years... and wonder why. One might also investigate the great rapidity with which theories change and evolve as they are adapted by various authors over time, or as they change between the contexts of differing publications. One way to measure this is to determine the "dynamic robustness" of the theories.

One point of congruence between modernist and postmodernist approaches is the need for *critical analysis*. This has been addressed above in some detail. In contrast, *reflexive* approaches are more postmodern in nature—involving forms of social construction. To make these approaches more valid, it is useful for the metatheoretical researcher to describe the process by which the original theory was created, and/or the process by which the new metatheory or metatheoretical perspective was developed. A more rigorous quantification of the reflexive approach may be difficult because of the inherent complexity of the process. Here, again, is an opportunity for a deeper more comprehensive exploration of reflexive approaches to metatheory.

Far from understanding such indeterminacies as an impediment, the postmodern perspective sees such looseness as an opening for *interpretation* by the researcher. To provide some sense of validation, some *introspection* is useful. Here, the researcher should provide readers with some indication of that introspective process. This may include surfacing one's inspirations, concerns, and excitements. The process of introspection may lead to spiritual insight. This is, as history has shown, rather difficult to validate from an outside perspective. And, from the inside view (one's own sense of spirituality) may not make a great difference to the larger community. If the researcher avoids such self-revelation, the analysis might be seen as primarily intuitive or artistic.

In the postmodern world, the prevailing view is that we should appreciate the *artistic* and the *intuitive* in all sciences—including theory and metatheory. We are also encouraged to recognize that each work of art has more validity in some contexts, than in others. So, for validating a theory or metatheory as an art form, it is important to consider the artistic tradition from which it emerged and the context in which it is being used. I find an example in my own experience as a life and business coach. To me, the process side of coaching follows a series of questions such as: What worked well? What might you do differently next time? The artistic aspect, to me, comes in my ability to engage the client in a way that makes him or her feel that this is all part of a very natural conversation and leads to profound revelations of self and life. If other members of that community share my view of the art of coaching, they can evaluate the extent to which my coaching was artistic, or merely mechanically following a set of rote questions. For evaluative purposes, it seems that the analyst should describe the artistic standards to which the theory conforms (or creates).

The modern view of science is further enhanced by the postmodern ideas of *humanistic* inquiry, which calls for an increased *awareness of the human condition*. These include a range of *value-based* inquiry including *naturalist, feminist,* and *neo-Marxist* approaches that seek to *educate, emancipate,* and improve the human condition. These approaches may be evaluated from a number of directions—most or all of which are useful for "decolonizing" or de-emphasizing privileged texts in favor of more localized insights.

One might ask of one's self, "What is my intention in this investigation?" Or, we might ask, "Who benefits from this research?" Or from the perspective of the possible participants, "To what extent does this research address our concerns?" These are, of course, a few of the many possible questions. Recursing briefly, it seems important to ask those who might be impacted about their opinions! Calás and Smircich, for example, question the faulty logic of modern knowledge and suggest a range of new logics to support research that will foster change and emancipation. When applied in practice, one effective approach was to avoid the idea of expert knowledge, and to engage co-researchers in our collective experiments of all kinds. Generally, we may consider theories more valid to the extent that they are more able to raise awareness, educate, and emancipate. Thus, again, two theories may be compared and the more preferable one advanced.

One problem with soft sciences is abduction without critical examination. A reader may be "taken" with an idea; and, as a result, may feel that a theory appears "too good to be false." As a result, researchers might believe their theories or values are of the highest importance because, perhaps, those values have not been examined or deconstructed. Susan Haack seeks to differentiate truth and meaning by investigating four possible relationships between the two. She accepts the idea that there is an overlap between the two—that sometimes truth and ethics are one and the same, while other times that does not appear to be the case. Here, however, she relies on that elusive idea of truth. In a brilliant work (that includes a thoughtful conversation on Haack's paper), Martyn Hammersley notes, "I am not convinced that the distinction between what is morally wrong and what is epistemologically wrong can be sustained". Rather than recounting millennia of thought, suffice to say that "truth" is an illusion that is sought by modernists and postmodernists alike. However, as "truth" seems so elusive, it seems an unlikely measure for validity and progress of metatheory.

Instead of a narrow, modernist science, Hammersley argues for link between epistemology, and morals, and calls for validation by a research community. This, in turn, seems to suggest that each and every paper should receive a through review by a close circle of readers. How do we accomplish such a thing? With all that is being written, who has time to do all that reading? Do we raise funds to pay for a large cadre of readers? It is an interesting possibility. Without funding, we might work to encourage readers to become more involved in the process—providing feedback to the authors and discussing the material more among themselves.

Then too, if knowledge is only valid if collaborative, should we require all that papers be co-authored? Or, is the act of learning, writing, review, reading, and conversation a sufficient level of collaboration? One might be tempted to say yes—it would certainly be easier to continue as we have—and to follow the traditions of modern academia. However, I feel that we are looking for something more. Instead of drawing a line and saying, "that side represents an unacceptable level of collaboration and this side is acceptable," we can pass beyond that dialectic and recognize that each written work may be represented on a scale of collaboration. And, reviewers, editors, and readers may rate that level of collaboration as a partial indicator for the collaborative-validity of the text.

For example, a book written by a single author would have a relatively low level of collaborative-validity compared with a paper written by two authors who received feedback from two reviewers. A book could claim a higher degree of validity where, for example, a team of authors determined among themselves the focus of the book, the focus of their individual chapters, and how those chapters would be linked. And, where those authors spent a month sitting around a table co-authoring and co-editing one another's work.

Many concerns of validity are absorbed in another postmodern approach, which is more closely related to practice than to theory. Specifically, *praxis-oriented* methods that develop and show the usefulness of specific practices and points of view, within a specific context have shown themselves to be fairly effective. For example, the *Action-Research* approach. These methods, it seems may best be evaluated from within the group that develops and applies them. One such approach is the Future Search process which includes the creation of a mind map—a shared sense of the situation. This mind map may be understood as a form of theory, because it may be used to improve understanding and guide action. This approach suggests we may evaluate theory creation, at least in part, by the level of collaboration involved in the construction process. We might say that having more people involved indicates a quantitative measure of collaboration, while their satisfaction with the mind map indicates a qualitative measure of success.

These approaches to evaluation suggest that the only one who can truly evaluate the success of one's own method is the person who employs it. This *phenomenological* approach has its own limits. For example, one person might believe that plants grow better with fertilizer, while another believes that music is more conducive to plant growth. If, as outside observers, a community sees these two people growing food in their gardens, and one is more successful than the other, we might take that success as an indicator of a more effective underlying theory. We might (and probably should) also inquire about the details of those methods. For example, are they sustainable? Do they lead to negative environmental impacts? And, importantly, is there another, as yet undiscovered explanation for the differences observed?

In the academic review process, one author might develop and apply a method which he or she sees as very effective. Yet, the larger community might find very different results. One test of many is to determine if a theory is *practical / workable.* This might include purposeful experiments or simply some application in daily life. A strict postmodern stance might give equal validity to all theories that are applied because each one seems workable to the person who is employing it. And, the simple fact that a person is using the theory and finding it workable is certainly one level of validation. Also, the more people who find a theory workable might lend another level of validity.

This leads to at least two other extensions. First, that within one's own mind, an individual should seek to evaluate multiple theories to find if one is more workable than another. This might be seen as similar to an individual seeking a religion or spiritual practice that works best for him or her. Second, is the idea that the practicality or workability of a theory might be evaluated through the lenses of other postmodern approaches. For example, We might look at two theories and determine that one is more workable than the other based on the amount of change or the amount of emancipation is provides.

Over the course of time and communication, the above methods of investigation, communication and participation may be broadly understood as a form of *social construction.* That perspective suggests rich and complex interactions that include theory, practice, intuition, art and inspiration. This opens yet another interesting recursion because, "the theory itself creates—it socially constructs—the terrain. A theory entails imposing interpretations (definitions, categories, and understandings) on behavior. Once we have a theory in mind, we pose questions that take those definitions, categories, and understandings for granted".

This perspective points to a strength and limitation of the post-modern perspective. First, we are looking at mirrors through mirrors—so we can never be certain about what we see. This may cause difficulty for those who seek some absolutist foundation upon which they can rest. This difficulty may even impel some to a useless level of skepticism where they believe that nothing is real (and, in consequence, nothing has meaning). It is unfortunate that one result of this stance is that some abandon the search for new and more wonderful insights. Instead, this constructionist perspective also indicates that we recognize our limitations. And, from there, we can go on to transcend them.

Implicit in many of these approaches is the idea that the development and/or assessment of theory and metatheory are part of a process. In the field of metatheory, the development of a specific theory or metatheory is (from one perspective) understood to occur through the iterative process of induction and deduction. And, there are many other process-based perspectives to suggest paths for creating and evaluating theory. These process-based approaches may also be applied to a science of metatheory—suggesting an iterative process of creation and testing. It seems important to the science of metatheory that we develop new tests and test those tests. Further, there is a great opportunity to identify the processes by which each of these methods of creation and validation lead to one another.

An important tool for building and evaluating theory and metatheory is creativity. Because, if we are stuck within the map of our theory, no matter what the apparent efficacy of that theory, no matter what level of validation it exceeds, we know that we will always benefit from new and more creative approaches. Each new approach adds to the existing range of approaches; and, suggests ways to integrate and extend them. Despite (or, perhaps, because of) the range of options reflected upon in this section:

> *It is far from evident that replacing conventional social science methodology with post-modern methods of interpretation and deconstruction constitutes any*

improvement in the social sciences. If adopted without modification, post-modern methodology leaves social science with no basis for knowledge claims and no rationale for choosing between conflicting interpretations.

Kitcher makes the important point that postmodern views, such as those presented in this section, should not be seen as existing in opposition to the views of modernity with the goal of eliminating the modernist view. Rather that these humanistic, artistic, and other diverse approaches actually serve to expand and change the modernist approaches. Changing the overall understanding of science to make it more rich, full, and inclusive. In the next section, I will investigate how we might understand a combination of modern and postmodern approaches. Not as a longer list of potential methodologies, but instead as existing in a carefully integrated relationship—an integral approach to science.

Integral Approaches to Metatheory

In moving metatheory toward a more scientific practice that includes modern and postmodern approaches, many forms of scholarship are possible. First, one might use any of the methods suggested here for creating and evaluating theory and metatheory. Another, more comprehensive approach, would be to use multiple methods. A more nuanced and sophisticated scholarship would involve a combination of methods that are mutually supporting—that is to say, integrated. This perspective is, in part, an answer to Fred Kofman's call to avoid ontological arrogance with our underlying ideology, and so encouraged us to accept alternative views. This approach is postmodern; and, the call also holds true for our ideology about postmodernism. We should be on the lookout for what is missing and how postmodernism might be improved. This is especially important as there is a lack of agreement on key issues—even within the postmodern community.

Between modern and postmodern there are many disconnects. For example, Critchley tells how the Vienna Circle sought a science free of emotion, religion, and metaphysics. While, in contrast, an integral approach is about exploring how they might be rejoined as in *The Marriage of Sense and Soul.* There, Wilber explores in some depth the importance of integrating science and religion. This is an appealing perspective because the traditional role of science is to find truth, while the traditional role of religion is to find meaning. This kind of integrative approach suggests that we are creating something that is more than the sum of the parts.

There are also overlaps between modern and postmodern approaches. Both seek knowledge and understanding, both call for critical analysis, and both appreciate the benefits applying theory ("testing" for modernists and "usefulness" or "workable" in postmodern terminology). What is missing between the two versions of science is a focus on how they might work together. This is an integral approach, which recognizes the benefits of both and seeks to bring them together to create a more complete and effective understanding. Kitcher suggests the need for overlapping and interlinked methods for measurement and Murray states:

> *Integral approaches give equal importance to the subjective and objective aspects of world. Seen through this lens science and technology are not divorced from questions of meaning, identify, aesthetics, and ethics.... It does so not mere by critiquing other theories but my proposing an integrative framework that coordinates these theories and also by incorporating subjective and Intersubjective matters of self, culture, and spirit.*

Let us move now to an exploration of approaches for a scientific advancement of metatheory that interlinks the best of modern and postmodern worlds. This approach is called triangulation which works under the assumption that it is desirable, to use a mix of approaches. This process is supported by a growing number of scholars. An excellent application of this approach may be found in Lewis who notes in an introductory explanation of her study:

> *The strategy—Meta-triangulation—entailed using paradigm lenses to construct alternative accounts of AMT implementation. The second section summarizes these accounts, which highlight varied vicious cycles during the change process. Third, I use paradox literature to build a meta-framework, depicting change as a multidimensional cycle, swirling around cognitive, action and institutional paradoxes. Paradigm lenses detail each paradox, revealing complex, systemic tensions between stability and change.*

The conclusion addresses implications for managing change paradoxes and future research.

A reading of Lyotard suggests triangulation using a combination of categorization, critical evaluation, and application. Koller suggests using a purposeful blend of approaches including contemplation (sprit), reason (mind), and body (senses). These kinds of triangulation suggest more than a range of coverage within a single sphere of thought, they also suggest an interrelationship between worlds. To expound on one such relationship, I draw on my recent work from the *Integral Review*.

Briefly, one may usefully and effectively view theory validation from three general points of view—or three worlds. Those three worlds are summed up as Theory, Facts/Data, and Meaning/Emotions. Within each world there are (at least) three levels of validation—with each level having more validity than the one below it. From an "outside" view, each theory may be seen to have greater or lesser validity within any one of the three worlds (see Table 3).

Another point is that there are two general ways to evaluate a theory or metatheory. One is to evaluate it from "within" one of the worlds; the other is to evaluate it from two or more perspectives "between" worlds. From an "outside" view, we can look at validation between the worlds. "Significantly, this distribution of worlds creates the opportunity to understand each of the three worlds in integral, co-causal relationship to one another". For example, a theory that is essentially a metaphor might have little or no validity in the world of theory. It might, if it were a compelling metaphor, have greater emotional/spiritual validity. If that metaphor were applied successfully to an organizational change effort, it might be seen to have greater perception/data based validity.

On another part of the spectrum, a theory from physics might have a high level of validity on the world of theory—and on the world of facts/data may prove very effective. On the other hand, on the world of meaning, the theory may be entirely uninspiring. One might see a well-written or inspirational, book as having great validity in the world of feeling and spirit, yet have little validity as a theory or in practical application. In contrast, a profundity of raw data might have great validity within its own world, but without

	Level 1	Level 2	Level 3
World One (Facts or data)	Uses objective data.	Uses objective data from multiple sources.	Future facts are predicted.
World Two (Meaning, emotions)	Makes sense to author.	Makes sense to editor, reviewers, and readers.	Consensus of expert opinion (this theory is preferred over other theoretical options).
World Three (Theory)	Includes logical arguments.	Theory is constructed of specific propositions.	Theory is constructed of co-causal propositions.

Table 3 *Dimensions of Validity*

a theory to make sense of it, it is merely noise—so has no validity within the world of feeling/spirit.

In the same way that great art calls for great reflection, a high level of validity in any one world seems to call for the generation of new understandings in each of the other worlds. And, we might imagine that a high level of validity in two worlds might create a very strong pull for increased validation in the third world. For example, having a great amount of data seems to call for a theory to make sense of it. Then too, a great deal of data also has a sensory component, so it calls for reflection and appreciation.

Popper's famous (and famously modernist) work suggests the best way to validate a theory is in the way it allows for predictions in perceptual data. A theory of physics, which fails to predict change, should be rejected. However, in the social sciences, the process is a highly complex issue of recursion. Changes in theory might cause changes in productivity but they may also alter an individual's values (e.g., one's personal feeling about what is important). So, an individual may no longer place the same value on the items being produced. Indeed, "individuals working on an assembly line might (based on an emerging mindset) change their minds and decide that the human producer is more important than the objects on the conveyor belt". There is a great opportunity for scholars to investigate the potential interrelationship of the many approaches presented in this article and this special issue, as a whole.

Advancing the Field

In *The Trouble With Physics: The Rise of String Theory, the Fall of Science and What Comes Next,* Lee Smolin reflects on his experiences as a physicist who was dedicated to the development of string theory—the creation of a single one, best, totalizing theory of physics. Smolin concluded that such a development is impractical and perhaps impossible. His experience turned him away from a staunch modernist perspective (and, may provide some insights for our exploration here). Smolin found that progress in science seems to take two forms—innovation and consensus. The two seem closely interrelated—we need both to advance.

This article has presented some basic ideas to clarify and encourage the continued emergence of metatheory as a science. To the extent that metatheoretical methodologies and processes seem to fit a fairly specific definition of science, it seems that we have the basic pieces in place to conduct a legitimate science of metatheory; although, more development and clarification of methodologies appears to be required. Further, if we are to be a science, we must also apply these rigorous methodologies—something the broader field of the social sciences has mainly avoided. There is reason to hope that we are moving toward a change—an improvement—including a more purposeful application of metatheoretical principles.

One may anticipate that metatheory, as a science, will not be immediately adapted by all social scientists. First, some may too deeply invested in the present paradigm. They might, for example, be interested in creating theory for publication instead of application. Second, by their seemingly successful experience, these academics may believe that they "know" how to create good theory—and so do not need to change their approaches. We might liken their theory-creation ability to the ability of a cook to bake a cake. They need no recipe because they have a great deal of experience in this department. Their students, however, have not yet learned the knack. And, the students may benefit most from learning how to follow a recipe. In short, we might achieve some success in advancing the practice of theorizing by developing and implementing a cross-disciplinary program to teach metatheory to graduate students. After all, they have not "grown up" with the fragmented fields of social theory, so may need additional tools to cope with the plethora of theories.

In the above sections, I have noted many opportunities and directions for future studies to advance scientific metatheory. Meehl states, "I view this disparity between a (purported) logical truism and scientific practice to be one of the most important and badly neglected metatheoretical puzzles." A great opportunity exists for meta-theorists to conduct cliometric analyses and so advance our understanding of theory, metatheory, and more. Another important opportunity exists for advancing the metatheory conversation through the application of metatheory. One place that this can occur with some ease is in the classroom. For example, professors could assign students to conduct purposefully metatheoretical analyses. Combined in a public database, the results might provide a useful source for metatheoretical analyses.

Reviewers of academic journals could aid in advancing the field by applying clear metatheoretical metrics when evaluating submissions. This way, each pass through the review process would result in better theory (rather than the creation of theory that is simply different). Similarly, those who review research proposals could also apply metrics to determine which theories are more likely to be effective than others.

The study of theory is appropriately referred to as metatheory (or metatheorizing) where, "The problem lies in how we differentiate between these different kinds of big pictures and the various ways of creating them. We have almost no way of formally distinguishing between metatheory as a scientific process of building metalevel conceptual frameworks or metatheory as grand storytelling or metatheory as philosophical musing. For the sake of all different forms of creating grand narratives this situation needs to change" (Edwards, personal communication, October 18, 2009). In this article, I hope that I have created a framework that we may use to address these issues.

Our nascent science is not at a point where we can apply labels with great confidence (although many have tried). It may be that we need to do some experimenting—we need to try out various methods of metatheorizing in quasi-controlled experimental settings and decide from the results what is our best tool for a particular project. We might find, in time, that certain metatheoretical approaches are useful for creating theory, while others are more useful for evaluating theory. Specifically, we need to have and to use a rigorous, scientific, and repeatable approach to metatheory if this field is to have any legitimacy; and, of greater importance, if we are to develop effective theories for addressing the many personal and societal issues of our age.

There are, of course, limits to the emerging science that I have sketched in this paper. I do not intend this to be the "last word" so much as a preliminary investigation of relevant components. My hope is not to be seen as some "absolute expert" so much as a "conversation-starter." As noted above, there are many opportunities presented in this paper for additional exploration. These include more nuanced investigations into each of the above-mentioned building blocks of modernist science, as well as investigations into how those blocks might be more carefully integrated with one another. Additionally, there are alternative definitions of science that might be explored and contrasted with the one presented above.

Reflections

In the development of this paper, I received a strong call from one reviewer to conform to the dictates of postmodernism, rather than modernism. Due to many personal stresses, I initially resisted that call. Taking time to breathe, and be more open, I worked to answer the call towards greater inclusiveness; I also worked to avoid the totalizing call of postmodernity and modernity, both. As a result, I was impelled on to a more integral approach that combines modernity and postmodernity.

It occurred to me that there is a key distinction between modernity and postmodernity. First, modern science sees progress as an accumulation of knowledge. Second, postmodernity seems to loosen the "direction" of what constitutes valid knowledge by accepting multiple forms of knowledge. Yet, postmodernity seems also to retain the idea that accumulation is important. This should come as no surprise, as accumulation serve to immediately validate any and every insight. It cannot be discarded without reducing the storehouse of knowledge—be it a modern or postmodern storehouse. What remains is the question of what to do with that storehouse of knowledge.

Integral approaches suggest that identifying interrelationships between existing forms of knowledge may create new knowledge. This takes us beyond simple accumulation and moves toward more effective approaches. This, in turn, implies the need for methods for creating theory and metatheory—where they are understood to be integrated knowledge, rather than simple accumulations. When one identifies relationships between previously unconnected bits of knowledge one tends to experience a sudden, perhaps spiritual, sense of wonderment or awakening. And, one is on the path to the creation of theory and metatheory.

Conclusion

It may be true, from a modern perspective, that we will benefit from validating theory and metatheory. It may be true, from a postmodern view, that there is no single approach to validating theory and metatheory. From an integral perspective, it seems reasonable that these forms of validation must be interrelated. In creating a science of metatheory, it seems we must recreate what it means to have a science.

At the risk of creating another layer of "meta-ness" it seems that the most useful approach will be to validate each theory and metatheory by using at least three forms of evaluation. And, importantly, those forms should be as different as possible. For example, it might be useful to evaluate a theory using a combination of spiritual, creative, and aesthetic measures. However, a more effective (and more challenging) approach would integrate more diverse points of view to provide very different measures (e.g., artistic, structural, and practical). In this, it may be seen that the greater diversity will lead to greater strength and the many will lead to the one—just as the success of one theory leads to greater diversity in application.

In this article, I have identified and investigated a variety of components needed for recognizing metatheory as a rigorous, and legitimate science from modern, postmodern, and integral perspectives. Those methods should be reflected, as rigorously as possible, in the multiple building blocks of science that point the way to advancement. Our ability to advance metatheory is founded on our ability to meet multiple requirements in an integrated way—not simply one or two. Implicit in this article is that idea that meta-theoreticians must be rigorous in their pursuit and application of these scientific ideals. We don't want metatheory to be flapping around in circles.

As meta-theoreticians strive to meet these goals, we will strengthen the withered wing of theory and so enable the bird of science to fly. Extending that metaphor slightly, we may recognize the need for a body—that connects those wings and provides a sense of purpose for that flight. For this body, we need to develop research communities whose members agree to conduct careful, rigorous, repeats of experiments and studies in metatheory (including building, testing, identifying the range of metatheories).

This new community will be reflected in our journals. This, in turn, implies that editors who are interested in supporting the emergence of a new science should stand ready to publish works that do not meet the more traditional "literary" standard of innovation, but instead rise to meet the equally high (and possibly more important) standard of scientific repetition. An interesting and useful collaborative effort might begin with a single metatheory, and ask seven authors to address that metatheory—each from the perspective of a single building block of science.

This article serves as a starting point for conversations on metatheory as a rigorous science, and how that science (as a process) might be improved for the development of better metatheory. Improvements in this field will improve our ability to advance our effectiveness as scholars working within and between other branches of the social sciences. In turn, those advances will have a profound effect on our ability to work for the betterment of our world.

Appendix: Definitions of Metatheory

Dominic Abrams and Michael Hogg

A metatheory is like a good travel guide—it tells you where to go and where not to go, what is worthwhile and what is not, the best way to get to a destination, and where it is best to rest a while. Metatheoretical conviction provides structure and direction, it informs the sorts of questions one asks and does not ask, and it furnishes a passion that makes the quest exciting and buffers one from disappointments along the way (p. 98).

A metatheory should provide an alternative framework for asking particular questions, not a complete explanation for all phenomena. The appropriate mission is not to convince others that the metatheory is right and that others are wrong, but to show how a particular metatheory can be useful to account for a specific class and range of phenomena (p. 100).

A strong metatheory helps to put the body parts together in a meaningful structure and then to theorize the links between those parts. In addition, identifying the metatheory behind a particular theory helps reveal potentially interesting and useful links to other theories (p. 100).

[Metatheory] encourages the integration of concepts across contexts (p. 100).

Jack Anchin

Unifying knowledge in any field of endeavor requires metatheory comprising a conceptual scaffolding that is sufficiently broad to encompass all of the specific knowledge domains distinctly pertinent to the field under consideration, that can serve as a coherent framework for systematically interrelating the essential knowledge elements within and among those domains, and that extends conceptual tendrils into other fields of study (p. 235).

Jack Anchin

Among vital purposes served by metatheory is its function as scaffolding for integrating more specific theories that conceptually and empirically map different aspects of the phenomena under study (p. 804).

Terese Bondas & Elisabeth O. C. Hall

Metatheory analysis is an examination of theories to determine the link between the theoretical perspective that frames each primary study and the methods, findings, and conclusions of the research (p. 115).

Richard Clarke

Metatheory is the 'theory of theory,' to be precise, the study of those underlying assumptions which shape particular theoretical perspectives.

Robert T. Craig

Metatheory is theory about theory.

Brenda Dervin

One major point here is that metatheory can be used in such a way that it releases research in always partial but still significant ways from implicit assumptions and draws these assumptions out into the light of day where they can be examined, interrogated and tested (p. 748).

Metatheory must be an inherently deconstructive enterprise.

David Faust

As data form the subject matter for theories, theories and other scientific products form a key subject matter for metatheory or meta-science, organized and directed by methods that, in large part, remain to be developed.

Deborah L. Finfgeld

[Metatheory is the] Analysis and interpretation of theoretical, philosophical, and cognitive perspectives; sources and assumptions; and contexts across multiple qualitative studies (p. 895).

Adam Maria Gadomski

A meta-theory M may represent the specific point of view on a certain class or set of theories T and this viewpoint generates meta-properties of T. Meta-properties are the consequence of the relation between M and T, but they are not the properties of any T application domain. More formally speaking, a theory T of the domain D is a meta-theory if D is a theory or a set of theories. For example, in computer science, the Theory of Data Bases Organization is a meta-theory for every specific (domain-dependent) theory of the data organization/structuring and management.

Willis Overton

A metatheory is a coherent set of interlocking principles that both describes and prescribes what is meaningful and meaning-less, acceptable and unacceptable, central and peripheral, as theory—the means of conceptual exploration—and as method—the means of observational exploration... a metatheory entails standards of judgment and evaluation. Scientific metatheory transcend... theories and methods in the sense that they define the context in which theoretical and methodological concepts are constructed. Theories and methods refer directly to the empirical world, while metatheories refer to methods themselves. There are many important features of metatheories, including the fact that they are ubiquitous—all theories and methods are formulated and operate within some metatheory—and the fact that they often reside quietly and unrecognized in the background of our day-to-day empirical science" (p. 154).

A metatheory is a coherent set of interlocking principles that both describes and prescribes what is meaningful and meaningless, acceptable and unacceptable, central and peripheral, as theory—the means of conceptual exploration—and as method—the means of observational exploration—in a scientific discipline. In other words, a metatheory entails standards of judgment and evaluation. Scientific metatheories transcend (i.e. 'meta') theories and methods in the sense that they define the context in which theoretical and methodological concepts are constructed. Theories and methods refer directly to the empirical world, while metatheories refer to the theories and methods themselves (p. 155).

Barbara Paterson, Sally Thorne, Connie Canam and Carol Jillings

Meta-theory is a critical exploration of the theoretical frameworks or lenses that have provided direction to research and to researchers, as well as the theory that has arisen from research in a particular field of study.

Meta-theory involves the analysis of primary studies for the implications of their theoretical orientations (p. 92).

George Ritzer

Metatheory is concerned... with the study of theories, theorists, communities of theorists, as well as with the larger intellectual and social context of theories and theorists (p. 188).

A metatheory is a broad perspective that overarches two, or more, theories.

Leslie Sklair

A metatheory is a set of assumptions about the constituent parts of the world and about the possibility of knowledge of them. The distinguishing characteristic of a metatheory is that it refuses to accept any burden of empirical proof by displacing the burden of empirical proof onto the theories that are logically deducible from it. An effective metatheory is one which manages to create a high degree of coherence between epistemology and the objects of knowledge (roughly, abstractions).... The hallmark of a metatheory in science is that it invites empirical proof (or, in some versions, refutation) by producing theories and hypotheses that can be tested.

Tendzin Takla and Whitney Pape

By metatheory we refer to the cluster of fundamental, but often implicit, presuppositions that underlie or embed a theory.

Sally Thorne, Louise Jensen, Margaret H. Kearney, George Noblit & Margarete Sandelowski

Metatheory [is] the study of the frames of reference used in a set of primary studies (p. 1357).

Jonathan Turner

... it is reasonable to conclude that metatheory should come after we have produced some theory. Metatheory is not about what assumptions and presuppositions sociology should have, but about the structure and implications of existent theories (p. 38).

... let me offer some proscriptions and prescriptions about how best to perform metatheorizing. First, here are some proscriptions or metatheoretical taboos (p. 39):

1. Avoid talking about theorists; instead, talk about theories.
2. Avoid discussions of intellectual context, place, and time; instead, discuss social processes denoted by concepts, models, and propositions.
3. Avoid debates over philosophical issues; instead, commit one's energies to the simple assumptions that there is a world out there and that it can be understood with concepts, models, and propositions.
4. Avoid commitments to ideologies; instead, develop concepts, models, and propositions that denote operative processes in the universe (there will always be someone to expose ideological biases without your help).
5. Ignore the particulars of history; instead, examine those more general and generic processes that cut across time and place (leave something for historians to do; or, if history is used, let it involve an empirical test or assessment of a theory or model).

Here is my list of prescriptions (pp. 40-41):

1. Evaluate the clarity and adequacy of concepts, propositions, and models.
2. Suggest points of similarity, convergence, or divergence with other theories.
3. Pull together existing empirical (including historical) studies to access the plausibility of a theory.
4. Extract what is viewed as useful and plausible in a theory from what is considered less so.
5. Synthesize a theory, or portions thereof, with other theories.
6. Rewrite a theory in light of empirical or conceptual considerations.
7. Formalize a theory by stating it more precisely.
8. Restate a theory in better language.

9. Make deductions from a theory so as to facilitate empirical assessment.

Walter Wallace

Synthetic metatheory sorts whole theories into two or more overarching categories. Analytic metatheory parses each theory into two or more components and then sorts these components into categories representing various types of assumptions, observable variables and causal relations among such variables. (p. 53)

Deena Weinstein and Michael Weinstein

Metatheory treats the multiplicity of theorizations as an opportunity for multiple operations of analysis and synthesis (p.140).

[Metatheory] has a concrete empirical referent, sociological theory, and studies its 'underlying structures'... As a... theory of theory, metatheory is simply that part of a general sociology of knowledge which happens to take sociology and sociological knowledge as its empirical referents. From the critic's perspective, there can be nothing objectionable about this kind of metatheory, for it opens up a substantive and empirical research agenda" (pp. 287-288).

Wikipedia

A metatheory or meta-theory is a theory whose subject matter is some other theory. In other words it is a theory about a theory. Statements made in the metatheory about the theory are called meta-theorems.

Shanyang Zhao

Metatheory is a subtype of metastudy that focuses on the examination of theory and theorizing.

References / Bibliography

Abes, Elisa S. "Theoretical Borderlands: Using Multiple Theoretical Perspectives to Challenge Inequitable Power Structures in Student Development Theory." *Journal of College Student Development* 50, no. 2 (2009): 141-57.

Abrams, Dominic, and Michael A. Hogg. "Metatheory: Lessons from Social Identity Research." *Personality and Social Psychology Review* 8, no. 2 (2004): 98-106.

Anchin, Jack C. "A Commentary on Henriques' Tree of Knowledge System for Integrating Human Knowledge." *Theory & Psychology* 18, no. 6 (2008): 801-16.

Anchin, Jack C. "A Commentary on Henriques' Tree of Knowledge System for Integrating Human Knowledge." . "Pursuing a Unifying Paradigm for Psychotherapy: Tasks, Dialectical Considerations, and Biopsychosocial Systems Metatheory." *Journal of Psychotherapy Integration* 18, no. 3 (2008): 310-49.

Argyris, Chris. "Double-Loop Learning in Organizations: A Theory of Action Perspective." In *Great Minds in Management: The Process of Theory Development*, edited by Ken G. Smith and Michael A. Hitt, 261-79. New York: Oxford University Press, 2005.

Bateson, Gregory. *Mind in Nature: A Necessary Unity*. New York: Dutton, 1979.

Bentz, Valerie M., and Jeremy J. Shapiro. *Mindful Inquiry in Social Research*. Thousand Oaks, CA: Sage, 1998.

Bondas, Terese, and Elisabeth O. C. Hall. "Challenges in Approaching Metasynthesis Research." *Qualitative Health Research* 17, no. 1 (2007): 113.

Bondas, Terese, and Elisabeth O. C. Hall. "Challenges in Approaching Metasynthesis Research. Qualitative Health Research." *Qualitative Health Research* 17, no. 1 (2007): 113-21.

Bonsu, Samuel K. "The Relationship between Customer Satisfaction and Economic Performance of the Firm: A Metatheoretical Review." Paper presented at the American Marketing Association. Conference Proceedings, Chicago, 1998, August.

Calás, Marta B., and Linda Smircich. "Organization Theory as a Postmodern Science." In *The Oxford Handbook of Organization Theory: Meta-Theoretical Perspectives*, edited by Haridimos Tsoukas and Christian Knudsen, 596-606. New York: Oxford University Press, 2005.

Campbell, Donald T. "Science's Social System and the Problems of the Social Sciences." In *Metatheory in Social Science: Pluralism and Subjectivities*, edited by Donald W. Fiske and Richard A. Shweder, 108-35. Chicago: University of Chicago Press, 1983.

Campbell, Nancy D. "Multiple Paths to Partial Truths: A History of Drug Use Etiology." In *Handbook of Drug Use Etiology: Theory, Methods, and Empirical Findings*, edited by Lawrence M. Scheier, 29-50. Washington, DC: American Psychological Association, 2009.

Chia, Robert. "Organization Theory as a Postmodern Science." In *The Oxford Handbook of Organization Theory: Meta-Theoretical Perspectives*, edited by Haridimos Tsoukas and Christian Knudsen, 113-40. New York: Oxford University Press, 2005.

Clarke, Richard L. W. "Metaphilosophy / Metatheory: Definitions." http://www.phillwebb.net/Topics/Metaphilosophy/MetaphilosophyDef.htm.

Colomy, Paul. "Metatheorizing in a Postpositivist Frame." *Sociological Perspectives* 34, no. 3 (1991): 269-86.

Commons, M. L., E. J. Trudeau, S. A. Stein, F. A. Richards, and S. R. Krause. "Hierarchical Complexity of Tasks Shows the Existence of Developmental Stages." *Developmental Review* 18 (1998): 238-78.

Craig, Robert T. "Metatheory." SAGE, http://sage-ereference.com/communicationtheory/Article_n244.html.

Critchley, Simon. *Continental Philosophy: A Very Short Introduction.* New York: Oxford University Press, 2001.

Deetz, Stanley. "Describing Differences in Approaches to Organizational Science: Rethinking Burrell and Morgan and Their Legacy." *Organization Science* 7, no. 2 (1996): 191-205.

Dervin, Brenda. "On Studying Information Seeking Methodologically: The Implications of Connecting Metatheory to Method." *Information Processing & Management* 35, no. 6 (November 1999): 727-50.

Dolan, Carl. "Feasability Study: The Evaluation and Benchmarking of Humanities Research in Europe." 51: Arts and Humanities research Council, 2007.

Dubin, Robert. *Theory Building.* Revised ed. New York: The Free Press, 1978.

Edwards, Mark. Organizational Transformation for Sustainability: An Integral Metatheory*. Routledge Studies in Business Ethics. New York: Routledge, 2010.

Edwards, Mark, and Russ Volkmann. "Integral Theory into Integral Action: Part 8." Integral Leadership Review, http://www.integralleadershipreview.com/archives/2008-01/2008-01-edwards-volckmann-part8.html.

Faust, David. "Why Paul Meehl Will Revolutionize the Philosophy of Science and Why It Should Matter to Psychologists." *Journal of Clinical Psychology* 61, no. 10 (2005): 1355-66.

Faust, David, and Paul Meehl. "Using Meta-Scientific Studies to Clarify or Resolve Questions in the Philosophy and History of Science." *Philosophy of Science* 69 (September 2002): S185-S96.

Finfgeld, Deborah L. "Metasynthesis: The State of the Art—So Far." *Qualitative Health Research* 13, no. 7 (2003): 893-904.

Fiske, Donald Winslow, and Richard A. Shweder, eds. *Metatheory in Social Science: Pluralisms and Subjectivities.* Chicago: University of Chicago Press, 1986.

Fuchs, Stephan. "Metatheory as Cognitive Style." *Sociological Perspectives* 34, no. 3 (1991): 287-301.

Gadomski, Adam Maria. "Meta-Knowledge Unified Framework: The Toga (Top-Down Object-Based Goal-Oriented Approach) Perspective." ENEA (Italian Agency for New Technologies, Energy and the Environment), http://erg4146.casaccia.enea.it/HID-server/Meta-know-1.htm.

Glaser, Barney G. "Conceptualization: On Theory and Theorizing Using Grounded Theory." *International Journal of Qualitative Methods* 1, no. 2 (Spring 2002 2002): 23-38.

Goulding, Christina. *Grounded Theory: A Practical Guide for Management, Business and Market Researchers.* Thousand Oaks, CA: Sage, 2002.

Gregor, Shirley. "The Nature of Theory in Information Systems." *MIS Quarterly* 30, no. 3 (September 2006): 611-42.

Haack, Susan. ""The Ethics of Belief" Reconsidered." In *Meta-Study of Qualitative Health Research: A Practical Guide to Meta-Analysis and Meta-Synthesis*, edited by Barbara L. Paterson, Sally E. Thorne, Connie Canam and Carol Jillings. Thousand Oaks, CA: Sage, 2001.

Hall, John R. *Cultures of Inquiry: From Epistemology to Discourse in Sociohistorical Research.* New York: Cambridge University Press, 1999.

Hammersley, Martyn. "Too Good to Be False? The Ethics of Belief and Its Implications for the Evidence-Based Character of Educational Research, Policymaking and Practice." Paper presented at the British educational research association annual conference, Heriot Watt University, Edinburgh, 11-13 September 2003 2003.

Hitt, Michael A., and Ken G. Smith. "Introduction: The Process of Developing Management Theory." In *Great Minds in Management: The Process of Theory Development*, edited by Ken G. Smith and Michael A. Hitt, 1-6. New York: Oxford University Press, 2005.

Hull, David L. *Science as a Process: An Evolutionary Account of the Social and Conceptual Development of Science.* Chicago: University of Chicago Press., 1988.

Kaplan, Abraham. *The Conduct of Inquiry: Methodology for Behavioral Science.* Chandler Publications in Anthropology and Sociology. Edited by Leonard Broom San Francisco: Chandler Publishing Company, 1964.

Kitcher, Philip. *The Advancement of Science: Science without Legend, Objectivity without Illusion.* New York: Oxford University Press, 1993.

Kofman, Fred. "Productivity Killers: Ontological Arrogance." Akialent, http://www.axialent.com/uploaded/press/Ontological%20Arrogance.pdf.

Koller, Kurt. "The Data and Methodologies of Integral Science." Integral World, http://www.integralworld.net/koller.html.

Landreine, Hope, and Elizabeth A. Klonoff. "Commentary [on Chapter 4 - Praxis in Health Behavior Program Development]." In *Handbook of Program Development for Health Behavior Research and Practice*, edited by Steve Sussman, 98-106. Thousand Oaks, CA: Sage, 2001.

Laszlo, Ervin. *Science and the Akashic Field: An Integral Theory of Everything.* 2 ed. Rochester, VT: Inner Traditions, 2007.

Law, John. "Collateral Realities." http://www.heterogeneities.net/publications/law2009collateralrealities.pdf.

Law, John. "Reality Failures." Paper presented at the CRESC Research Residential on Knowledge Failure, Derby, UK, 25-26 February, 2010 2010.

Leong, Siew Meng. "Metatheory and Meta-methodology in Marketing: A Lakatosian Reconstruction." *Journal of Marketing* 49, no. 4 (Fall 1985): 23-40.

Lewis, Marianne W. "Systemic Paradoxes of Organizational Change: Implementing Advanced Manufacturing Technology." In *Cybernetics and Systems Theory in Management: Views, Tools, and Advancements*, edited by Steven E. Wallis. Hershey, PA: IGI Global, 2009.

Lewis, Marianne W., and Andrew J. Grimes. "Meta-triangulation: Building Theory from Multiple Paradigms." *Academy of Management Review* 24, no. 4 (October 1999): 627-90.

Lyotard, Jean-Francois. *The Postmodern Condition: A Report on Knowledge*. Translated by Geoff Bennington and Brian Massumi. Theory and History of Literature, Volume 10. Vol. 10, Minneapolis, MN: University of Minnesota Press, 1984. 1979.

MacIntosh, Robert, and Donald MacLean. "Conditioned Emergence: A Dissipative Structures Approach to Transformation." *Strategic Management Journal* 20, no. 4 (April 1999): 297-316.

Meehl, Paul E. "Cliometric Metatheory: Ii. Criteria Scientists Use in Theory Appraisal and Why It Is Rational to Do So." *Psychological Reports* 91, no. 2 (2002): 339-404.

Meehl, Paul E. *Psychological Reports* 71, no. 2 (1992): 339-467.

Mintzberg, Henry. "Developing Theory About the Development of Theory." In *Great Minds in Management: The Process of Theory Development*, edited by Ken G. Smith and Michael A. Hitt, 355-72. New York: Oxford University Press, 2005.

Morgeson, Frederick P., and David A. Hofmann. "The Structure and Function of Collective Constructs: Implications for Multilevel Research and Theory Development." *Academy of Management Review* 24, no. 2 (April, 1999): 249-65.

Murray, Tom. "Collaborative Knowledge Building and Integral Theory: On Perspectives, Uncertainty, and Mutual Regard." *Integral Review* 2, no. 1 (2006): 210-68.

Ofori-Dankwa, Joseph, and Scott D. Julian. "Complexifying Organizational Theory: Illustrations Using Time Research." *Academy of Management Review* 26, no. 3 (July 2001 2001): 415-30.

Orwell, George. *1984*. New York: Signet, 1949.

Ostroff, Cheri, and David E. Bowen. "Moving Hr to a Higher Level: Hr Practices and Organizational Effectiveness." In *Multilevel Theory, Research, and Methods in Organizations: Foundations, Extensions, and New Directions*, edited by Katherine J. Klein and Steve W. J. Kozlowski, 211-66. San Francisco: Jossey-Bass, 2000.

Overton, Willis F. "A Coherent Metatheory for Dynamic Systems: Relational Organicism-Contextualism." *Human Development* 50 (2007): 154-59.

Paterson, Barbara, Sally Thorne, Connie Canam, and Carol Jillings. *Meta-Study of Qualitative Health Research: A Practical Guide to Meta-Analysis and Meta-Synthesis*. London: Sage, 2001.

Popper, Karl. *The Logic of Scientific Discovery*. Translated by Karl Popper, Julius Freed and Lan Freed. New York: Routledge Classics, 2002. 1959 (English), 1935 (German).

Ritzer, George. *Explorations in Social Theory: From Metatheorizing to Rationalization*. London: Sage, 2001.

Ritzer, George. *Metatheorizing*. Newbury Park, CA: Sage, 1992.

Ritzer, George. "Metatheorizing in Sociology." *Sociological Forum* 5, no. 1 (1990). 3-15.

Ritzer, George. "Metatheory." Blackwell, http://www.sociologyencyclopedia.com.

Ritzer, George. "Sociological Metatheory: A Defense of a Subfield by a Delineation of Its Parameters." *Sociological Theory* 6, no. 2 (1988): 187-200.

Roe, Emery. *Taking Complexity Seriously: Policy Analysis, Ttriangulation and Sustainable Development*. New York: Kluwer Academic, 1998.

Rosenau, Pauline Marie. *Post-Modernism and the Social Sciences: Insights, Inroads, and Intrusions*. Princeton, NJ: Princeton University Press, 1992.

Ross, Sara Nora, and Nancy Glock-Grueneich. "Growing the Field: The Institutional, Theoretical, and Conceptual Maturation of "Public Participation," Part 3: Theoretical Maturation." *International Journal of Public Participation* 2, no. 1 (June 2008): 14-25.

Rousseau, Denise M., Joshua Manning, and David Denyer. "Chapter 11: Evidence in Management and

Organizational Science: Assembling the Field's Full Weight of Scientific Knowledge through Syntheses." *The Academy of Management Annals* 2, no. 1 (2008): 475-515.

Scheff, Thomas J. *Emotions, the Social Bond, and Human Reality*. Cambridge, UK: Cambridge University Press, 1997.

Science Council. "Defining Science." The Science Council, http://www.sciencecouncil.org/DefiningScience.php.

Shotter, John, and Haridimos Tsoukas. "Theory as Therapy: Towards Reflective Theorizing in Organizational Studies." Paper presented at the Third Organizational Studies Summer Workshop: 'Organization Studies as Applied Science: The Generation and Use of Academic Knowledge about Organizations,' Crete, Greece, 7-9 June 2007 2007.

Skinner, Quentin. "Introduction." In *The Return of Grand Theory in the Human Sciences*, edited by Quentin Skinner, 1-20. New York: Cambridge University Press, 1985.

Sklair, Leslie. "Transcending the Impasse: Metatheory, Theory, and Empirical Research in the Sociology of Development and Underdevelopment." *World Development* 16, no. 6 (1988): 697-709.

Smith, Linda Tuhiwai. *Decolonizing Methodologies: Research and Indigenous Peoples*. London: Zed Books, 1999.

Smolin, Lee. *The Trouble with Physics: The Rise of String Theory, the Fall of a Science, and What Comes Next*. Orlando, FL: Houghton Mifflin Harcourt, 2006.

Southerland, John W. *Analysis, Administration, and Architecture*. New York: Van Nostrand, 1975.

Steiner, Elizabeth. *Methodology of Theory Building*. Sydney: Educology Research Associates, 1988.

Stinchcombe, Arthur L. *Constructing Social Theories*. Chicago: University of Chicago Press, 1987. 1968.

Sussman, Steve, and Alan Sussman. "Praxis in Health Behavior Program Development." In *Handbook of Program Development for Health Behavior Research and Practice*, edited by Steve Sussman, 79-97. Thousand Oaks, CA: Sage, 2001.

Takla, T. N., and W. Pape. "The Force Imagery in Durkheim: The Integration of Theory, Metatheory, and Method." *Sociological Theory* 3 (1985): 74-88.

Thorne, Sally, Louise Jensen, Margaret H. Kearney, George Noblit, and Margarete Sandelowski. "Qualitative Metasynthesis: Reflections on Methodological Orientation and Ideological Agenda." *Qualitative Health Research* 14, no. 10 (2004): 13-42.

Treviño, A. Javier. "Sociological Theory at the Crossroads." *Contemporary Sociology* 32, no. 3 (may, 2003 2003): 282-88.

Tsai, Stephen D., Hong-Quei Chiang, and Scott Valentine. "An Integrated Model for Strategic Management in Dynamic Industries: Qualitative Research from Taiwan's Passive-Component Industry." *Emergence* 5, no. 4 (2003): 34-56.

Turner, Jonathan H. "Developing Cumulative and Practical Knowledge through Metatheorizing." *Sociological Perspectives* 34, no. 3 (1991): 249-68.

Turner, Jonathan H. "The Misuse and Use of Metatheory." *Sociological Forum* 5, no. 1 (1990): 37-53.

Van de Ven, Andrew H. *Engaged Scholarship: A Guide for Organizational and Social Research*. New York: Oxford University Press, 2007.

Wacker, J. G. "A Definition of Theory: Research Guidelines for Different Theory-Building Research Methods in Operations Management." *Journal of Operations Management* 16, no. 4 (1998): 361-85.

Wallace, Walter. "Metatheory, Conceptual Standardization and the Future of Sociology." In *Metatheorizing*, edited by George Ritzer, 53-68. Newbury Park, CA: Sage, 1992.

Wallis, Steven E. "The Complexity of Complexity Theory: An Innovative Analysis." *Emergence: Complexity and Organization* 11, no. 4 (2009): 26-38.

Wallis, Steven E. *Intelligent Complex Adaptive Systems*, edited by Ang Yang and Yin Shan, 1-25. Hershey, PA: IGI Publishing, 2008.

Wallis, Steven E. "From Reductive to Robust: Seeking the Core of Institutional Theory." *Under submission - available upon request* (2009).

Wallis, Steven E. "A Sideways Look at Systems: Identifying Sub-Systemic Dimensions as a Technique for Avoiding an Hierarchical Perspective." Paper presented at the International Society for the Systems Sciences, Rohnert Park, California, July 13, 2006 2006.

Wallis, Steven E. "The Structure of Theory and the Structure of Scientific Revolutions: What Constitutes an Advance in Theory?". In *Cybernetics and Systems Theory in Management: Views, Tools, and Advancements*, edited by Steven E. Wallis. Hershey, PA: IGI Global, 2009.

Wallis, Steven E. "A Study of Complex Adaptive Systems as Defined by Organizational Scholar-Practitioners." Fielding Graduate University, 2006.

Wallis, Steven E. *Integral Review* 6, no. 3 - Special Issue: "Emerging Perspectives of Metatheory and Theory" (2010): 73-120.

Wallis, Steven E. "Validation of Theory: Exploring and Reframing Popper's Worlds." *Integral Review* 4, no. 2 (December 2008): 71-91.

Weinstein, Deena, and Michael A. Weinstein. "The Postmodern Discourse of Metatheory." In *Metatheorizing*, edited by George Ritzer, 135-50. Newbury Park, CA: Sage, 1991.

Weisbord, Marvin Ross, and Sandra Janoff. *Future Search: An Action Guide to Finding Common Ground in Organizations and Communities*. 2 ed. San Francisco: Berrett-Koehler, 2000.

Wikipedia. "Metatheory." http://en.wikipedia.org/wiki/Metatheorem.

Wilber, Ken. *The Marriage of Sense and Soul: Integrating Science and Religion*. New York: Broadway Books, 1999.

Wallis, Steven E. *A Theory of Everything: An Integral Vision for Business, Politics, Science, and Spirituality*. Boston, MA: Shambhala, 2001.

Zhao, Shanyang. "Metatheory." SAGE, http://sage-ereference.com/socialtheory/Article_n193.htm.

Footnotes

1. Yet, in theories of physics, we see no such bounding. For example, Ohm's I=E/R relates to volts, amps, and ohms of resistance. Those are the aspects of the theory, and they do not apply to other things such as color or emotion. Therefore, it seems to make more sense to suggest that a theory is self-bounding—based on its constituent propositions.

2. Dubin notes that the process of theory construction may be inescapable as man creates models of the sensory world then uses those models to comprehend the world. Or, put another way, a theorist is one who observes part of the world and seeks to find order Christina Goulding, *Grounded Theory: A Practical Guide for Management, Business and Market Researchers* (Thousand Oaks, CA: Sage, 2002)., p. 5-6. While it may be inescapable, that does not mean it is hopeless or useless. For example, we cannot negate gravity, but we can use it to our advantage (e.g., downhill skiing, accelerating space probes through a "slingshot effect," enjoying the benefits of rainfall, and more).

3. Imagine a poet who is using a word processor. Each time she types a word; the computer inters a random word in its place. Can poetry be made this way? No—the very process of creativity requires reliable tools. During the scientific revolution, each scientist had his own theory—a very creative situation. Today, each person still has the opportunity to have his or her own theory. Yet, there is one prevailing (perhaps dominating) set of laws for electricity. While some might see those laws as infringing on more creative interpretations, I would suggest that those laws enable creative work in the design of electronic devices such as cell phones and computers. And, in turn, those devices enhance creativity, art, and communication in myriad ways.

Chapter 6
The Structure of Theory and the Structure of Scientific Revolutions—What Constitutes an Advance in Theory?

Wallis, S. E. (2010). The structure of theory and the structure of scientific revolutions: What constitutes an advance in theory? In S. E. Wallis (Ed.), Cybernetics and systems theory in management: Views, tools, and advancements (pp. 151-174). Hershey, PA: IGI Global.

Abstract: *From a Kuhnian perspective, a paradigmatic revolution in management science will significantly improve our understanding of the business world and show practitioners (including managers and consultants) how to become much more effective. Without an objective measure of revolution, however, the door is open for spurious claims of revolutionary advance. Such claims cause confusion among scholars and practitioners and reduce the legitimacy of university management programs.*

Metatheoretical methods, based on insights from systems theory, provide new tools for analyzing the structure of theory. Propositional analysis is one such method that may be applied to objectively quantify the formal robustness of management theory. In this chapter, I use propositional analysis to analyze different versions of a theory as it evolves across 1,500 years of history. This analysis shows how the increasing robustness of theory anticipates the arrival of revolution and suggests an innovative and effective way for scholars and practitioners to develop and evaluate theories of management.

Introduction

As scholars, we seek to improve our understanding of management practices. An important part of this process is how we advance our theories. While an advance in understanding might be understood as relating to individual perception, advances in theory relate to the development of formal structures that are communicable, testable, and useable across our discipline. The question of what actually constitutes an advance in theory is still open, and new answers to that question are only now emerging. For example, it has been claimed that a theory of greater complexity should be considered as one that is more advanced (Ross & Glock-Grueneich, 2008). Another approach claims that improved theories are those that combine multiple theoretical lenses (Edwards & Volkmann, 2008). Still another approach suggests that theories of greater structure may be considered more advanced (Wallis, 2008b).

For scholars outside this growing metatheoretical conversation, the standard method for advancing a theory is to determine if that theory works in practice. However, each theorist seems to claim that his or her theory is best, so this is not a very useful measure. Investigating the Faust-Meehl Strong Hypothesis for Cliometric Metatheoretical investigations, Meehl notes that many authors claim their theories are good because they are parsimonious. However, Meehl (2002: 345) notes, this claim is misused, and represents a weak claim for successful theory.

Popper (2002) suggests that the best theories are those that are falsified. Yet, this level of testing seems to represent too high a hurdle for social scientists (Wallis, 2008e). Few theorists even attempt to falsify their own theories, or encourage others to do so. Some authors, in claiming that they have developed an advanced theory, invoke the spirit of Thomas Kuhn and his description and discussion of paradigmatic revolution.

Drawing on centuries of hindsight, Kuhn (1970) developed the idea of scientific paradigms; each of which includes laws, theories, application and instruments which combine to support "coherent traditions of scientific research" (Kuhn, 1970: 10). A paradigmatic revolution is said to occur when the traditions of a science change in significant ways. For example, moving from the Ptolemaic view of the solar system (where the Earth is at the center, surrounded by nested crystal spheres on which are embedded stars, planets, etc.) to a Copernican view where the sun is at the center. Revolutions also result in major improvements to the effectiveness of practitioners. With modern physics, it is possible to have communication satellites, while under the Ptolemaic paradigm, no such achievement would be possible.

Some authors in the field of management claim that their theories are not only effective and useful, but have achieved the status of paradigmatic revolution—ushering in a new age of management, presumably as great as the shift in thinking between Ptolemy and Copernicus. For example, after the development of Total Quality Management (TQM) by Ishikawa, a Kuhnian revolution was claimed. It was argued of TQM that, "All of these characteristics and underlying philosophies point to fundamental changes in the rules of business--a paradigm shift" (Amsden, Ferratt, & Amsden, 1996). While some authors claim revolution, others lend legitimacy to such claims. For example, Clarke & Clegg (2000: 45) refer to a proliferation of paradigms and describe over twenty publications that claim significant paradigmatic changes. They closely investigate some claims of paradigmatic revolution including, "*Transition From Industrial To Information Age Organization*." On the other hand, some authors are content to strongly imply a revolution, as would be found in a shift toward more spiritual management practices (Steingard, 2005). Still others do not make such claims, but explicitly seek revolution in their field (e.g., Stapleton & Murphy, 2003).

The nature of these claims seems to suggest that management science, as with the broader social sciences, does not have a shared understanding of what constitutes a Kuhnian revolution, or even the advance in theory needed for such a revolution. This lack of advance is reflected in management studies where the field is disparaged as being fragmented (Donaldson, 1995) by academics and where practitioners have little interest in the theories of academia (Pfeffer, 2007). In short, these "paradigm wars" lead to a loss of legitimacy from philosophers and practitioners (McKelvey, 2002).

The responsibility for these spurious claims may rest upon Kuhn's shoulders. While he wrote convincingly his focus, "leaves largely intact the mystery of how science works" (Nickles, 2009). In a sense, Kuhn described that houses of theory were built, but did not describe the method of construction. We can say that Kuhn's approach missed the mark in two important and closely related ways. The first was his focus on empirical data as a tool for advancing revolution. That focus, we will briefly explore in this section. Second, in looking at data, Kuhn missed the opportunity to focus on theory. That focus we will investigate (and remedy) in the remainder of this chapter.

For the first focus, Kuhn (1970) highlighted the idea that collecting facts is critical to the advancement of science. For example, he suggests that Coulomb's success in developing a revolutionary theory of electrostatic attraction (EA) depended on the construction of a special apparatus to measure the force of electric charge (Kuhn, 1970: 28). Kuhn provides this kind of specific example for the development of empirical data, but does not provide close descriptions of how scientists used that data to develop their theories. This pursuit of the empirical is exemplified by Popper (2002: 113) who suggests that scientists should begin with relatively arbitrary propositions. Then, they should move to the more serious work of deductive testing and falsification. In his view, it seems that the development of theory is relegated to a secondary status, while objective analysis reigns supreme. This empirical approach has colored the social sciences from the outset—and for good reason.

Social scientists of the early 19th century might have experienced an appreciable envy of their counterparts in physics, who were then reveling in the newfound success of their science. It was as if social scientists had looked up one day to see their counterparts living in comfortable homes of brick—safe behind solid walls of useful theory—while the social scientists languished outside in the cold. Those early social theorists recognized the benefit of having solid walls, but were unsure about the process of building a house of theory. It must have seemed to them that the house of physics was built using factual bricks of empirical analysis. Comte, for example, is said to have developed theories using a positivist, or empirical approach (Ritzer, 2001). However, in the words of

Poincaré, "...a collection of facts is no more a science than a heap of stones is a house" (Bartlett, 1992). When social scientists used that empirical approach, however, the results were disappointing. Instead of solid walls, they had only piles of bricks.

By the middle of the twentieth century, it was becoming clear that social theory was not very useful in practice (Appelbaum, 1970; Boudon, 1986). As a result, three general remedies emerged. One remedy was for scientists to focus on smaller scale systems (Lachmann, 1991: 285). This approach led to the development of organizational studies and management science, as found in the writings of Lewin, McGregor, and others (Weisbord, 1987). Another remedy focused on investigations that were essentially a-theoretical. These "epistemologies of practice" (Schön, 1991) suggested the need for investigation, reflection, and action instead of the act of creating formal theories (Burrell, 1997; Shotter, 2004). Finally, the failure of a social science based on empirical investigation prompted the call for still more empirical investigation—a call that continues today (e.g., Argyris, 2005).

Unfortunately, the results generated by those alternatives do not seem to be any more useful than their predecessors. For example, despite the popularity of these approaches, studies have shown that Total Quality Management (TQM) fails at least 70 percent of the time (MacIntosh & MacLean, 1999), organization development culture change efforts seem to fail about the same rate (Smith, 2003) and Business Process Engineering (BPR) should not be considered a viable approach (Dekkers, 2008). Other authors echo this concern. For example, Ghoshal (2005) suggests that management theory as taught in MBA programs is a contributing factor to serious issues such as the Enron collapse which leads to concerns about the viability of management science in academia (Shareef, 2007). *The essential idea, that social scientists could engage in empirical observation and use the resulting data to create useful theories, appears to have been flawed.*

Kuhn reported that houses of theory were built, and that they were built from empirical bricks, but he did not explain the process by which the bricks were assembled. If we are to understand how to build solid houses of theory, it seems that our focus should be directed toward understanding theory. Only by looking between the bricks can we learn how they are put together. Only by investigating how these houses of theory are assembled, regardless of what empirical building blocks are used, can we find how they are built, and how we may advance management theory.

In this chapter I will investigate one method for objectively measuring theory to ascertain if this method may be used as a path for the advancement of management theory. In a metaphorical sense, I will identify the previously unknown techniques of bricklayers responsible for the well-built house of physics and, from that perspective, suggest how management scientists might build solid houses of useful theory. In this, we will seek to answer Kuhn's question, "Why should the enterprise sketched above move steadily ahead in ways that, say, art, political theory, or philosophy, does not?" (Kuhn, 1970: 160). Several perspectives may prove useful in this investigation. The first comes from developments in systems theory and the closely related field of complexity theory, specifically mutual causality—the idea that everything is interrelated. When applied with academic rigor, this idea of interrelatedness allows the objective analysis of the structure of theory.

Moving forward, we first review background information that identifies important similarities between theories of physics and theories of the social sciences. These similarities allow us to conduct analyses on one form of theory and draw inferences to another. Next, as part of understanding both forms of theory, we will look at the broader context of the growing conversation on metatheory, with a focus on the structure of theory, which may be analyzed in an objective way described in terms of formal robustness. That understanding of theory leads into the main thrust of the chapter where EA (electrostatic attraction) theory is analyzed in various forms as it evolved over time as it evolved over time—moving from antiquity (where the theory merely described curiosities), into modernity (where robust theories supported paradigmatic revolution). By comparing the developmental path of EA theory with the present structure of some theories of management, inferences about the present state of management theory can be made along with suggestions for accelerating the advancement of management theory toward more effective application in practice.

Background

The Common Ground Between Physical and Social Theories

The contrast between the physical sciences and the social sciences may framed in terms of complexity. In classical physics it is generally considered possible to develop predictive theories or laws to explain and forecast the workings of the natural universe because the physical universe is relatively stable and predictable. In contrast, complexity theory suggests that social systems exhibit inherent complexities, understood as non-linear dynamics (Olson & Eoyang, 2001; Wheatley, 1992). Therefore, in the social world, prediction (and the creation of predictive theory) is considered problematic or impossible. This point of view is not a strong one, however, because a statement such as, "It is not possible to create useful theories in social systems" is itself a theory that makes a prediction about a social system. The self-contradictory nature of that position renders it questionable. Therefore, we cannot rule out the possibility of predictive theories within the social sciences (Fiske & Shweder, 1986).

Moreover, the present chapter is not about theory creation, so many of these concerns may be set aside. The goal here is to measure the similarities between existing theories. This is an important distinction because the a-theoretical camp has not shown that theories of management *cannot exist* because, indeed, they do. They have only shown that within the current paradigm of management science we don't have the ability to create *effective* theory, which is reasonable since the evidence shows we don't. The validity of the analysis in this chapter rests on the similarity between theories of physics and management, leading us to ask: How might the two be compared?

The essential commonality between theories explored in the present chapter is found in the *structure* of the theory as seen in the interrelated nature of the propositions contained within the theory. Theories of physics and theories of management both contain interrelated propositions. For an example, physicist Georg Ohm developed the proposition (for a simple electrical circuit) that an increase in resistance and an increase in current would result in an increase in voltage. As an example from management theory, Bennet & Bennet (2004) suggest that (in an organization) more individuals and more interactions will result in more uncertainty. This similarity between propositional structures in theories of physics and theories of management provides a basis for comparison. Thus, structural inferences from one set of theories may be applied to another set of theories with some level of reliability.

Of course, other aspects are not held in common between theories of physics and theories of management—specifically where theories from each discipline are used to describe relationships among different things. For example, Newton's laws describe the motion of planetary bodies in the context of our solar system, while management theory describes some relationships between humans in the context of the workplace. In this chapter we are focusing on the theories themselves, not the things described by those theories. Therefore, as demonstrated above, the comparison between structures of theories should hold true. *Metaphorically, we are not trying to differentiate between the bricks of physics and the rough stones of management; we are looking at the mortar that can serve equally well to hold them all together.*

Further, we are not testing the process by which the theories were created, such as the formal process of grounded theory (Glaser, 2002) or more intuitive methods (Mintzberg, 2005). Neither will we consider the falsifiability (Popper, 2002) of those theories per se, although this analysis does suggest some insights and opportunities for further investigations along those lines. Because this is essentially a metatheoretical investigation, we begin with a brief explication of metatheory.

Metatheory

In previous decades, the term metatheory meant the speculative construction of one theory from two or more theories (Ritzer, 2001). This understanding of metatheory is being superseded by scholars who use the term to describe investigations into the structure, function, and construction of theory (Wallis, 2008b, 2008e) as well as the more carefully considered construction of overarching theory (Edwards & Volkmann, 2008) and investigations into the validity of theory as related to the complexity of theory (Ross & Glock-Grueneich, 2008), as well as the investigations of other authors in the present volume.

The "theory of theory" or metatheoretical conversation draws on insights from Kuhn (1970), Popper (2002), and Ritzer (2001), as well as methodologies from Stinchcombe (1987), Dubin (1978), and Kaplan (1964). More recently, notable scholars such as Weick (2005), Van de Ven (2007), Starbuck (2003) and others have summarized our present understanding of theory development and called for a new look at theory. The goal of this conversation is to develop a better understanding of theory—including how theory is created, structured, tested, and applied—in order to engender better theories and support the development of improved applications.

Theory may be understood as, "an ordered set of assertions about a generic behavior of structure assumed to hold throughout a significantly broad range of specific instances" (Southerland, 1975: 9). Theory is of key importance to practitioners as, "practice is *never* theory-free" (Morgan, 1996: 377) and to scholars because "there are *no* facts independent of our theories" (Skinner, 1985: 10) (emphasis, theirs). While Burrell (1997) suggests that theory has failed, others call for better theory (e.g., Sutton & Staw, 1995). This investigation falls into the later camp. This topic is of critical importance to our field because, "What constitutes good, useful, or worthy theory in our field remains up in the air and cannot be resolved through empirical validation alone." (Maanen, Sørensen, & Mitchell, 2007: 1153). This echoes Nonaka's (2005) argument that the creation of theory requires more than the traditional, positivist approach of seeking objective facts.

While there has been a great deal of conversations around the creation of theory (e.g., K. G. Smith & Hitt, 2005) and the testing of theory (Lewis & Grimes, 1999; Popper, 2002), these conversations have not shown efficacy in advancing theory. Although, some academicians seem to imply that the creation of more theory is taken as a reasonable measure of success. For example, one web page notes that an accomplished professor emeritus, "...is the sole author of six books. He is author or co-author of over 100 papers" (GMU, 2009). While these are certainly impressive numbers, there was no mention of the value of the work or its application in the world—the value sits upon a shelf. Another popular method for determining accomplishment is counting citations. These methods (and others) might indicate some sort of popularity (Wallis,

2008a), but do not seem to indicate any way to advance a theory. Indeed, in a recent HERA study, the author admits that there does not seem to be any reliable method for evaluation (Dolan, 2007). Another approach is called for.

Structure of Theory

Kaplan (1964: 259-262) suggests six forms of structure for theoretical models. His forms of structure include a literary style (with an unfolding plot), academic style (exhibiting some attempt to be precise), eristic style (specific propositions and proofs), symbolic style (mathematical), postulational style (chains of logical derivation), and the formal style. The formal style avoids "reference to any specific empirical content" to focus instead on "the pattern of relationships."

Dubin's (1978) approach is similar, in that he suggests how theories of the highest "efficiency" are those that express "the rate of change in the values of one variable and the associated rate of change in the values of another variable" (Dubin, 1978: 110). Such relationships might be understood as structural in the sense that they represent causal and co-causal relationships between events. Because those relationships are well explained, we might expect such a theory to be more useful in practice, allowing the practitioner to use the theory to understand or predict those changes. In short, we might expect that theories with a higher level of structure to be more effective in practice.

Because theory indicates changes between multiple interrelated events, the structure of theory might be understood as a system. Briefly, the history of systems theory might be best seen in Hammond (2003) while the breadth of the theory might be best seen from the systemic perspective of Daneke (1999). The application of systems theory to management is provided in Stacey, Griffin, & Shaw (2000) while Steier (1991) draws on an understanding of cybernetics to explore reflexive research and social construction. More relevant to the present study, advances in complexity theory, systems theory, and cybernetics suggest that a systemic perspective might provide a useful lens for viewing management theory (e.g., Yolles, 2006).

The systems perspective may be applied to a conceptual system. In this case, the present methodology focuses on the systemic relationship between propositions in a body of theory measured in terms of "robustness" (Wallis, 2008b). It should be noted, by way of clarification, that the understanding and use of robustness as used here is different from a more common understanding, where robustness might be understood as strong, resistant to change, longevity, or widely distributed. Rather, *robustness refers to the specific and objective measure of the relationship between propositions in a theory.*

The idea of a robust theory comes from physics and mathematics, and represents a theory with complete internal integrity. For example, Ohm's law of electricity (E=IR) is considered to be a robust theory because it is amenable to algebraic manipulation—that is to say, this formula is equally valid if written as I=E/R, which means it is equally valid whether it is used to find volts (E), amps (I), or resistance (R). When I undertook to understand the structure of theory, I drew on insights from dimensional analysis, systems theory, Hegelian dialectic, and Nietzsche's insights into the co-definitional relationships between the dimensions of those dialectics. These, and other ideas, I combined to develop the process of Reflexive Dimensional Analysis (Wallis, 2006a, 2006b). That methodology was further refined (Wallis, 2008b) to develop a method of propositional analysis that could be used to objectively determine the robustness of a theory.

By way of background, a causal proposition describes a relationship between aspects—where each aspect relates to some observable or conceptual phenomena. For an abstract example, a proposition might be represented as, "A causes B" (or, "changes in A cause changes in B"). A co-causal proposition might be, "Changes in A and B cause changes in C." Such co-causal propositions are described as concatenated (Kaplan, 1964; Van de Ven, 2007).

Van de Ven suggests that concatenated concepts may be difficult to justify and suggests, instead, that the more commonly used chain of logic is the better way to construct a theory. A chain of logic might be understood as explaining how changes in A are caused (or explained) by changes in B that are caused by changes in C. Conversely, Stinchcombe (1987), suggests that such a chain is less effective because any intermediary terms (B, in this abstract example) are redundant and so do not represent a useful addition to the theory. Further, any such chain must ultimately rest on some unspoken assumption. Extending the abstract example, the chain of steps (A, B, C, etc.) continues until the argument reaches a point where everyone agrees that some foundational claim is "true" (perhaps Z, in this case). However, insights developed from Argyris' "ladder of inference" suggest that those underlying assumptions are not necessarily reliable guides (Senge, Kleiner, Roberts, Ross, & Smith, 1994: 242-243). In short, relying on unspoken assumptions may lead to folly as easily as wisdom.

The robustness of a theory may be objectively determined in a straightforward manner (for an in-depth example, see Wallis, 2008b). First, the body of theory is investigated to identify all clear propositions. Those propositions are then compared with one another to identify overlaps and redundant aspects are dropped. Second, the propositions are investigated for conceptual relatedness between the aspects described in the propositions. Those propositions that are causal in nature are conceptually linked with aspects of the theory that are resultant (each aspect may then be understood as a dimension representing a greater or lesser quantity of some aspect or phenomena). Those resultant aspects that are described by two or more causal aspects are understood to be concatenated and are considered to be more complex, more complete, and more useful than aspects that are not as complex, or as well structured.

Third, the number of concatenated aspects in the theory are divided by the total number of aspects in the theory to provide a ratio—a number between zero and one. This ratio is the robustness of the theory and represents the degree to which the theory is structured. A value of zero represents a theory with no robustness, as might be found in a bullet point list of concepts with no interrelationship between them. A theory with a value of one suggests a fully robust theory; an example is Ohm's E=IR. Because of the successful application of robust theories in math and physics, it may be expected that *a robust theory of the social science can be reliably applied in practice, and will be more easily falsifiable in the Popperian sense* (Wallis, 2008c). Metaphorically, bricks that are directly mortared to other bricks would be highly robust (as found in a structured wall or home), while bricks that are scattered about would not be robust at all. A pile of bricks would be somewhere in the middle (as a pile might have slightly more structure than scattered bricks, though far less structure than a home).

For an abstract example of determining robustness, consider a theory of five aspects (A, B, C, D, and E), each representing differentiable concepts or phenomena. The causal relationships between these aspects are suggested by two propositions: (1) A causes B; and (2) More C and more D results in more E. Of these, only E is concatenated because there are two aspects of the theory that are causal to E. Therefore, the robustness of this theory is 0.20 (the result of one concatenated aspect divided by five total aspects).

This method of propositional analysis allows us to examine a theory and assign a relatively objective measure of that theory's structure. With this method of measurement in hand, we can apply the yardstick of robustness to theories across time. And, importantly, we can determine if the robustness of the theory is increasing, decreasing, or merely wandering.

Forward to Revolution

According to Kuhn, a paradigmatic revolution is a situation where, "the older paradigm is replaced in whole or part by an incompatible new one" (Kuhn, 1970: 92). Kuhn also suggests that a scientific revolution occurs when a new paradigm emerges, one that provides a better explanation, answers more questions, and leaves fewer anomalies. Such a revolution in management is expected to bring more effective theories and practices to managers. However, Kuhn's description of revolution is problematic because it does not describe how much better the explanation must be. Nor does it describe exactly what reduction in anomalies must occur for a paradigmatic change to be considered revolutionary. There is no method of measurement. This issue was highlighted recently when Sheard (2007) framed the conversation as a contrast between superficial and profound revolution and asked, "Who decides what is 'profound'?" Yet, in his reflections, Sheard shied away from establishing a metric for delineating revolution stating that revolutions, "may be qualitatively sensed, but are not amenable to any ratio of distinction" (Sheard, 2007: 136).

This lack of distinction between superficiality and profundity may lead to spurious and conflicting claims for the existence of revolutionary change. Taking an example from social entrepreneurship theory, many authors agree that the act of social entrepreneurship is an important part of that theory (e.g., Austin, Stevenson, & Wei-Skillern, 2006; Bernier & Hafsi, 2007; de Leeuw, 1999; Guo, 2007; Mort, Weerawardena, & Carnegie, 2003). Yet, Fowler (2000) suggests that the focus is not so much the act of social entrepreneurship as it is the social value proposition created by that act. Is this difference between Fowler and others revolutionary? Following Kuhn's example of historical evaluation, it is impossible to know without centuries of perspective; which, in turn, renders the concept of paradigmatic revolution useless for any conversation around contemporary issues. So scholars and practitioners continue to claim revolutionary improvement without any measure of what that means. In short, claims of revolutionary theory are unsupportable because there is no clear understanding of what constitutes a revolution.

While scholars earn their pay by arguing points such as these, managers are rewarded for effectively applying those theories in practice. Academia does not seem to be producing any useful tools for today's managers; indeed, "Management research produced by academics does not fare particularly well in the marketplace of ideas that might be adopted by managers" (Pfeffer, 2007: 1336). Pfeffer goes on to cite the work of Mol & Bikenshaw (forthcoming) who suggest that of the 50 most important innovations in management, none of them originated in academia. He also cites Davenport & Prusak (2003) as noting that business schools "have not been very effective in the creation of *useful* business ideas" (emphasis theirs). Obviously, this does not bode well for the social sciences in general or business schools, in particular.

In short, the current paradigm of social sciences in the social sciences (in general) and management science (in particular) has not produced anything that managers can reliably apply for great effectiveness, let alone anything that might be considered revolutionary. Meanwhile, business students must wonder about the value of their tuition and managers must spend their lives without knowing what, if any, theory to apply for successful practice. This leads us to consider a challenging possibility: If we can identify a quantifiable link between management theory and paradigmatic revolution, we may be able to evaluate the usefulness of theory in terms of its potential for revolutionary implementation. That, in turn, would allow us to predict, and or instigate, a revolution in management theory.

A Working Hypothesis

An important part of a systemic perspective is to avoid looking at "things" because an improved understanding may be gained by looking at the relationships between them (Ashby, 2004; Harder, Robertson, & Woodward, 2004).The present chapter follows that suggestion by investigating the co-causal relationships between aspects within a theory in terms of that theory's robustness. Instead of looking to our empirical bricks of data, we will focus on the theoretical mortar that binds them together. The present hypothesis suggests that *revolution is enabled by the structure of the theory rather than some notion of objective data*. To investigate this idea, I will revisit Kuhn's work—investigating it from a metatheoretical perspective.

Within his descriptions of paradigm revolution, Kuhn provides several examples of revolutionary theorists and their fields of study. Among others, he mentions Coulomb (and his work in advancing electrical theory), Newton (mechanical motion), and Einstein (relativity). Each of those scientists developed the final theory—advancing his field to paradigmatic revolution. The theories developed by these exemplars of scientific revolution are similar in at least one important way. Each has a robustness of 1.0 (on a scale of zero to one). For example, Newton's F=ma has three aspects; each one concatenated from the other two. Therefore the robustness of Newton's formula is 1 (the result of three concatenated aspects divided by three total aspects).

Is it only coincidence that all these revolutionary theories have perfect robustness? If so, it seems an odd sort of combination, especially given the great differences between some areas of study. For example, who would imagine, a priori, that a theory involving electricity and a theory involving planets would have the same structure? Yet, the robustness of both theories is the same (Robustness = 1.0) as are many effective theories in physics. Such a relationship between structure and efficacy should not be too surprising because (as noted above) theories with a higher level of structure have long been expected to be more advanced. The problem, in physics and management, has been the lack of a standardized method for measuring the structure of a theory. Therefore, instead of seeking steadily higher levels of structure, management theorists were content to reach a level of logical structure deemed acceptable by the editorial/review process; there was no incentive to advance beyond that mark.

So, rather than dismiss this relationship out of hand, the following analysis suggests that there is some sort of connection between the structure of a theory and the usefulness of that theory in practice. Simply put, it is suggested that *the robustness of a theory may be understood as a key indicator of a scientific revolution.* The implication of this insight is very significant. If there is an objective approach to identifying revolutionary theory when it emerges, the development of that theory can be accelerated, thus enabling a revolution in management theory within decades instead of centuries, and, importantly, providing the attendant benefits to humanity and our understanding of our social world.

If the robustness of theory were, as hypothesized, a valid indicator of paradigmatic revolution, we would expect to see the development of a theory from low robustness to high robustness over some length of time. This expectation raises a question: How does the robustness of those theories change over time?

In choosing a body of theory to study, it should be stated here that no paradigmatic revolution has occurred in management theory. Perhaps this lack is why managers express frustration in their need for usable knowledge (Czarniawska, 2001) of the sort that would be provided by useful theories. Because of this lack, we cannot draw upon management theory for the present analysis. Therefore, we must investigate the history of some other body of theory. Because Kuhn used the example of electrical theory as part of a revolutionary paradigm, that area of theory appears to offer a reasonable area of study. And, as noted above, because the analysis is limited to the structure of the theory, the inferences drawn from EA (electrostatic attraction) theory may be transferred to the structure of management theory. Thus, for this analysis, I will investigate the evolution of EA theory and determine the level of robustness during different stages of the development of this theory. Mimicking Kuhn, for data I will draw on "The Development of the Concept of Electric Charge: Electricity from the Greeks to Coulomb" (Roller & Roller, 1954).

Analysis

In the present section, the method of propositional analysis will be applied to differing versions of the theory of electrostatic attraction as found in Roller & Roller, as that theory was understood at different points through history. The present history begins with a revolution in thinking that occurred when ancient Greeks began to explain, rather than simply describe, phenomena they encountered.

Roller & Roller present Plutarch as an example of thinking about this time. Following the example of Roller & Roller, this particular analysis includes magnetism, because, in ancient times, both magnetism and what we now understand as electrostatic attraction were believed to represent the same phenomena.

Around the year 100 CE, Plutarch wrote that lodestones (naturally occurring magnets) exhale, thus pushing the air, which would then push objects of iron. Amber behaved the same, except that amber needed rubbing to encourage it to exhale. The exhalations of amber would then push the air, which then would effect small objects (such as hair) instead of iron (Roller & Roller, 1954: 3).

Deconstructing Plutarch's theory into its essential propositions, it may be said that rubbing (amber) creates exhalations; exhalations push air; and air pushes small objects. Magnets exhale, exhalations push air, air pushes iron. In this, it may be seen that there are seven aspects of the theory (Rubbing, Amber, Magnet, Air, Iron, Push, Small objects). Magnets are causal to Push, which is causal to the movement of Air and moving Air causes Push, which is causal to the movement of Iron. In this sequence, there are four aspects and each one is the result of only one other aspect. None of them are concatenated. Indeed, Exhalations are synonymous with Push, as both appear to be a general representation of some form of force.

Instead of concatenation, the relationship between these aspects may be understood as linear (Stinchcombe, 1987: 132). Where, for an abstract example, it may be said that A changes B, which changes C. In such a relationship, Stinchcombe notes, the concept of B is redundant. Therefore, it may be seen that Plutarch's theory contained a redundant term by including the concept of air in the model. Redundant terms detract from the robustness of the model. Similarly, Occam's razor suggests the need for parsimony in theory construction. In these older versions of the theory, the linear relationships between multiple aspects of the theory are examples of a lack of parsimony and therefore a weakness in the structure of the theory.

In contrast to linear relationships between some aspects of the model, Rubbing and Amber, together, cause Exhalations (Push); that Push is causal to the movement of Air, which is causal to the movement of Small objects. Here, the Push may be understood as a concatenated aspect of the theory because it is caused by a combination of Rubbing and Amber.

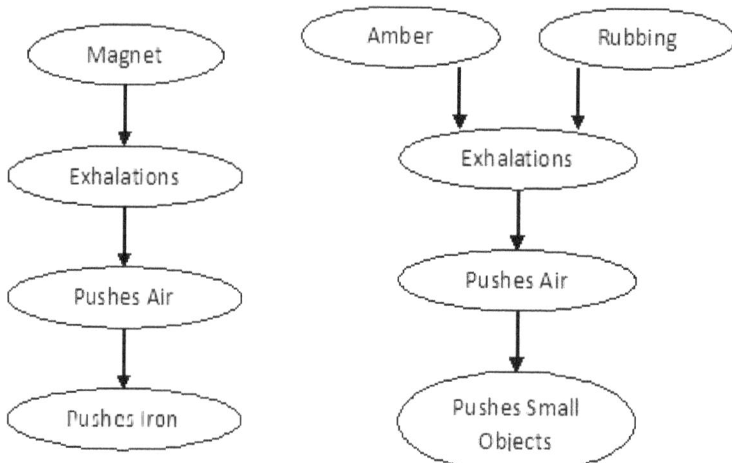

Figure 1 *Plutarchean Electrostatic Attraction Theory*

In determining the level of robustness of this version of theory by propositional analysis, it should be noted there are seven aspects—only one of which is concatenated. The others have simple, linear, causal relationships. Therefore, the robustness of Plutarch's theory may be set at 0.14 (the result of one concatenated aspect divided by seven total aspects).

The next clear demarcation of theory surfaces in 1550. Jerome Cardan theorized that the rubbing of amber produced a liquid. And, that dry objects (such as chaff) would move toward the amber as they absorbed the liquid (Roller & Roller, 1954: 4-5). Here, there are six aspects (Rubbing, Amber, Liquid, Movement, Absorption, and Dry objects). And, as with Plutarch's model, only one is concatenated (Rubbing and Amber together produce Liquid). Therefore, the robustness of Cardan's theory is 0.17 (the result of one divided by six).

Although Cardan suggests a liquid instead of Plutarch's exhalations of air, the theories are similar in their structure and level of robustness. The relationship between these theories stands as an example of how theories may appear to change because the terminology has changed. That change in terminology may be accompanied by changes in underlying assumptions, philosophical justifications, or even simple changes in speculation. Yet, looking at the theory through the metatheoretical lens of robustness, it becomes clear that no useful change has occurred at all.

Roller & Roller note that the scientific revolution took hold about this time. More scientists began to investigate electrical phenomena. Those scientists began to develop new insights and a new vocabulary to relate the results of their experiments. About this time, every important experimenter had his own theory (Kuhn, 1970: 13). Continued experimentation led to the discovery that objects besides amber would exhibit what is now called electrostatic attraction when they were rubbed. About 1600, Gilbert called this class of objects, "electrics."

According to Roller & Roller (1954: 10-11) Gilbert's version of theory adds heat to the theory and suggests that rubbing and heat (but only heat from rubbing), and electrics, causes the release of an invisible liquid which then causes the attraction of dry objects. A moist barrier would interfere with that attraction. Additionally, Gilbert held that amber, glass, and gems all belonged to a class of material called "wet." So, he concluded, that wet things were electric things. Yet, he found that wet things that softened or melted (such as wax and ice) would not attract. Similarly, electrics with impurities could not be rubbed to develop EA. Therefore, more aspects and more propositions were added to the theory suggesting that electrics were the result of "wet" things with higher melting points and lower levels of impurities.

Gilbert's theory of magnetism suggested that each magnet possesses a "form" and that form would awaken a similar form in particles of iron that was near to the magnet. Indeed, he noted that the nearer the iron was to the magnet, the more mutual attraction would occur. Here, attraction may be understood as a synonym for movement.

In total, Gilbert's theory included 14 aspects (Rubbing, Heat, Electrics (objects that exhibit EA), Impurities, Melting point, Attraction, Liquid, Dry objects, Moist barriers, Magnets, Form, Awakening, Iron, Proximity). Of these, Electrics may be said to be concatenated because they are formed with more Liquidness and fewer Impurities. Similarly, Attraction is concatenated because it is generated by more Electrics, more Rubbing, and more Heat. Additionally, it may be said that Magnets and greater Proximity results in more Awakening. Therefore, Gilbert's version of electrical theory has a robustness of 0.21 (the result of three concatenated aspects divided by 14 total aspects).

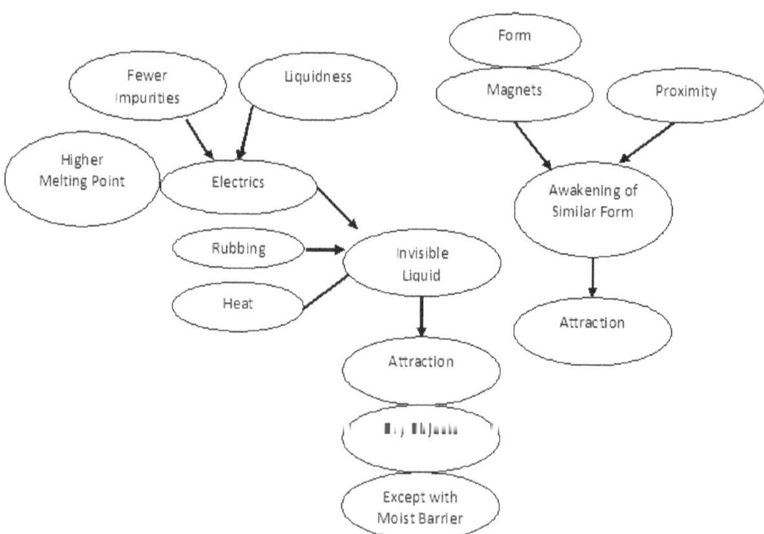

Figure 2 *Gilbertian version of Electrostatic Attraction Theory*

In his time, Gilbert might have claimed that his own theory of attraction explained more than Plutarch's theory. The moist barrier, for example, was not a part of Plutarch's theory. Because Gilbert's theory explains more than previous theories, today's management theorist might be tempted to claim that it represents a scientific revolution. Yet, Kuhn does not suggest that a revolution occurred until much later. This example suggests how today's general academic understanding of what constitutes a revolution is unclear, and how that lack of clarity opens the door for false claims of revolutionary advancement.

With continued experimentation, scientists generated more innovative terminology. By the mid 18th century, the "two-fluid theory" had emerged (Roller & Roller, 1954: 47-48); described as:

- There are two kinds of electric fluids (vitreous & resinous).
- Unelectrified objects contain equal amounts of the two fluids.
- Rubbing an object removes one of the two electric fluids.
- More imbalance results in greater strength of electrification.
- Touching objects will cause fluid to flow from one object to another, and so de-electrify the objects as the levels of electric fluid come into balance.
- Objects with similar fluids repel one another when they are near.
- Objects with differing fluids attract one another when they are near.

In this theory, there are eleven aspects (Vitreous fluid, Resinous fluid, Objects, Rubbing, Balance, Electrification, Repulsion of objects, Attraction of objects, Flow of fluid, Touching, and Nearness). The relationships leading to concatenated aspects may be summarized as:

- Rubbing and Objects decreases Balance thus increasing Electrification.

- Touching and Electrification and Objects cause Flow which then increases Balance and so decreases Electrification.
- Objects and Nearness and Balance cause Attraction.
- Objects and Nearness and lack of Balance cause Repulsion.

Note that many aspects are described in terms of linear relationships. For example, Flow increases Balance which decreases Electrification. It may be seen that the four aspects of Balance, Flow, Attraction, and Repulsion are concatenated by their relationship to the other aspects. Therefore, the robustness of this theory is 0.36 (the result of four concatenated aspects divided by eleven total aspects).

The existence of many theories, and many aspects, meant that more new theories were called for thus creating, "a synthesis that serves not only to reconcile the contradictory features, but to provide explanations of a wider range of phenomena" (Roller & Roller, 1954: 81).

In what may be considered the final stage of development of this theory, from the perspective of Roller & Roller, Charles Coulomb developed his theory about 1785. Coulomb focused on only three aspects. Force, Electric charge, and Distance. He determined that the Force was equal to the Distance (squared) divided by the Charge. He also found that the Distance (squared) is equal to the Force multiplied by the Charge. And, the Charge is equal to the distance (squared) divided by the Force.

Therefore, it may be understood that each of the three aspects of Coulomb's theory is concatenated from the other two. Therefore, the robustness of Coulomb's theory is 1.0 (the result of three concatenated aspects divided by three total aspects). One benefit of developing a theory with a robustness of 1.0 was that mathematical techniques could now be used in conjunction with the experimental process. The result of this combination was revolutionary.

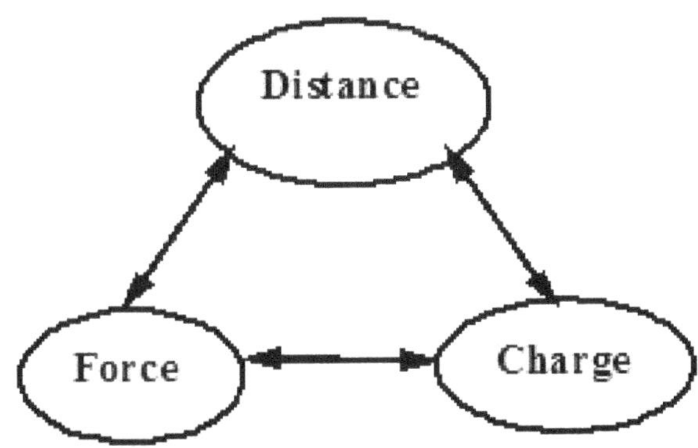

Figure 3 *Coulomb's Theory of Electrostatic Attraction*

Discussion

The above investigation used information from Roller & Roller (1954) to benchmark the development of EA theory at five points across hundreds of years. This objective analysis used the total number of aspects for each theory, and the concatenated aspects of those theories to identify the formal robustness of each of the five theories. That information is summarized in Table 1.

Year	Total Number of Aspects	Number of Concatenated Aspects	Robustness	Name of theorist or theory
100	7	1	0.14	Plutarch
1550	6	1	0.17	Cardan
1600	14	3	0.21	Gilbert
1750	11	4	0.36	Two Fluid theory
1785	3	3	1.0	Coulomb

Table 1 *Summary of Aspects and Robustness of Theories*

The graph in figure 4 shows the variation in the total number of aspects and the number of concatenated aspects over time. Note the pre-revolutionary surge during the scientific revolution. This period of increased experimentation led scientists to suggest more aspects and to identify more relationships between those aspects. This may be understood as the time when the "scientist in crisis" is generating "speculative theories" (Kuhn, 1970: 87). Those theories may lead to a "critical mass" of ideas (Geisler & Ritter, 2003) and generate communities of practice (Campbell, 1983: 127). This peak lends validity to the idea that theory which is more complex may be considered more advanced (Ross & Glock-Grueneich, 2008).

Also, note the change in focus over time. Early on, the body of theory included magnetics and electrostatics because both were thought to represent the same essential effect. As time and experimentation progressed, the field of study narrowed to the study of electrostatic attraction, only. Yet, despite the narrowed focus, the number of aspects that may be said to define theory continued to increase.

The same phenomenon appears to be occurring within the social sciences. In the field of management theory, the fragmentation of the field has been reported as problematic (Donaldson, 1995); yet, that fragmentation might also be understood as a narrowing of the focus—with each fragment representing a more focused sub-field.

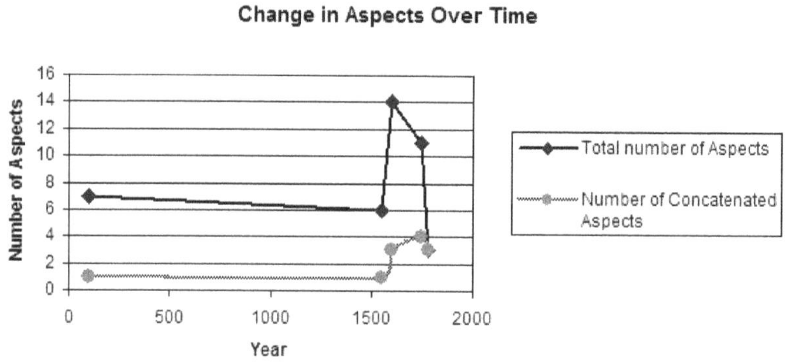

Figure 4 *Change in Aspects Over Time*

In Figure 5, the robustness of theories is plotted over time. Note the rapid increase in robustness on the right hand of the chart. This rise corresponds with Kuhn's description of scientific revolution culminating in paradigmatic change. The asymptotic change suggests an "event horizon" or a "phase shift" beyond which a theory may be considered a law. After this point in paradigm change, the focus would not be on developing new forms of theory, rather the focus would be on conducting empirical analysis to verify and/or falsify the theory. Once verified, the focus would shift to the application of theory in practice as a useful tool. This kind of change has not been found in management theory.

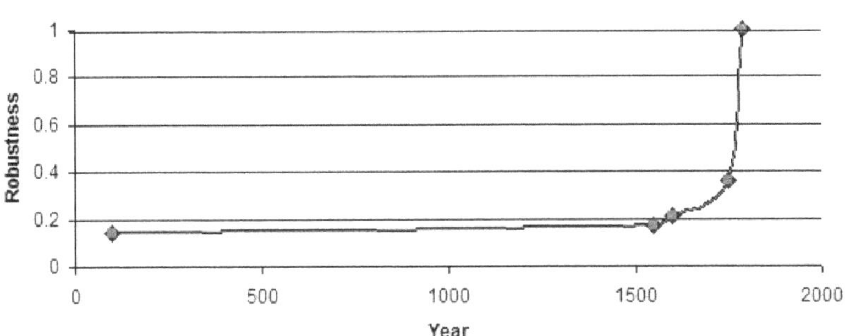

Figure 5 *Robustness Over Time*

The asymptote at the right-hand side of figure 5 is also suggestive of a "power curve," a vertical (or nearly so) line that stands as a diagrammatic indicator that something significant has occurred, such as a quantum increase in the capacity of a system (Kauffman, 1995). In this case, the system under consideration is the structure of a theory. The theory at the top of the curve has significant capacity for enabling action, where theories at the bottom of the curve have very limited capacity. In short, theory with higher robustness has greater capacity to support paradigmatic revolution than theories of lower robustness.

The relationship between robustness and paradigmatic revolution suggests that the robustness of a theory may be used as a milestone for marking the objective development of a theory and progress toward revolution. Similarly, the robustness of a theory may also be considered something of a predictor for such a revolution.

In addition to the relatively passive approach of tracking changes in theories developed by others, a more activist approach may also be inferred. That is to say, if a theorist works toward the purposeful creation of robust theory, she may be able to purposefully spark a paradigmatic revolution. With that opportunity, there are also limitations. For example, we should probably disregard a speculative theory that seems to represent a robust relationship if it has no basis in reality.

To conclude this section, the advancement toward robustness appears to be a useful indicator and potential instigator of paradigmatic revolution. The test of robustness might be understood as a validation of theory in the Popperian sense, although it is validation with considerably more rigor than Popper appears to have considered in his time. While validation is useful and necessary, there still remains the need to falsify those theories, as suggested by Popper (2002). In the following two sub-sections, I will investigate the implications of advancing management theory toward robustness. The first will relate more to academicians, who might be more interested in advancing management theory. The second will relate more to practitioners, who might be more interested in the application of robust theory.

Whither Management Theory?

To date, no comprehensive test of robustness has been conducted of the field of management theory. However, some indicator of the field may be inferred from tests of robustness performed on bodies of theory from related sub-fields. In figure 6, the robustness of each of eight theories are indicated on the curve from figure 5—the development of EA theory. By showing those theories in relation to the time required to develop EA theory to robustness, we may infer how much time might be required for each body of theory to reach a useful level of robustness. This is, of course, working under the assumption that the current paradigm of management science might follow the same trajectory as the EA theory—a question that is very much open for consideration.

Briefly, those eight studies are related to management theory as follows. Social entrepreneurship theory is part of the general rubric of management theory for the simple reason that management is sometimes understood in terms of an entrepreneurial activity. A study of social entrepreneurship theory found a robustness of 0.13 (Wallis, 2009d). Integral theory has been applied to a wide range of disciplines in the social sciences (including management). A study of integral theory found a robustness of 0.10 (Wallis, 2008d). A study of a structure of ethics found a robustness of 0.15 (Wallis, 2009f). Organizational learning theory, has been found to have a robustness of 0.16 (Wallis, 2009c). Peak performance theory has a robustness of 0.17 (Wallis, 2009e). Institutional theory is a little better with a robustness of 0.31 (Wallis, 2009b). Higher levels of robustness are found in studies related to systems theory. A study of complexity theory, as it has been applied to organizations, finds a robustness of 0.56 (Wallis, 2009a). And, a study of Complex Adaptive Systems (CAS) theory as it relates to management and organizations finds a robustness of 0.63 (Wallis, 2008b).

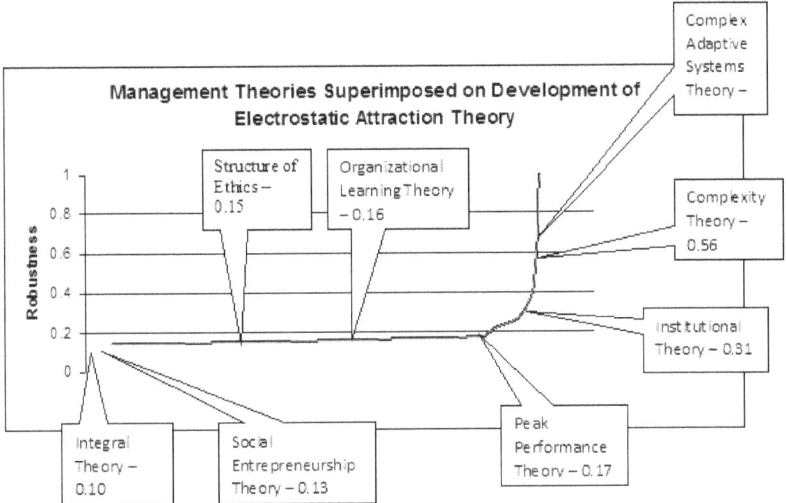

Figure 6 *Robustness of Some Management Theories*

It should be noted that the areas of theory with the highest levels of robustness are those that are more closely related to systems theory (i.e., complexity theory and CAS theory). Therefore, it may be suggested that future studies of management should utilize approaches based in systems theory, cybernetics, and complexity theory. More studies of robustness are required in and of the field of management theory to confirm this idea.

From figure 6, we may begin to answer the question, "How long until management theory experiences a true paradigmatic revolution?" Most theories of management noted here have a robustness at or below Cardan's theory of EA (Robustness = 0.17). When Cardan presented his theory in 1550, Coulomb's revolutionary theory was a distant 235 years away. When the two-fluid theory of electricity emerged in 1750 (R = 0.36), Coulomb's discovery was a relatively modest 35 years in the future. Some management theories today, specifically in CAS theory and complexity theory, have levels of robustness that exceed the two fluid theory, suggesting those theories might achieve revolutionary status much sooner. *Accelerating management theories toward robustness (and related revolution in improved efficacy) is important because of the depth and breadth of problems faced by practitioners.*

Linking Theory and Practice

Managers, consultants, CEOs, and all kinds of practitioners might read this chapter and ask, "What's in it for me?" In brief, then, this study suggests that theories of higher robustness are expected to be more effective in application. Further, we may now purposefully advance a theory toward robustness and so enable a paradigmatic revolution. With a paradigm revolution, improvements in theory are associated with highly effective improvements in practice.

To encapsulate an example from history, the low robust theories of EA were associated with explaining mere curiosities, such as the way that hair is attracted to amber because of exhalations. In management practice, this is like a performance management guru claiming that a stitch in time saves nine. Perhaps it is true in some metaphorical sense, but making it work in practice is entirely in the hands of the employee. Worse, there is no way to tell if these kinds of claims are actually true. This means that managers are likely relying on false information. In short, managers might be better off with no theory at all.

The EA theories of medium robustness, developed during the scientific revolution, were associated with public displays of new scientific insights—creating controlled shocks and showing how electrostatics might work under varying conditions; interesting, but of uncertain value. In management, this might be conceptually similar to conflict resolution or a facilitated organizational change process. There are successes and failures. The successes are widely touted while the failures are quietly ignored. Either way, the credit or blame might be applied to the consultant, the theory, both, or some other factor (such as a sudden change in the economy).

At the end of the scientific revolution, when the understanding of EA reached a robustness of one, it had become a very useful theory. A recent search of the US patent database revealed over 7,000 patents that draw upon the principle of EA. In management terms, using a highly robust theory would be like having a way to accurately predict the behavior of your employees, the actions of the market, or the best strategic plan to follow. Obviously, we have none of those—yet.

As management theory advances toward paradigmatic revolution, we can expect to see the same kinds of evolution in practical application. Early theories will explain curiosities while later theories may be used for effective application. What will the future look like when we have robust theories of management? We can no more predict such a future than Plutarch could predict the invention of cell phones or laptop computers. Rather than attempting to predict an incomprehensible future, this section will serve to suggest what practitioners might do in the near term to improve their use of theory and to support more effective practice.

As we know too well, the emergence of a new management guru is accompanied by the arrival of thick hardback books gracing the bookstore shelves, their covers proclaiming some "new way" to work. Their pages are packed with wisdom-filled anecdotes and careful explanations detailing exactly why the reader should adopt this new point of view. Essentially, each new book represents a new theory, a new lens that provides the thoughtful reader with a new way to see the world. Along with that new view comes the implication and/or description of new actions, policies, and behaviors the practitioner should enact to achieve success. However, implementing a new system is a difficult and expensive process. Because, as the limited success of TQM and other methods have shown, past benefits are no indicator of future success. The difficulty here is that practitioners really have no certain way of determining if that guru's approach is likely to be successful, or not. The present chapter provides some remedy, suggesting that practitioners should choose the theory with the highest level of robustness (avoiding choosing theories based only on their popularity), because, as suggested above, theories that are more robust have more capacity to support more effective action.

Because robust theories are not immediately available, a more thoughtful approach is called for. It is important that the practitioner not be a "consumer" of theory following the faddish dictates of each emerging guru. Rather, the practitioner must become something of a researcher, possibly working in parallel with academicians and in thoughtful interaction with management texts.

An important idea here is that practitioners need to measure what is occurring in the workplace (we have no choice in this, we do it automatically) while at the same time, our sense of measurement (what we measure and how we measure it) must change. We must learn to understand the invisible things, such as morale (Dubin, 1978; van Eijnatten, van Galen, & Fitzgerald, 2003). In the development of the physical sciences, Kuhn suggests that the emergence of a new paradigm with its robust theory is accompanied by the development of new instruments for measurement (Kuhn, 1970: 28). However, in the study of management, no instruments currently exist for a CEO to measure (for example) the morale of her corporation. Indeed, the only instruments available to practitioners are the practitioners themselves. This "self as instrument" (McCormick & White, 2000) includes the self-identification of phenomenological reactions. *An important consideration here is how to calibrate ourselves as an instrument for effective measurement.*

Bateson (1979) suggests that the process of calibration is enhanced by the use of "double description" where multiple streams of information are combined to suggest a new, third form of information that is more useful than the previous two. Other examples of double description include binocular vision (where the extra sense of depth is added), and synaptic summation (where neuron A and neuron B must both fire to trigger neuron C). In example after example, Bateson shows how these double descriptions create an extra dimension of understanding. Importantly, this is an understanding that is of a different (and higher) logical type.

A parallel may be drawn between the structure of Bateson's approach and the structure of theory. As described above, the idea of a concatenated aspect of theory may be found in a proposition describing how aspect A and aspect B combine to understand aspect C. This understanding of aspect C through the understanding of aspects A and B suggests a greater level of understanding—a higher logical type. The parallel between the idea of double description and the idea of concatenated theory suggests that applying the two ideas in parallel might provide more benefits than either one of them alone. In short, this similarity suggests the practitioner might improve the process of self-calibration using robust theory as a guide. The validation of this idea will require additional studies.

Limitations and Opportunities for Future Research

This study is limited because it examines the development of theory in only a single field—electrostatic attraction. Future studies of this type might investigate other forms of theory. For example, studies might be conducted on the evolving robustness of theories of motion, thermodynamics, and relativity. Such studies will provide additional insight into the advancement of management theory.

The present study is also limited because it presents the insights of a single researcher. Future studies may investigate the validity of the present methodology by engaging multiple researchers in parallel studies of the same body of data. It is anticipated that, procedural errors aside, the results of multiple researchers will be similar to those presented here. This kind of study would add greatly to the validity of the metatheoretical conversation and so support the advancement of theory in a useful direction.

Future research involving the testing or development of theory should include propositional analysis as an objective test for the formal robustness of theory. That way, scholars and practitioners will have a method of effectively evaluating the theory and its potential for advancement, calibration, and application. This "R" level should be indicated in the abstract of each publication.

Future studies might also investigate the robustness of management theories over time and investigate the link between robust theory and calibration. These studies will be critical to supporting paradigmatic revolutions in management. More generally, managerial scholars may best be served by abandoning (at least temporarily) the tight focus on empirical research. Rather than employ methods based on empirical perception and so-called facts, scholarship should instead focus on metatheoretical efforts to identify robust relationships between concatenated aspects of theory. Only after we have developed highly robust theories will it make sense to engage in empirical research.

Of course, such an approach is more easily said than done. As with the development of electrical theory, we might expect many arguments to emerge around the specific meaning of each aspect of the theory. However, as long as we keep in mind that each aspect should be understood only in terms of two or more other aspects of the theory, we will have a compass indicating the direction toward success in the field of management theory.

Conclusion

Although management science does not have theories that are highly effective in practice, we do have a boom in theory creation. In some sense, the increase in theory creation parallels the global information boom. While the business world is well acquainted with the difficulties (and opportunities) of that vast amount of information (Wytenburg, 2001), we academicians can readily retreat to our disciplinary niches (or create new ones), and so insulate ourselves from information overload. Instead of retreating, I suggest that we advance; and, in so doing, I suggest that we recalibrate our views of the world.

When we look at the vast number of management theories, we should not see a fragmented and chaotic field. Instead, from a metatheoretical vantage point, we should see a field that is rich in resources from which we can build more robust theories. Rather than focus on empirical data from direct observation, *we should use existing theories as data* to investigate and advance management theory using rigorous metatheoretical methodologies. Given that we have the pursuit of robustness as a viable direction for positive and objective advancement and that we have a huge field of theory to draw from, *my hope is that we can advance effective theories into practice on the order of years, rather than centuries.*

In the present chapter, I used propositional analysis (a metatheoretical methodology founded on principles of systems theory) to objectively investigate the development of theory over time in terms of its formal robustness. In this, I expanded our understanding of paradigmatic revolutions by developing a more detailed understanding of the role played by the structure of theory in advancing a science toward revolution. Specifically, by explicating how the achievement of fully robust theory appears to be an integral aspect of paradigmatic revolution. Importantly, with this new understanding of formal robustness, we have the opportunity to measure, and purposefully advance management theory toward paradigmatic revolution and improved efficacy.

What does it take to be an Einstein? What does it take to be a stonemason who can take a pile of bricks and turn it into a well-built house? What does it take to create a revolutionary theory that, in turn, radically alters the fabric of management life? The study presented in this chapter suggests how scholars and practitioners in the field of management may anticipate great professional success by developing and applying management theory of appropriate robustness. Without purposeful advancement, management theory can expect to remain moribund for decades or centuries—with dire implications for practitioners and management programs in universities.

Bibliography

Amsden, R. T., Ferratt, T. W., & Amsden, D. M. (1996). TQM: Core paradigm changes. *Business Horizons, 39*(6), 6-14.

Appelbaum, R. P. (1970). *Theories of Social Change*. Chicago: Markham.

Argyris, C. (2005). Double-loop learning in organizations: A theory of action perspective. In K. G. Smith & M. A. Hitt (eds.), *Great minds in management: The process of theory development* (pp. 261-279). New York: Oxford University Press.

Ashby, W. R. (2004). Principles of the self-organizing system. [Classical Papers]. *Emergence: Complexity and Organization, 6*(1-2), 103-126.

Austin, J., Stevenson, H., & Wei-Skillern, J. (2006). Social and commercial entrepreneurship: Same, different, or both? *Entrepreneurship Theory and Practice, 30*(1), 1-22.

Bartlett, J. (1992). *Familiar Quotations: A Collection of Passages, Phrases, and Proverbs Traced to their Sources in Ancient and Modern Literature* (16 ed.). Toronto: Little, Brown.

Bateson, G. (1979). *Mind in nature: A necessary unity*. New York: Dutton.

Bennet, A., & Bennet, D. (2004). *Organizational Survival in the New World: The Intelligent Complex Adaptive System*. Burlington, Massachusetts: Elsevier.

Bernier, L., & Hafsi, T. (2007). The changing nature of public entrepreneurship. *Public Administration Review, 67*(3), 488-503.

Boudon, R. (1986). *Theories of Social Change* (J. C. Whitehouse, Trans.). Cambridge, UK: Polity Press.

Burrell, G. (1997). *Pandemonium: Towards a Retro Organizational Theory*. Thousand Oaks, California: Sage.

Campbell, D. T. (1983). Science's social system and the problems of the social sciences. In D. W. Fiske & R. A. Shweder (eds.), *Metatheory in social science: Pluralism and subjectivities* (pp. 108-135). Chicago: University of Chicago Press.

Clarke, T., & Clegg, S. (2000). Management Paradigms for the New Millennium. *International Journal of Management reviews, 2*(1), 45-64.

Czarniawska, B. (2001). Is it possible to be a constructionist consultant? *Management Learning, 32*(2), 353-266.

Daneke, G., A. (1999). *Systemic Choices: Nonlinear Dynamics and Practical Management*. Ann Arbor: The University of Michigan Press.

de Leeuw, E. (1999). Healthy cities: Urban social entrepreneurship for health. *Health Promotion International, 14*(3), 261-269.

Dekkers. (2008). Adapting organizations: The instance of Business process reengineering. *Systems Research and Behavioral Science, 25*(1), 45-66.

Dolan, C. (2007). *Feasibility Study: The Evaluation and Benchmarking of Humanities Research in Europe*: Arts and Humanities research Council.

Donaldson, L. (1995). *American Anti-Management Theories of Organization: A Critique of Paradigm Proliferation*. New York: Cambridge University Press.

Dubin, R. (1978). *Theory building* (Revised ed.). New York: The Free Press.

Edwards, M., & Volkmann, R. (2008). Integral theory into integral action: Part 8. Retrieved 11/03/08, 2008, link.

Fiske, D. W., & Shweder, R. A. (Eds.). (1986). *Metatheory in social science: Pluralisms and subjectivities*. Chicago: University of Chicago Press.

Fowler, A. (2000). NGDOS as a moment in history: Beyond aid to social entrepreneurship or civic innovation? *Third World Quarterly, 21*(4), 637-654.

Geisler, E., & Ritter, B. (2003). Differences in additive complexity between biological evolution and the progress of human knowledge. *Emergence, 5*(2), 42-55.

Ghoshal, S. (2005). Bad management theories are destroying good management practices. *Academy of Management Learning & Education, 4*(1), 75-91.

Glaser, B. G. (2002). Conceptualization: On theory and theorizing using grounded theory. *International Journal of Qualitative Methods, 1*(2), 23-38.

GMU. (2009). John Nelson Warfield. *Faculty Expertise Database*, 2009, link.

Guo, K. L. (2007). The entrepreneurial health care manager: Managing innovation and change. *The Business Review, Cambridge, 7*(2), 175-178.

Hammond, D. (2003). *The Science of Synthesis: Exploring the Social Implications of General Systems Theory*. Boulder, Colorado: University Press.

Harder, J., Robertson, P. J., & Woodward, H. (2004). The spirit of the new workplace: Breathing life into organizations. *Organization Development Journal, 22*(2), 79-103.

Kaplan, A. (1964). *The conduct of inquiry: Methodology for behavioral science*. San Francisco: Chandler Publishing Company.

Kauffman, S. (1995). *At Home in the Universe: The Search for Laws of Self-Organization and Complexity* (Paperback ed.). New York: Oxford University Press.

Kuhn, T. (1970). *The Structure of Scientific Revolutions* (2 ed.). Chicago: The University of Chicago Press.

Lachmann, R. (Ed.). (1991). *The Encyclopedic Dictionary of Sociology* (4 ed.): The Dushkin Publishing Group.

Lewis, M. W., & Grimes, A. J. (1999). Meta-triangulation: Building theory from multiple paradigms. *Academy of Management Review, 24*(4), 627-690.

Maanen, J. V., Sørensen, J. B., & Mitchell, T. R. (2007). The interplay between theory and method. [Introduction to special topic forum]. *Academy of management review, 32*(4), 1145-1154.

MacIntosh, R., & MacLean, D. (1999). Conditioned emergence: A dissipative structures approach to transformation. *Strategic Management Journal, 20*(4), 297-316.

McCormick, D. W., & White, J. (2000). Using One's Self as an Instrument for Organizational Diagnosis. *Organization Development Journal, 18*(3), 49-63.

McKelvey, B. (2002). Model-centered organization science epistemology. In J. A. C. Baum (Ed.), *Blackwell's Companion to Organizations* (pp. 752-780). Thousand Oaks, California: Sage.

Meehl, P. E. (2002). Cliometric metatheory: II. Criteria scientists use in theory appraisal and why it is rational to do so. [Monograph Supplement]. *Psychological Reports, 91*(2), 339-404.

Mintzberg, H. (2005). Developing theory about the development of theory. In K. G. Smith & M. A. Hitt (eds.), *Great minds in management: The process of theory development* (pp. 355-372). New York: Oxford University Press.

Morgan, G. (1996). *Images of Organizations*: Sage.

Mort, G. S., Weerawardena, J., & Carnegie, K. (2003). Social entrepreneurship: Towards conceptualisation. *International Journal of Nonprofit and Voluntary Sector Marketing, 8*(1), 76-88.

Nickles, T. (2009). Scientific Revolutions. *Stanford Encyclopedia of Philosophy* Retrieved 04/23/2009, 2009, from http://plato.stanford.edu/entries/scientific-revolutions/

Nonaka, I. (2005). Managing organizational knowledge: Theoretical and methodological foundations. In K. G. Smith & M. A. Hitt (eds.), *Great Minds in Management: The Process of Theory Development* (pp. 373-393). New York: Oxford University Press.

Olson, E. E., & Eoyang, G. H. (2001). *Facilitating Organizational Change: Lessons From Complexity Science.* San Francisco: Jossey-Bass/Pfeiffer.

Pfeffer, J. (2007). A modest proposal: How we might change the process and product of managerial research. *Academy of Management Journal, 50*(6), 1334-1345.

Popper, K. (2002). *The logic of scientific discovery* (K. Popper, J. Freed & L. Freed, Trans.). New York: Routledge Classics.

Ritzer, G. (2001). *Explorations in social theory: From metatheorizing to rationalization.* London: Sage.

Roller, D., & Roller, D. H. D. (1954). *The Development of the Concept of Electric Charge: Electricity from the Greeks to Coulomb* (Vol. 8). Cambridge, Mass: Harvard University Press.

Ross, S. N., & Glock-Grueneich, N. (2008). Growing the field: The institutional, theoretical, and conceptual maturation of "public participation," part 3: Theoretical maturation. [Editorial]. *International Journal of Public Participation, 2*(1), 14-25.

Schön, D. A. (1991). *The Reflective Turn: Case Studies In and On Educational Practice.* New York: Teachers Press.

Senge, P., Kleiner, K., Roberts, S., Ross, R. B., & Smith, B. J. (1994). *The Fifth Discipline Fieldbook: Strategies and Tools for Building a Learning Organization.* New York: Currency Doubleday.

Shareef, R. (2007). Want better business theories? Maybe Karl Popper has the answer. *Academy of Management Learning & Education, 6*(2), 272-280.

Sheard, S. (2007). Devourer of our convictions: Populist and academic organizational theory and the scope and significant of the metaphor of 'revolution'. *Management and Organizational history, 2*(2), 135-152.

Shotter, J. (2004). Dialogical Dynamics: Inside the Moment of Speaking. Retrieved June 81, 2004, 2004, from http://pubpages.unh.edu/~jds/thibault1.htm

Skinner, Q. (1985). Introduction. In Q. Skinner (Ed.), *The return of grand theory in the human sciences* (pp. 1-20). New York: Cambridge University Press.

Smith, K. G., & Hitt, M. A. (eds.) (2005). *Great Minds in Management: The Process of Theory Development*: Oxford University Press.

Smith, M. E. (2003). Changing an organization's culture: Correlates of success and failure. *Leadership & Organization Development Journal, 24*(5), 249-261.

Southerland, J. W. (1975). *Systems: Analysis, administration, and archetectura*: Van Nostrand.

Stacey, R. D., Griffin, D., & Shaw, P. (2000). *Complexity and Management: Fad or Radical Challenge to Systems Thinking.* New York: Routledge.

Stapleton, I., & Murphy, C. (2003). Revisiting the Nature of Information Systems: The Urgent Need for a Crisis in IS Theoretical Discourse. *Transactions of International Information Systems, 1*(4).

Starbuck, W. H. (2003). Shouldn't organization theory emerge from adolescence? *Organization, 10*(3), 439-452.

Steier, F. (1991). *Research and Reflexivity.* London: Sage Publications.

Steingard, D. S. (2005). Spiritually-Informed Management Theory: Toward Profound Possibilities for Inquiry and Transformation. [Essay]. *Journal of Management Inquiry, 14*(3), 227-241.

Stinchcombe, A. L. (1987). *Constructing social theories.* Chicago: University of Chicago Press.

Sutton, R. I., & Staw, B. M. (1995). What theory is not. [ASQ Forum]. *Administrative Science Quarterly, 40*(3), 371-384.

Van de Ven, A. H. (2007). *Engaged scholarship: A guide for organizational and social research.* New York: Oxford University Press.

van Eijnatten, F. M., van Galen, M. C., & Fitzgerald, L. A. (2003). Learning dialogically: The art of chaos-informed transformation. *The Learning Organization, 10*(6), 361-367.

Wallis, S. E. (2006a, July 13, 2006). *A sideways look at systems: Identifying sub-systemic dimensions as a technique for avoiding an hierarchical perspective.* Paper presented at the International Society for the Systems Sciences, Rohnert Park, California.

Wallis, S. E. (2006b). *A Study of Complex Adaptive Systems as Defined by Organizational Scholar-Practitioners.* Unpublished Metatheoretical Dissertation, Fielding Graduate University, Santa Barbara.

Wallis, S. E. (2008a). Emerging order in CAS theory: Mapping some perspectives. [Research/Review]. *Kybernetes, 38*(7).

Wallis, S. E. (2008b). From reductive to robust: Seeking the core of complex adaptive systems theory. In A. Yang & Y. Shan (eds.), *Intelligent Complex Adaptive Systems* (pp. 1-25). Hershey, PA: IGI Publishing.

Wallis, S. E. (2008c). From Reductive to Robust: Seeking the Core of Complex Adaptive Systems Theory. In A. Yang & Y. Shan (eds.), *Intelligent Complex Adaptive Systems.* Hershey, PA: IGI Publishing.

Wallis, S. E. (2008d). The integral puzzle: Determining the integrality of integral theory, link.

Wallis, S. E. (2008e). Validation of theory: Exploring and reframing Popper's worlds. *Integral Review, 4*(2), 71-91.

Wallis, S. E. (2009a). The complexity of complexity theory: An innovative analysis. *Emergence: Complexity and Organization, 11*(4), 26-38.

Wallis, S. E. (2009b). From reductive to robust: Seeking the core of institutional theory. *Under submission - available upon request.*

Wallis, S. E. (2009c). Seeking the robust core of organizational learning theory. *International Journal of Collaborative Enterprise, 1*(2), 180-193.

Wallis, S. E. (2009d). Seeking the robust core of social entrepreneurship theory. In J. A. Goldstein, J. K. Hazy & J. Silberstang (eds.), *Social Entrepreneurship & Complexity.* Litchfield Park, AZ: ISCE Publishing.

Wallis, S. E. (2009e). Theory of peak performance theory. *Under submission - available upon request.*

Wallis, S. E. (2009f). Towards a robust systemization of Gandhian ethics. *Under submission - available upon request.*

Weick, K. E. (2005). The experience of theorizing: Sensemaking as topic and resource. In K. G. Smith & M. A. Hitt (eds.), *Great Minds in Management: The Process of Theory Development* (pp. 394-413). New York: Oxford University Press.

Weisbord, M. R. (1987). *Productive Workplaces: Organizing and Managing for Dignity, Meaning, and Community*: Jossey-Bass.

Wheatley, M. J. (1992). *Leadership and the New Science.* San Francisco: Barrett-Koehler.

Wytenburg, A. J. (2001). Bracing for the future: Complexity and computational ability in the knowledge era. *Emergence, 3*(2), 113-126.

Yolles, M. (2006). Knowledge cybernetics: A new metaphor for social collectives. *Organizational Transformation and Social Change,* 3(1), 19-49.

Chapter 7 How to Choose Between Policy Proposals— A Simple Tool Based on Systems Thinking and Complexity Theory

Wallis, S. E. (2013). How to choose between policy proposals: A simple tool based on systems thinking and complexity theory. Emergence: Complexity & Organization, 15(3), 94-120.

Abstract: *Complexity and systems approaches can be applied for the creation and evaluation of policy proposals. However, those approaches are difficult to learn and use. Therefore, those conceptual tools are not available to the general public. If citizens were able to analyze policies for themselves with relative ease, they would gain a powerful tool for choosing and improving policy. In this paper, I present a relatively simple method that can be used to measure the structure (complexity and co-causal relationships) of competing policies. I demonstrate this method by conducting a detailed comparison of two economic policies that have been put forth by competing political parties. The results show clear differences between the policies that are not visible through other forms of analysis. Thus, this method serves as a "David's sling"—a simple tool that can empower individuals and organization to have a greater influence on the policy process.*

Introduction

Our world faces a growing list of concerns including war, poverty, drugs, crime, environmental crisis, and economic collapse. The way that we understand and organize ourselves to engage these problems is through the creation and application of policy. Those policies are based on expert analysis. However, a growing body of research suggest that policy experts are unable to develop effective policy for dealing with these issues (Rhodes, 2008). Indeed, it could be said that our global problems are a result of our shared lack of policy competence.

Because we lack policy competence, we cannot develop a single, shared, policy that is inarguably effective. As a result, each political party creates their own policy and claims that their version of policy will solve our problems (if we elect them into office). This has the result of placing the voting public in the very difficult situation of trying to choose between policies without a good understanding of how such a decision might best be made. In this paper, I will present and demonstrate a tool that will empower voters to make more effective policy choices.

An economic policy serves as a map, reflecting each party's understanding of the economy and how it may be improved. In the present election cycle, as I write this article, each party has posted their economic policy on their official website. The voters must choose between those policies by voting for the party's candidates. However, for choosing policies, we have no tool more effective than intuition. And, as any gambler can tell you, intuition is not very reliable.

More information is often said to be helpful in making that kind of decision. Yet, acquiring information is a "cognitively taxing task" (Redlawsk, 2004: 595). So-called 'factual' evidence is problematic because of underlying assumptions (MacGillivray & Gallagher, 2012). And, even when we do have the information, the amount of information obtained by voters may not have a large effect on voters' decisions (Carmines & Stimson, 1980). Certainly, that information-based approach has not proved useful thus far.

Indeed, what we have here is a "wicked" problem (Rittel & Webber, 1973). For such problems the normal, linear, approaches of science will not provide useful solutions. Our lack of ability to develop and choose policy means that the adoption of policy is based (at least in part) on the cultural and religious norms of a society (Simmons & Elkins, 2004). That is to say, in many instances, tradition and habit may be more influential than knowledge and logic. This is bad news if our tradition has become one of making poor economic policy.

Without a reliable tool for choosing between competing policies, debate becomes divisive instead of constructive. The sides become polarized. Society becomes fragmented and democracy comes one step closer to failure.

Systems thinking (ST) and complexity theory (CT) have been suggested as ways to gain a better understanding of policy problems (Dennard, Richardson & Morçöl, 2008a; Morçöl, 2010). Yet, systems thinking is difficult to learn (Nguyen *et al.*, 2012). In short, "Complexity thinking is hard" (Tait & Richardson, 2011: v) and complexity science has not been effective at creating tools for practitioners.

For example, in a recent special issue on "Complexity and Public Policy" (S. Landini & S. Occelli, eds.) Morçöl (2012) suggests that policy makers have been unable to address the problems associated with urban sprawl, in part because of the confusing multiplicity of theories on the topic. While he explores the topic form a CT perspective, he also admits that there is no coherent version of CT. Why then should policy be analyzed form one fuzzy/conflicted view instead of another? Verweij (2012) steps back from the use of theory and presents his version of complexity as "sensitizing concepts" to provide hints as to where managers might look for insights in organizing large scale projects. Again, it is unclear how or why one set of sensitizing ideas should be better than another. This is not to say that the mentioned papers are not interesting and informative. Indeed, they represent good scholarship in a reputable academic journal. This is only to highlight the difficulty of our challenge.

Any individual would be better able to investigate and understand these issues if he or she possessed an advanced degree in systems or complexity. However, that level of education is out of reach for most people. Instead, the method presented in this paper is relatively simple. It will not require a PhD in systems sciences. Using this approach, an individual with reasonable education and intelligence will be able to evaluate two policies and objectively decide which one is more likely to work effectively in practical application. This is quite an exciting development for a number of reasons.

First, scholars in complexity and systems thinking may find this an interesting approach because complexity and systems perspectives tend to focus on observations of world systems. In this article, conversely, I apply CT to study conceptual systems, which are used to understand world systems. This double perspective acts as a lens on top of a lens—as in a microscope that allows us to see much more than ever before and start a revolution of insight (Dent, 1999; Wallis, 2010b). For scholars, this approach bridges the gap between theory and practice. Using this method, scholars can measure the structure of a policy with some empirical accuracy. Those results can be linked with objective results in the real world. In short, a new stream of research is now available.

Second, this method is useful for a wide variety of policy fields. It can be used to analyze and improve policy in areas such as drug abuse, military, economic, foreign, domestic, and others. In that way, this method is a tool for addressing real-world problems. A small improvement in policy can have profound results in practice. For example, a one percent improvement in the U. S. military policy could save over six billion dollars per year with no loss of functional efficacy (Wallis, 2011).

Third, this approach is appealing because it is a non-partisan approach. The evaluations are based on fundamental building blocks of logic. Partisan party philosophies are not involved. Additionally, this approach opens a new path toward bridging and integrating conflicting policies. This, in turn, opens the door to healing the rifts that have grown in our society.

The fourth, and perhaps the most obvious benefit, is that this approach will be of interest to practitioners. These are the policy makers, analysts, politicians, public administrators and others who can benefit from a new perspective on policy. This will allow them to create more effective policy.

Our current political system encourages fragmentation and disenfranchisement. This is because the factions with the most power have the largest advertising budget, so we hear their message more often. That situation may have shaped the "voice" of some policy makers—encouraging them to present a message or story that is simple, repetitive, and loud. The typical voter, in contrast, has no such voice and is buffeted by the conflicting claims of the political parties.

Having no decision making tool but our unreliable intuition, members of the voting public are easily drawn into argument and conflict. The approach presented in this paper is important because it serves as a "David's sling:" a simple tool that can be applied by the disenfranchised to overcome the rhetoric of the powerful. By providing a new understanding of policy and a new vocabulary, the nature of the political dialog will change for the better.

When correctly applied, the tool presented here is very empowering in the same way that, "Systems based facilitation can level the intellectual playing field" (Daniels & Walker, 2012: 114). Thus, this approach may be used to increase the "systems intelligence" (Jones & Corner, 2012) allowing people to, "act intelligently even in the absence of objective knowledge" (Hämäläinen and Saarinen, quoted in Jones & Corner, 2012).

This approach is different from other approaches to policy. For example, those that promote specific aspiration, such as freedom. From one perspective we can compare the definition of aspiration with the definition of policy. While a policy is a map, an aspiration is a single location on the map. So, an aspiration cannot stand in for a policy. Indeed, without the complete policy map, the understanding of how the world works, there is no sure way to reach to goal. The destination is not the map.

By itself, an aspiration is an example of reductive thinking. Essentially, a lone aspiration is disconnected from other complex co-causal relationships. It says that one thing is important but does not recognize what the costs are to achieve that benefit. Nor does it recognize the large number of unanticipated outcomes. Nor does it show how other things may be important or otherwise connected.

Second, from a metapolicy perspective, if we say that one aspiration (such as freedom) is a valid way to determine the validity of a policy, then we must accept all aspirations as valid measures. Then we are right back to the start—with no objective way to decide which policy is best. Third, aspirations are based on assumptions that have not been analyzed according to an objective, rigorous, approach. That is why policies based on aspirations are "problematic" (Wæver, 2011: 467) one example is the failed "war on drugs." That aspiration without understanding seems to have caused more problems than it has solved (e.g., Baum, 1996).

Another consideration for choosing policy relates to the issues involved. The methods presented in this paper do not consider what topics "should" be addressed. That seems to be a choice best left to the policy makers. However, it should be noted that no policy problem exists in isolation. The interconnected nature of issues seems to suggest that we should address as many issues as possible using collaborative approaches while applying integrated knowledge management strategies (e.g., Bammer, 2008; Meek 2008; Runhaar, Dieperink & Driessen, 2008).

More important than stretching toward a single aspiration or single issue, the present approach employs a more systemic perspective. When looking at theories, there seems to be a general consensus that better theories are more systemic in nature to maximize sustainability (e.g., Dubin, 1978; Friedman, 2003). Further, a policy is like a theory in that both are a kind of conceptual construct that may be used to understand and engage the world around us. Therefore, it is possible to apply the same idea to policies—and see that the better policy is the one that is more systemic.

Previous scholars have suggested that logics should be used to evaluate policies (e.g., Ball, 1995; Gasper & George, 1997; Hambrick, 1974). However, their logics are related to the reification of empirical data and seeking to make a policy argument a convincing one. Those previous scholars did not provide a rigorous, quantitative, way to determine the how systemic a theory or policy might be. Recent advances have suggested how this might be done (more on this below).

In complexity theory, the idea of co-causality has emerged from concepts such as autopsies (Maturana & Varela, 1972) that have been used to understand how organizations exist as a self-organizing systems. Hofstadter (1980) wrote on self-reference and recursion to show how these new insights might be used to better understand the nature of systems; from the physical to the organizational. Hofstadter (1980) provides an interesting reference point by suggesting that each point in a system refers to itself indirectly.

In the same way that social relations are multi-causal and complex (White, 1994), so too the relationships between concepts within a policy are best understood as complex and multi-causal. Briefly, if we assume that everything in our world is interconnected it makes sense to understand that world using a policy that is made of interconnected concepts. "And, critically, that there is a correlation between the quantifiable structure of a policy and the effectiveness of that policy in practical application" (Wallis, 2011: 14).

To summarize this introduction, this is a new approach based in CT & ST with many apparent benefits. This approach enhances our capacity as a society to create an innovation spur that will encourage and accelerate the development of more effective policies through more constructive collaboration. This acceleration can occur in private analysis, public debate, and academic literature. Lamborn (1997) and deLeon (1999) suggest that a systems approach might prove useful for understanding policies. The present article answers their call by presenting a systemic view of the policies themselves.

Policies, Discussions, and Decisions

The policy process is often said to include the following stages of development: Agenda Setting; Policy Formation and Legitimization; Implementation; and Evaluation (Sabatier, 1999: 6). It is generally agreed that more studies are needed to understand the overall process, and to better understand each stage of the process. While all stages are important, this paper will focus our attention on the legitimization of a policy. That may be understood as making, "An authoritative choice among those specified alternatives" (Kingdon, 1997: 3).

Here, we will talk about a policy as a set of ideas that serve as a map. This map represents the way the world is understood. As a map, the policy shows what a nation must do to get some desired results. More formally a policy is, "A cognitive structure (like a theory) representing how a community or organization understands the world, thus enabling them to take specific actions to achieve

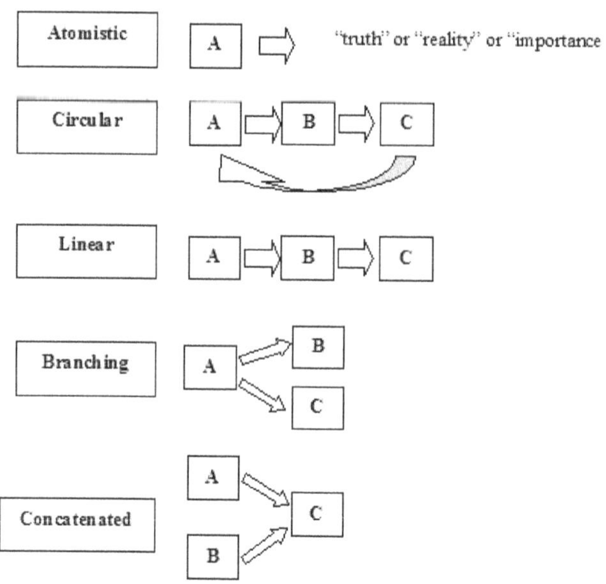

Figure 1 *Five Forms of Logic*

their goals" (Wallis, 2011: 102). A policy is not a goal (such as an aspiration for freedom, or a desire for world peace) nor is it an action (such as the act of building more solar power systems). A policy is an understanding of the way the world works so that we can better understand what actions we might take to achieve our goals.

Historically, policies have been created through a process of recognizing problems, engaging in careful analysis, followed by political wrangling (Kingdon, 1997). While that process may seem reasonable, the results have not been impressive. A growing body of research questions the policy process and the effectiveness of the resulting policies (e.g., Albritton, 1994; deLeon, 1999; John, 2003; Lamborn, 1997; Sabatier, 1999; Schmidt, Scanlon & Bell, 1979; Scott Jr., 2010; White, 1994; Wroughton & Kaiser, 2008).

In short, we seem incapable of developing effective policy in any sort of reliable way (Rhodes, 2008; Wroughton & Kaiser, 2008). If the process is so fraught with failure, and the political wrangling dominated by large powerful parties, how can the average voter decide or have a voice in that process or decide between policies?

To decide between policies, one way of looking at them has been to choose the policy that seems the most logical. But what is logic? Most people would answer that question by saying something is logical when it makes sense. On the surface, that view may seem reasonable. When we look more closely, however, there are serious problems.

What happens, for example, when two people look at the same policy and have different conclusions about how sensible it is? Too often, each person begins to see the other as unreasonable and illogical. The two may argue; and, in doing so, start down the road to open conflict. A more rigorous and objective approach is needed.

In this section, I present five fundamental structures of logic with examples of how these logics have been used (and misused) in policies and policy debates. This presentation of logics is significant in two ways. First, it shows that some logics are more useful than others for creating policies. Second, by understanding the five forms of logic, we can learn to objectively measure them. This gives us a very useful tool for cutting through the Gordian knot of policy confusion. The five forms of logic are presented in abstract form in Figure 1.

An Atomistic logic is essentially an unsupported truth claim. That is much like saying "A is true" or "A is important" or "A is real." By themselves, Atomistic claims are not good for explaining or proving anything at all. One example of an Atomistic logic would be the Libertarian claim, "All persons are entitled to keep the fruits of their labor" (Platform, 2012). Such an entitlement would lead to arguments with others who would claim that taxation is necessary. Unfortunately, these kinds of claims are frequently included in policy conversations. When U.S. foreign policy is too focused on a single facet or issue instead of taking into account the nuances of history, language, and culture, we end up with highly problematic unintended consequences (Daniels, 2012).

To make an Atomistic logic a little more useful, it must be backed up with other claims. This leads us to Linear logics. A Linear logic uses simple explanations. For example, the Boston Tea Party economic policy (Principles, 2012) suggests that more government intervention erodes free markets, which in turn causes economic decline.

This kind of logic (abstractly) is very similar to claims of "proof" that say, A is true because of B and B is true because of C" and so on until Z. The problem is that we can never be sure what "Z" is. It remains a hidden assumption—it is never explained. It is as if a long ladder of logics is resting with its bottom rungs in a murky pond. Those hidden assumptions, as with Atomistic logics, may easily lead to arguments and worse.

A Circular logic is one where a change in any aspect will lead back to itself. This is a tautology and is of little worth. For example, Figure 1 shows how, more A will cause more B, which will cause more C, and that will lead to more A. Or, in short, one could say that more A leads to more A. The use (or, more accurately, the misuse) of a Circular logic is generally understood and frowned upon.

The Branching structure of logic is a little more complex. This is where "more A causes more B *and* more C. When they are perceived as positive, Branching logics may be characterized as a form of "silver bullet" thinking. This is where one (supposedly simple) cause leads to multiple beneficial effects.

An example is the famous "Just say no" speech where President and Nancy Regan spoke about the war on drugs. Their message was a powerful one. Simply by saying no to drugs (causal action), a wide range of benefits (resultant effects) would occur. Everyone would be safe. The dignity and health of workers would be preserved. America would be free. Our values and institutions would be strong, children would be clean-eyed and clear minded, leading lives that are exciting, rewarding, stimulating, full of trust, love, confidence, and hope (Speech, 1984).

The approach may have seemed logical (and sensible) to many listeners the time. Indeed, the effort received broad bi-partisan support and considerable funding. However, the hard results did not match the stirring rhetoric. Instead of dignity, freedom, and clean-eyed children, there were many unanticipated consequences including soaring expenses, rising drug potency, increased criminal activity, overcrowded prisons, clogged courts, and miscarriages of justice (Baum, 1996).

Branching logic (like Atomistic, Linear, and Circular logics) gives the appearance of making sense. However, those forms of logic are not functionally useful when they are applied to understanding and taking effective action in the real world. The fifth structure of logic is much more useful. That is the Concatenated form of logic. That is where changes in A and changes in B cause changes in C.

The strength of the Concatenated logic can be seen in the brilliant work of George Bateson (1979). Bateson called this approach a "dual description." The basic idea here is that any two descriptions are better than one. Whenever two perspectives are combined, a new (and better) understanding emerges. Or, for a biological example, one eye cannot discern depth. Having two eyes gives two perspectives, which the brain integrates to create a third perspective, one with the added understanding of depth.

Delving deeper into philosophy, the Concatenated logic can be seen in the classic Hegelian dialectic (e.g., Appelbaum, 1988). In that framework, thesis and antithesis lead to synthesis. More recently, and more related to CT, the idea of a Concatenated aspect is similar to the idea of emergence in that something new may be seen or understood.

With this understanding of the relative values of the logics involved, we gain a new ability. We can deconstruct a policy into its constituent logics. By identifying what logical "building blocks" have been used to create the policy, we gain a new perspective on how well the policy is built. By counting those logics, we can actually see how logical a policy is. By counting the logics in two policies, we can compare them and decide which one is more logical in an objective sense rather than an intuitive sense. This approach also shows which policy is more co-casual and non-linear in structure.

This kind of approach also has some conceptual roots in Ashby's law of requisite variety. His, "law holds that for a biological or social entity to be adaptive, the variety of its internal order must match the variety imposed by environmental constraints" (Boisot & McKelvey, 2010: 421). While this is certainly an interesting idea there is no measure of how complex a policy must be if it is to be effective in practical application.

White (1994) claims that our economic paradigm is flawed because there are "multiple views of social reality and policy problems and no definitive way to adjudicate among them" (p. 862). To be sure, there are existing methods that have been proposed as potential candidates for evaluating policy. However, the usefulness of those older methods such as "rational comprehensive approach to planning" remain in question (Dennard, Richardson & Morçöl, 2008b: 7).

Similarly, argument continues as to the relative usefulness of other, older methods such as root and branch analysis, and simply "muddling through" (Scott Jr., 2010). While such approaches are related to a "rich literature... applications prove impossible" (Scott Jr.. 2010: 7). Also, the process of "program evaluation" began in the early sixties. While this approach was supposed to improve the application of policies by evaluating and understanding the results of their implementation, it has since become clear that, "Program evaluation has not led to successful policies or programs" (Schmidt, Scanlon, & Bell, 1979: 1).

The new approach presented below is based on previous research, that shows which policies are more likely to be effective in practical application (Wallis, 2010a; Wallis, 2011). Studies using this method have suggested that policies containing greater complexity tend to be more successful than policies that are simpler (Wallis, 2011). For example, in an analysis of military policies in 1870, the Prussian policy was found to be more than twice as complex than the military policy of the French. In the Franco-Prussian war, which the French fully expected to win, the Prussians won with surprising ease (Wallis, 2011).

This new approach also suggests that policies with more concatenated structures of logic will be more effective in practical application. Consider, for example, a comparison of two economic policies implemented in the 1980s. One policy contained nothing but Linear and Atomistic logics. The other policy contained 20% Concatenated aspects. The more concatenated policy led to lower rates of unemployment and a more stable economy (Wallis, 2011).

In the next section, we will learn how to calculate the interrelatedness of the concepts within the policy as well as the complexity of the policy, thus gaining a new tool and perspective for policy analysis and comparison.

Propositional Analysis

In this section, I present a method called Propositional Analysis (PA). PA is a six-step process for analyzing the logical structures of policies. This analysis is used to determine the number of concatenated aspects and what percent of the concepts in a policy are well understood. Following that presentation, I present the economic policies of the Republican Party and the Democratic Party and use PA to analyze each of them. That process is a demonstration of how PA can be used to compare two policies. By learning this method, individuals will be able to analyze policies and decide which ones are better than others with some level of objectivity.

Let's begin by defining a proposition as, "A declarative sentence expressing a relationship" (Van de Ven, 2007: 117). Each proposition explains how changes in one thing are related to changes in others. For example, a proposition might say that higher employment leads to more consumer spending. Knowing what a proposition is makes it possible to identify one and to analyze it properly to determine its structure.

In the past, scholars have tended to agree that conceptual systems (such as policies) are more effective when they exhibit a higher level of structure (e.g., Dubin, 1978). Until recently, however, there was no formal, empirical approach to measure the structure. And, as the saying goes, if it cannot be measured, it cannot be managed.

Recently, PA was developed as a means to measure the systemic nature of policy with some level of objectivity (Wallis, 2010d; Wallis, 2011). PA is used to measure what fraction of a policy is well understood (compared with the fraction that is not so well understood). That level of understanding has also been described as measuring the extent to which the policy model exists as a co-causal system. PA has been used in many studies including organizational theory (Wallis, 2009b), complexity theory (Wallis, 2009a), ethics (Wallis, 2010c), policy (Wallis, 2010d), and others.

Originally, PA was used to analyzing the logical structure of theories (Wallis, 2008a; Wallis, 2009a). While there is not enough space here for a detailed explanation, research has shown that theories with more structure (more Concatenated logics, more co-casual relationships) were more useful in practical application (Wallis, 2010a). For example, an ancient version of electrostatic attraction theory did not show a high level of interrelationship between the concepts. As a result, the theory was not very useful (perhaps most applicable as a conversation starter at ancient Roman cocktail parties). In contrast, the modern theory of electrostatic attraction shows a high level of interrelation between the concepts. And, as a result, that theory is highly useful for designing the many electronic devices we find so useful today (although, admittedly, it is not so amusing at cocktail parties).

Later, PA was applied to study the logical structures of policies (Wallis, 2010d). Very recently, studies into the logical structure of policy yielded some interesting fruit. Policies that seemed logical when they were made, quickly fell apart in practical application (Wallis, 2011).

In one study, as noted in the previous section, two economic policies were compared. One worked quite well in application. It proved useful for reducing unemployment, supporting economic recovery, and stabilizing the economy. The other economic policy actually caused more economic problems than it cured. That economy became less stable than before, and unemployment increased. The differences in those policies were not clear to those who made them. However, the differences in success could have been predicted if they had used PA.

Propositional Analysis is a six-step process.

1. Identify the logical propositions within a policy text (found in a publication or speech).
2. Diagram the causal relationships between the concepts/aspects within the propositions.
3. Combine those smaller diagrams where they overlap to create a larger, integrated, diagram.
4. Identify and count the concatenated aspects.

5. Count the total number of aspects to determine the complexity of the policy.
6. Calculate the robustness of the policy by dividing the number of Concatenated aspects by the total number of aspects.

In the following subsections, I will analyze the economic policies of the Republican and Democratic parties. I chose to study economic policies because the economy is a particularly important topic at this time. To make as fair a comparison as possible, both policies were drawn from the official party websites. To the greatest extent possible, I use the words of the authoring parties with only small changes for clarity. I am not conducting these analyses to promote one side or the other. My goal here is to provide objective examples of how to apply PA. That way, others can learn to analyze policies and choose for themselves.

Republican Economic Policy

The Republican Party economic policy is presented as:

> *We believe in the power and opportunity of America's free market economy. We believe in the importance of sensible business regulations that promote confidence in our economy among consumers, entrepreneurs and businesses alike. We oppose interventionist policies that put the federal government in control of industry and allow it to pick winners and losers in the marketplace.* (http://www.gop.com/index.php/issues/issues/)

Step 1: Identify the logical propositions within a policy.

This step sifts through the statements to get a very clear sense of what is being said in the policy. Remember that we are looking for causal relationships. The first sentence reflects how the Republican belief in the power and opportunity of America's free market economy. In that sentence, there is no causal relationship. There is only an Atomistic statement that America's free market economy has power and opportunity.

Proposition 1 (P1)—There is power and opportunity of America's free market economy.

The second sentence provides a causal relationship. Here, I rephrase it slightly to make the understanding more clear from a structural perspective.

Proposition 2 (P2)—More sensible business regulations cause more confidence in our economy among consumers, entrepreneurs and businesses alike.

The third sentence presents a more complicated although still Linear relationship:

Proposition 3 (P3)—More interventionist policies cause more federal government control of industry which (in turn) causes the more federal government picking of winners and losers in the marketplace

Step 2: Diagram the causal relationships between the concepts/aspects.

In this step, the goal is to diagram the propositions. In Figure 2, I place each distinct aspect of the policy into a separate box. Arrows represent the causal relationships. This step helps to make the relationships more clear.

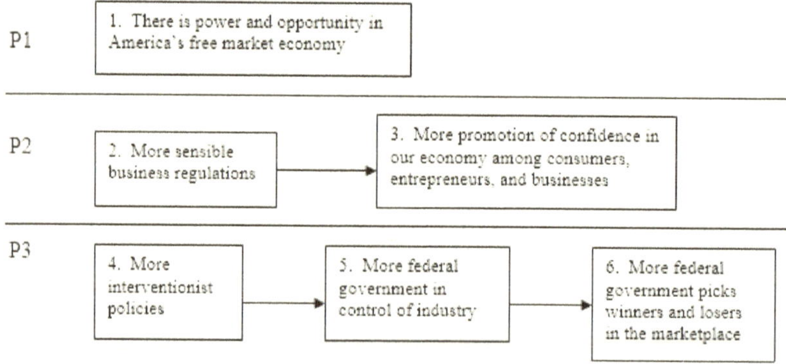

Figure 2 *Republican Party Economic Policy*

Step 3: Integrate those smaller diagrams where they overlap to create a larger, combined, diagram.

At this stage of the analysis, we look at each of the boxes to see if any are very similar (or perfectly identical). If two or more aspects are close enough so that we can think of them as representing the "same thing," they can be said to overlap. For an abstract example, if one proposition says "more A causes more B" and another proposition says "more C causes more B" we can put the two together into a single logic structure where more A and more C cause more B. This process helps to clarify the policy.

In the case of the Republican Party economic policy, there are no identical aspects. There are some that are close, however. For example, #1 says that the free market is powerful. And, #6 talks about the marketplace. However, #6 is about picking winners and losers in the market, not directly about the power of the free market.

Therefore, because there are no overlaps, the diagram for the policy remains the same as in Figure 2.

Step 4: Identify and count the Concatenated aspects.

Recalling the discussion from the previous section, a Concatenated logic is where two causes merge to create a single result. That resultant aspect is the Concatenated aspect. In Figure 2, it can be seen that aspect #1 is an Atomistic logic. The others exist in Linear relationships. Therefore, there are zero concatenated aspects in this policy.

Step 5: Count the total number of aspects to determine the complexity of the policy.

Because we have placed all of the aspects in boxes (and numbered them) it is a simple process to count the number of aspects in the policy. Here, there are six aspects so the complexity of the policy is six.

Step 6: Calculate the robustness of the policy by dividing the number of Concatenated aspects by the total number of aspects.

Because there were zero Concatenated aspects, the robustness of the policy is zero (the result of zero divided by six).

To summarize and conclude this analysis, the Republican Party economic policy has a complexity of six and a robustness of zero.

The complexity is an indication of the range of ideas or possibilities that are accepted as being relevant to the economy. The robustness of zero indicates that none of those aspects are very well understood. That is to say, there is vanishingly small level of systemic integration so there is little chance of the policy being effective in practical application.

Democratic Economic Policy

The Democratic economic policy is:

> *Democrats are moving forward with a "Made in America" economic plan to strengthen American industries and create jobs for American workers by:*
>
> - *Ending tax loopholes that let corporations hide profits overseas, and investing those dollars in small businesses that create jobs in America;*
> - *Providing tax cuts to small businesses and expanding lending so that businesses can create new jobs;*
> - *Investing in a clean-energy economy, and providing tax credits to spark manufacturing of windmills, solar panels, and electric cars here at home; and*
> - *Putting Americans to work rebuilding roads, bridges, rails, and ports, strengthening our economy and our infrastructure across all 50 states.*
>
> *Democrats stand for the values of hard work and responsibility, and we know that as a country we are most successful when we invest in our people—middle-class families and small business owners—who can grow our economy from the bottom up. Together, we have begun to lay a new foundation for growth, building an economy that works for all Americans.* (http://www.democrats.org/issues/economy_and_job_creation)

Step 1: Identify the logical propositions within a policy.

As with the Republican Party economic policy, the policy is deconstructed into its constituent propositions.

Proposition 1 (P1)—Fewer tax loopholes that let corporations hide profits overseas leads to more investing those dollars in small businesses that in turn leads to more creation of jobs - leading to more strengthening of American industries and more jobs for American workers.

Proposition 2 (P2)—More tax cuts to small businesses and more lending to businesses will result in the creation of new jobs - leading to more strengthening of American industries and more jobs for American workers.

Proposition 3 (P3)—More investment in a clean-energy economy and more providing of tax credits will lead to more manufacture of windmills, solar panels, and electric cars here at home - leading to more strengthening of American industries and more jobs for American workers.

Proposition 4 (P4)—More rebuilding roads, bridges, rails, and ports, will lead to leading to more strengthening of infrastructure and more strengthening of American industries and more jobs for American workers.

Proposition 5 (P5)—Value hard work

Proposition 6 (P6)—Value responsibility

Proposition 7 (P7)—More investment in our people (middle-class families and small business owners) causes more success, which (in turn) causes more growth our economy from the bottom up.

Step 2: Diagram the causal relationships between the concepts/aspects.

In this step, again, I simply take the propositions and place each aspect in a box and connecting them with causal arrows to create a visual representation of the policy.

Step 3: Integrate those smaller diagrams where they overlap to create a larger, combined, diagram.

Propositions P1, P2, P3, and P4 are all casual to aspect #1. Therefore, they have some overlap and can be integrated into a single

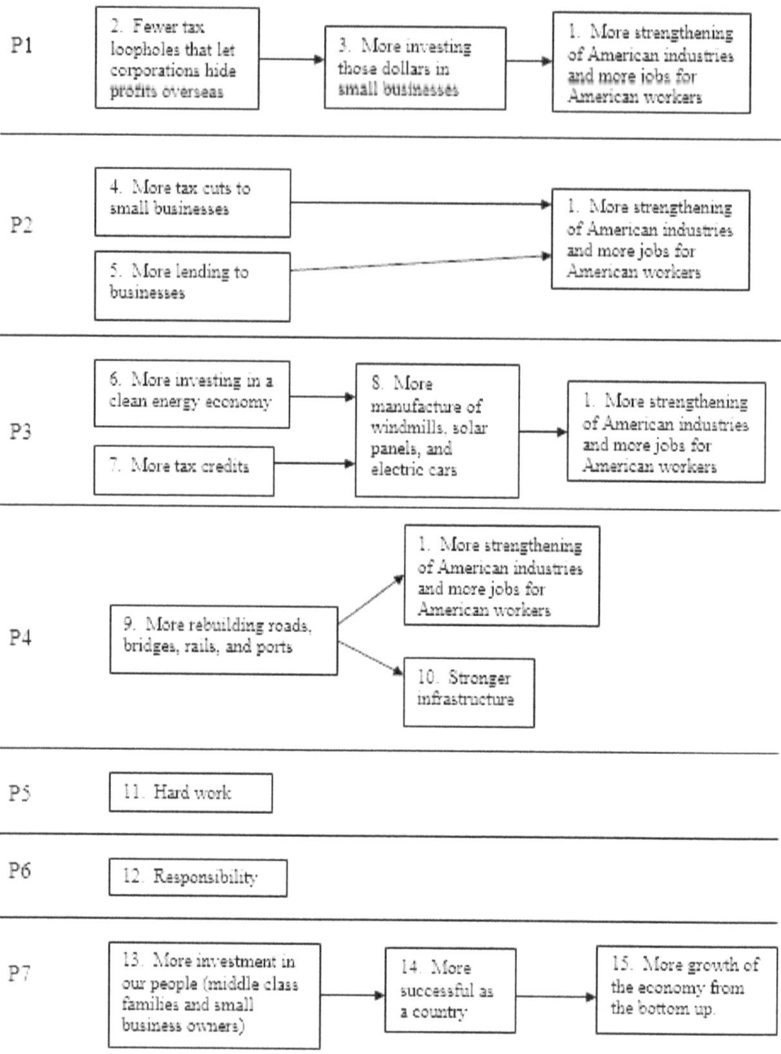

Figure 3 *Propositions of Democratic economic policy in diagrammatic form*

diagram. The other propositions do not connect with any of the other aspects. Bringing these multiple smaller diagrams together gives an integrated diagram as seen in Figure 4.

Step 4: Identify and count the Concatenated aspects.

Aspect #1 and aspect #8 have more than one causal influence. Therefore, they count as Concatenated aspects. They are better

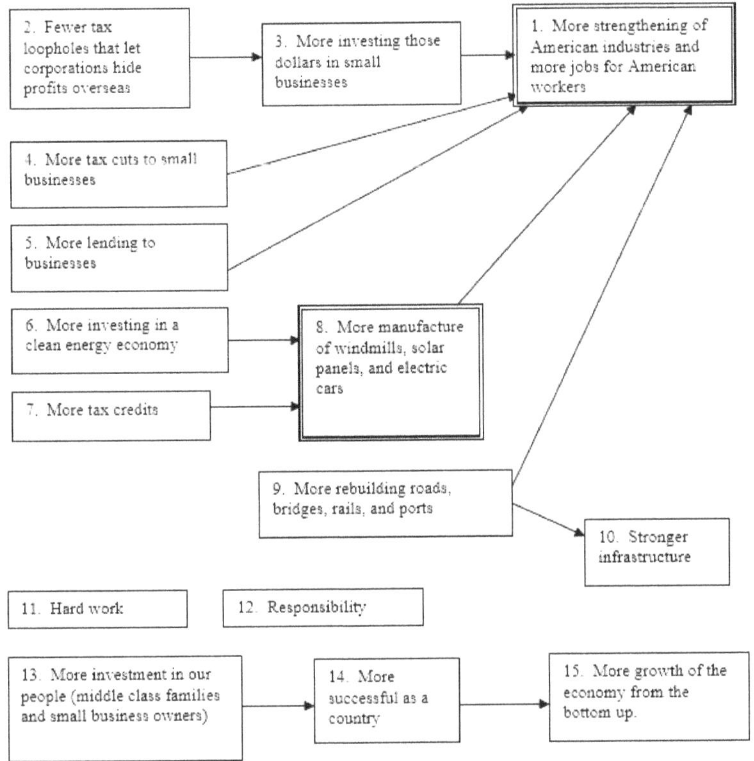

Figure 4 *Integrated Diagram of Democratic Economic Policy*

understood than the other aspects in the policy that are Atomistic or Linear.

Step 5: Count the total number of aspects to determine the complexity of the policy.

Looking at Figure 4, it seems clear that there are 15 aspects (one aspect in each box). Therefore, the Democratic economic policy has a complexity of 15. That is an indicator of the conceptual breadth of this policy

Step 6: Calculate the robustness of the policy by dividing the number of Concatenated aspects by the total number of aspects.

From step 4, there are two Concatenated aspects. From step 5, there are 15 total aspects. To calculate the robustness of the policy, we simply divide two by 15. Therefore, the robustness of this policy is 0.13 (the result of two divided by 15). This is an indicator of how effective the policy might be in practical application.

Comparisons of Policies and Limitations of this Process

In this subsection, I will briefly compare the Democratic and Republican economic policies. I will also discuss some strengths and limitations of the PA approach to comparing policies.

The complexity indicates the conceptual breadth of a policy. That is to say, the more complex the policy, the more aspects it contains. With more complexity comes more ideas and the inclusion of more economic issues. The opposite is also true. If a policy does not mention employment, that policy can hardly claim to show how to improve the level of employment.

In some sense, the greater the complexity of a policy, the more likely it is to address the "right" issues and therefore to provide a better map and path to resolving issues. The choice as to which issues are the "right" ones depends on how much effort the policy makers are willing to invest in the policy process. Because everything in the world is connected, we cannot really say that we can exclude anything from the analysis. One approach for addressing this problem is to involve as many stakeholders as possible. This kind of collaborative approach increases the ability of the policy makers to include more diverse points of view (and to gain more participants for implementation).

In contrast to the breadth of complexity, the depth of a policy is shown by the robustness. This tells us what percentage of concepts in the policy are well understood. The robustness also corresponds with other insights into the structure of a policy. These include the level of systemic integration among the aspects, its structure, and the level of co-causality of the aspects within the policy as a system unto itself. It also indicates the likelihood of that policy being successful if it is effectively implemented in practical application. A policy with a robustness closer to one will be more effective than a policy with a robustness that is closer to zero.

The following Figure 5 provides a graphic comparison of the two policies analyzed here.

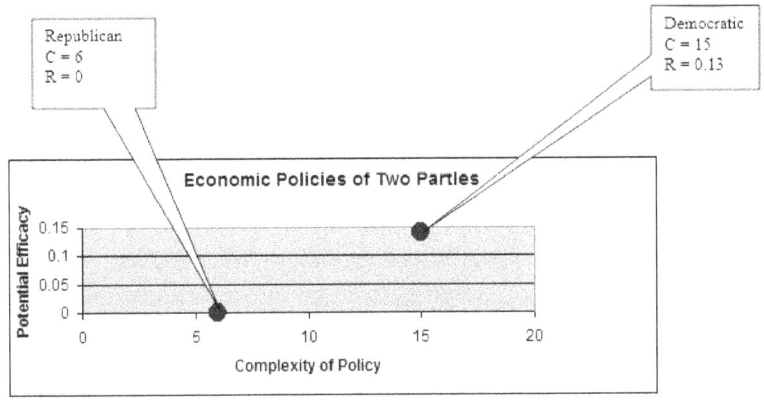

Figure 5 *Comparison of two economic policies*

As discussed above, studies into policy suggest that the more complex policies will be more effective in practical application than policies that are less complex. Similarly, policies that are more robust will be more effective than those that are less robust. Here, it is graphically clear that the Democratic economic policy is superior to the Republican policy in both complexity and robustness. These measurements suggest that the Democratic policy is preferential to the Republican one.

Any good analysis requires multiple perspectives (Roe, 1998; Wallis, 2008b). So other considerations (in addition to PA) should be taken into account. There is not sufficient space in this article to cover all those considerations in depth. However, they should be mentioned to provide a more complete picture of what is needed to choose between policies.

First, the study presented in this article assumes that the policies are based on some kind of observation and data. Do sensible regulations really promote confidence in the economy? Will closing tax loopholes actually result in those dollars being invested in small business? Policies are more believable when the propositions are supported by academic studies. For the purpose of this study, I will presume that both political parties have "done their homework" before promoting their policies.

Second, we must consider if the policy will be implemented as planned. Even the best policy in the world is useless if it is not acted upon. We won't really know until one party or the other is elected to the relevant positions of power. So, for that indicator of policy success, one should probably look at the track record of the parties' politicians. In short, one would need to determine how good each has been about keeping their campaign promises.

Third, it is worth considering that these policies are taken from the official party websites. While this may provide a fair comparison, it is entirely possible that each party has another policy on file somewhere. We could conduct better analyses if we had access to those versions of policy. This concern relates to transparency. Are the political parties telling us what they really believe, or are they telling us what we want to hear?

The fourth area of concern regards the conceptual relevance of each policy. Does the policy address issues that are relevant to the reader? If a reader is interested in building confidence, it might make more sense to focus on the Republican policy. If the reader were interested in job growth, the Democratic policy would seem to be more useful.

Other considerations of this sort are covered in a number of excellent policy texts (e.g., Kingdon, 1997; Sabatier, 1999). But these are very difficult methods, and beyond the reasonable reach of most voters.

Summary and Conclusion

A policy is understood as a map, a theory of how the world works. A policy is not a specific goal or an action to achieve that goal. The broader policy process includes: Agenda Setting; Policy Formation and Legitimization; Implementation; and Evaluation (Sabatier, 1999: 6). Failure at any stage of this process will result in a failed policy. Success at one stage, "does not necessarily imply success in others" (Kingdon, 1997: 3). This paper has focused on the legitimization of a policy including the process of choosing between competing policies.

Because existing methods have not proved useful in developing existing policies, this paper has demonstrated a new method for choosing between policies based on an analysis of the logics that are internal to the text of each policy. This method makes it possible to determine how good a policy map we have so that we may choose our goals more effectively and take action with more assurance of achieving policy success.

Nations are like gamblers. They place their policy bets and hope for good results. To some extent, each nation can afford to make some policy mistakes. We must wonder, however, how many mistakes we can afford to make before our luck runs out. For some individuals, and nations, it already has.

This problem is of particular concern when those mistakes are so expensive in terms of ecological, human, and financial costs. Today, we have a global economic crisis. We also have a growing ecological crisis and looming crises of military engagement and international relations. All of these are the result of poorly placed policy wagers.

Sadly, policy is still made the old fashioned way. Individuals and factions throw ideas against the wall during analyses and debates. That approach leads to incoherent policies consisting of Atomistic and Linear logic structures. In such policies, concepts are poorly understood because they are founded on vague assumptions. Their Atomistic, Linear, Branching, and Circular structures of logic lead to arguments over interpretations and implementation—further fragmenting society. Is it any wonder why so many policies fail?

It is not enough, however, to replace the existing fuzzy lens of policy theory with an equally fuzzy lens of complexity theory. When a policy works effectively, we cannot simply claim that its success is due to the application of complexity thinking as in (Lehmann, 2012). Nor can we simply replace an existing concept such as "shared values" with a new concept such as "vortex" as in (Sturmberg, 2012).

Methods of evaluating policy based on the rational comprehensive approach to planning, root and branch analysis, and muddling through have not proved useful. We would all benefit if we used more CT and ST approaches applied to the process of policy creation, evaluation, and implementation. Such efforts would change the way we think about policy and change the nature of the policy conversation. Despite its weaknesses, CT and ST do show great promise. However, it is not practical to provide a graduate education to our entire population. Propositional Analysis provides a relatively simple tool based in complexity theory and systems thinking. More information about how to use this approach may be found at http://projectfast.org

Studies using PA have shown that policies are more likely to be effective in practical application when they are more complex and more robust. That tool enables users to analyze, integrate, and understand policy on a new and deeper level than ever before. This is particularly important because, "it is rare for scientists and citizens to speak a common language or even find forums where both voices are validated" (MacGillivray & Gallagher, 2012: 68). PA, therefore, serves to empower people at all levels to have a greater understanding of policies and therefore have a more influential voice in the policy process. That influence will serve as an innovation spur to accelerate the improvement of policy and may even be useful for integrating multiple policy perspectives among multiple stakeholders. Using the David's sling of PA, the disempowered gain a new tool that will help them to shape the policy conversation for the benefit of all.

The methods presented in this paper, used to measure the structure (complexity and robustness) of policy will allow us to choose policies that are good and satisfying. The structure of a policy is not the only consideration for policy makers. It is also important to consider the reliability of the information used to create the theory. Another consideration is the implementation of the policy. The best policy in the world means little if it is not followed, or if it is not implemented with sufficient resources. Ideally, such considerations should be evident within the policy, itself.

In short, if we use more effective methods for developing policy, we can be more successful. Our policy bets will pay off more often. Of equal importance, this paper has provided a David's sling—a tool that disenfranchised people can use to help make their voices heard.

References

Albritton, Robert B. 1994. "Comparing policies across nations and over time." Pp. 159-176 in *Encyclopedia of policy studies*, edited by Stuart S. Nagel. New York: Marcel Dekker.

Appelbaum, Richard P. 1988. *Karl Marx*: Sage.

Ball, William J. 1995. "A pragmatic framework for the evaluation of policy arguments." *Review of Policy Research* 14:3-24.

Bammer, Gabriele. 2008. "Integrating policy analysis and complexity: Developing the new specialization of integration and implementation sciences." Pp. 249-262 in *Complexity and policy analysis: Tools and concepts for designing robust policies in a complex world*, edited by Linda Dennard, Kurt A. Richardson, and Goktug Morçöl. Goodyear, Arizona: ISCE Publishing.

Bateson, Gregory. 1979. *Mind in nature: A necessary unity*. New York: Dutton.

Baum, Dan. 1996. *Smoke and mirrors: The war on drugs and the politics of failure*. Waltham, MA: Little, Brown &Co.

Boisot, Max, and Bill McKelvey. 2010. "Integrating modernist and postmodernist perspectives on organizations: A complexity science bridge." *Academy of Management Review* 35:415-433.

Carmines, Edward G., and James A. Stimson. 1980. "The two faces of issue voting." *The American Political Science Review* 74:78-91.

Daniels, Peter. 2012. "A systems perspective on U.S. foreign policy in the Middle East: A propositional analysis." *E:CO - Emergence: Complexity & Organizations* 14.

Daniels, Steven E., and Greg B. Walker. 2012. "Lessons from the trenches: Twenty years of using systems thinking in natural resource conflict situations." *Systems Research and Behavioral Science* 29:104-115.

deLeon, Peter. 1999. "The stages approach to the policy process: What has it done? Where is it going?" Pp. 19-32 in *Theories of the Policy Process*, edited by Paul A. Sabatier. Boulder, Colorado: Westview Press.

Dennard, Linda, Kurt A. Richardson, and Goktug Morçöl (eds.). 2008a. *Complexity and policy analysis: Tools and concepts for designing robust policies in a complex world*. Goodyear, Arizona: ISCE Publishing.

Dennard, Linda, Kurt A. Richardson, and Goktug Morçöl (eds.). 2008b. "Editorial." in *Complexity and policy analysis: Tools and concepts for designing robust policies in a complex world*, edited by Linda Dennard, Kurt A. Richardson, and Goktug Morçöl. Goodyear, Arizona: ISCE Publishing.

Dent, Eric B. 1999. "Complexity science: A worldview shift." *Emergence* 1:5-19.

Dubin, Robert. 1978. *Theory building*. New York: The Free Press.

Friedman, Ken. 2003. "Theory construction in design research: Criteria: Approaches, and methods." *Design Studies* 24:507-522.

Gasper, Des, and R. Varkki George. 1997. "Analyzing argumentation in planning and public policy: Assessing, improving and transcending the Toulmin model." Pp. 40 in *Working Paper Series, No. 262*: Institute of Social Studies, the Hague, Netherlands.

Hambrick, Ralph S. Jr. 1974. "A guide for the analysis of policy arguments." *Policy Sciences* 5:469-478.

Hofstadter, Douglas R. 1980. *Gödel, Escher, Bach: An Eternal Golden Braid*: Vintage Books / Random House.

John, Peter. 2003. "Is there life after policy streams, advocacy coalitions, and punctuations: Using evolutionary theory to explain policy change?" *Policy Studies Journal* 31:481-498.

Jones, Rachel, and James Corner. 2012. "Stages and dimensions of systems intelligence." *Systems Research and Behavioral Science* 29:30-45.

Kingdon, John W. 1997. *Agendas, alternatives, and public policies*: Pearson Education.

Lamborn, Alan C. 1997. "Theory and politics in world politics." *International Studies Quarterly* 41:187-214.

Lehmann, Kai. 2012. "Dealing with violence, drug trafficking and lawless spaces: Lessons from the policy approach in Rio de Janeiro." *E:CO - Emergence: Complexity & Organizations* 14:51-66.

MacGillivray, Alice E., and Krista G. Gallagher. 2012. "A policy paradox: Social complexity emergence around an ordered science attractor." *E:CO - Emergence: Complexity & Organizations* 14:67-85.

Maturana, Humberto R., and Francisco J. Varela. 1972. *Autopoiesis and cognition: The realization of the living*. Hingham, MA: Kluwer.

Meek, Jack W. 2008. "Partnerships and metropolitan governance: An adaptive systems perspective." in *Complexity and policy analysis: Tools and concepts for designing robust policies in a complex world*, edited by Linda Dennard, Kurt A. Richardson, and Goktug Morçöl. Goodyear, Arizona: ISCE Publishing.

Morçöl, Goktug. 2010. "Issues in reconceptualizing public policy from the perspective of complexity theory." *E:CO* 12:52-60.

Morçöl, Goktug. 2012. "Urban sprawl and public policy: A complexity theory perspective." *E:CO - Emergence: Complexity & Organizations* 14:1-16.

Nguyen, Nam C., Doug Graham, Helen Ross, Kambiz Maani, and Ockie Bosch. 2012. "Educating systems thinking for sustainability: Experience with a developing country." *Systems Research and Behavioral Science* 29:14-29.

Platform. 2012. "Libertarian Party Platform."

Principles. 2012. "Boston Tea Party Principles."

Redlawsk, David P. 2004. "What voters do: Information search during election campaigns." *Political Psychology* 25.

Rhodes, Mary Lee. 2008. "Agent-based modeling for public service policy development: A new framework for policy development." in *Complexity and policy analysis: Tools and concepts for designing robust policies in a complex world*, edited by Linda Dennard, Kurt A. Richardson, and Goktug Morçöl. Goodyear AZ: ISCE Publishing.

Rittel, Horst W. J., and Melvin M. Webber. 1973. "Dilemmas in a general theory of planning." *Policy Sciences* 4:155-169.

Roe, Emery. 1998. *Taking complexity seriously: Policy analysis, triangulation and sustainable development*. New York: Kluwer Academic.

Runhaar, Hens A. C., Carel Dieperink, and Peter P. J. Driessen. 2008. "Policy analysis for sustainable development: Complexities and methodological responses." Pp. 197-213 in *Complexity and policy analysis: Tools and concepts for designing robust policies in a complex world,* edited by Linda Dennard, Kurt A. Richardson, and Goktug Morçöl. Goodyear, Arizona: ISCE Publishing.

Sabatier, Paul A. (Ed.). 1999. *Theories of the policy process.* Boulder, Colorado: Westview Press.

Schmidt, Richard E., John W. Scanlon, and James B. Bell. 1979. *Evaluability assessment: Making public programs work better.* Washington, D.C.: Department of Health, Education, and Welfare - Project Share.

Scott Jr., Roland J. 2010. "The science of muddling through revisited." *E:CO* 12:5-18.

Simmons, Beth A., and Zachary Elkins. 2004. "The globalization of liberalization: Policy diffusion in the international political economy." *American Political Science Review* 98:171-189.

Speech. 1984. "Just say no: Words to the nation." in *Public address by President Ronald Regan and Nancy Regan.*

Sturmberg, Joachim. 2012. "Health care policy that meets the patient's needs." *Emergence: Complexity & Organizations* 14:86-104.

Tait, Andrew, and Kurt A. Richardson. 2011. "Guest editorial: From theory to practice." *E:CO* 13:v-vi.

Van de Ven, Andrew H. 2007. *Engaged scholarship: A guide for organizational and social research.* New York: Oxford University Press.

Verweij, Stehan. 2012. "Management as system synchronization: A case of the Dutch A2 Passageway Mastricht Project." *E:CO - Emergence: Complexity & Organizations* 14:17-37.

Wæver, Ole. 2011. "Politics, security, theory." *Security Dialogue* 42:465-480.

Wallis, Steven E. 2008a. "From reductive to robust: Seeking the core of complex adaptive systems theory." Pp. 1-25 in *Intelligent complex adaptive systems,* edited by Ang Yang and Yin Shan. Hershey, PA: IGI Publishing.

Wallis, Steven E. 2008b. "Validation of theory: Exploring and reframing Popper's worlds." *Integral Review* 4:71-91.

Wallis, Steven E. 2009a. "The complexity of complexity theory: An innovative analysis." *Emergence: Complexity and Organization* 11.

Wallis, Steven E. 2009b. "Seeking the robust core of organisational learning theory." *International Journal of Collaborative Enterprise* 1:180-193.

Wallis, Steven E. 2010a. "The structure of theory and the structure of scientific revolutions: What constitutes an advance in theory?" Pp. 151-174 in *Cybernetics and systems theory in management: Views, tools, and advancements,* edited by Steven E. Wallis. Hershey, PA: IGI Global.

Wallis, Steven E. 2010b. "Techniques for the objective analysis and advancement of integral theory." Pp. 17 in *Integral Theory Conference 2010: Enacting an Integral Future.* Pleasant Hill, CA.

Wallis, Steven E. 2010c. "Towards developing effective ethics for effective behavior." *Social Responsibility Journal* 6:536-550.

Wallis, Steven E. 2010d. "Towards the development of more robust policy models." *Integral Review* 6:153-160.

Wallis, Steven E. 2011. *Avoiding policy failure: A workable approach*. Litchfield Park, AZ: Emergent Publications.

White, Louise G. 1994. "Values, ethics, and standards in policy analysis." Pp. 857-878 in *Encyclopedia of policy studies*, edited by Stuart S. Nagel. New York: Marcel Dekker.

Wroughton, Lesley, and Emily Kaiser. 2008. "Financial storm tips world toward recession: IMF." edited by Tom Hals: Reuters.

Chapter 8
Existing and Emerging Methods for Integrating Theories Within and Between Disciplines

Wallis, S. E. (2014). Existing and emerging methods for integrating theories within and between disciplines. Organizational Transformation and Social Change, 11(1), 3-24.

This paper is based on a presentation at the 56th annual meeting of the International Society for Systems Sciences (ISSS). July 15-22, 2012, at San Jose State University, California.

Abstract: *Our conceptual systems (including theories, models, policies, schema, etc.) all help us to understand the world around us. For highly complex situations such as those found in natural systems and service systems, it is important to understand them from an interdisciplinary perspective because these real-world systems do not respect the boundaries of any single discipline. While many conceptual systems exist, they have not proven highly effective for understanding issues that are the focus of their disciplines. Still fewer conceptual systems have been developed that cut across disciplinary boundaries—and they have not been shown to be any more effective than their mono-disciplinary companions. This paper investigates emerging and existing methods for creating and integrating theories within and between disciplines. This includes "soft" methods (ad hoc, cherry picking, and intuitive) as well as "rigorous" (Formal Grounded Theory (FGT), Reflexive Dimensional Analysis (RDA), and Integrative Propositional Analysis (IPA)). The present paper demonstrates that soft methods are relatively easy to use, but they do not produce conceptual systems of great or lasting value. In contrast, it is proposed that the rigorous methods are more likely to yield conceptual systems that are measurably more systemic, more useful and more* *effective for understanding and engaging the highly complex systems of our world.*

Introduction

If we are to enact change in society and organizations, we must have better conceptual tools. That is to say, we need better conceptual systems for better understanding organizations, society, and change. In this paper, I investigate how we might accelerate our ability to create more effective conceptual systems by integrating conceptual systems across disciplines.

For natural systems or service systems most research may be categorized as inductive or deductive. While these are good for "normal" science, more interesting revolutions in science may occur when a deep thinker considers two conceptual systems and seeks to compare, contrast, and combine them. Galileo and Einstein began with this kind of approach. Because of the paradigmatic revolutions they triggered, we all lead much richer lives. Were they unique in their ability to seek and find new insights from existing conceptual systems? Or, is this an approach that we all may use? In this paper, we will investigate multiple methods for integrating conceptual systems to determine which methods might be more useful. The results suggest that more rigorous methods provide a more useful and more systemic approach to integrating conceptual systems.

Although natural systems and social systems are often considered to be desperate disciplines, we may legitimately investigate conceptual systems of the natural and social spheres using the same tools because both sets of conceptual systems rely on causal propositions (Wallis, 2010a). "A proposition is a declarative sentence expressing a relationship among some terms." (Van de Ven, 2007: 117). When that proposition expresses a causal relationship, "two or more aspects [concepts] are related so that a change in one causes a change in one or more others. A causal relationship is often expressed as a proposition, hypothesis, or a diagram" (Wallis, 2011: 100). For example, "More A causes more C."

In short, although conceptual systems of service systems and theories of natural systems are from different disciplines, they are made of the same propositional "building blocks." Therefore, it is possible to integrate them on the level of conceptual systems.

As scholars, we like to think that we are good at evaluating conceptual systems and deciding the extent to which those conceptual systems might be useful or effective for research and/or practice. Unfortunately, while we are able to evaluate the work of others with some level of objectivity, we have less objectivity when evaluating ourselves (Dunning, Heath & Suls, 2004). Worse, those who are low-performers do the worst job of self-evaluation (Ehrlinger, Johnson, Banner, Dunning & Kruger, 2008). In the social sciences, in general, we (as a community) are low performers. By that I mean no disrespect for any individual scholar. Instead, I mean that the social sciences, in general, appear to be "low performers."

For example, public policies frequently fail (Wallis, 2011), organizational theory seems to have failed (Burrell, 1997), psychology is not held in high regard by other professionals (e.g., Kovera & McAuliff, 2000), and social change theory does not seem to be effective (Appelbaum, 1970; Boudon, 1986). Indeed, the promise of the social sciences is "largely unfulfilled" (Spicer, 1998). This creates an additional level of challenge because we are trying to integrate the low-performing theories of the social sciences with the more reliable theories of the natural sciences—although they too are sometimes held in low favor (Smolin, 2006).

The systems community strives to take a new view—but there is no guarantee that our views will be any more effective than previous ones. Mainly, our community looks at natural systems and/or service systems using a systemic view to better understand the world around us. Relatively few scholars study conceptual systems (Wallis, 2013). This is an important concern because our conceptual systems determine how we understand and engage the world around us. Those conceptual systems are our theories, models, mental models, schema, policies, and so on. In the present paper, I will use the term "conceptual system" to refer to theories, although I may occasionally use other terms to refer to more specific forms of conceptual systems.

Like previous schools of thought, most systems research may be broadly categorized as either inductive (begin with data and move toward creating a conceptual system) or deductive (begin with a conceptual system and then conduct experiments to test that conceptual system) (Hitt & Smith, 2005). These are well and good for "normal" science because they lead to the creation of conceptual systems, data, and insights.

In contrast to those approaches of normal science, revolutionary advances in science are sometimes caused when deep thinkers consider two (or more) conceptual systems and seek to compare, contrast, and combine them.

In this paper we will identify and discuss a range of existing and emerging methods for integrating conceptual systems within and between academic disciplines. These include soft approaches such as "ad-hoc," "cherry picking," and "intuitive" methods. Soft methods have been used throughout the history of the social sciences without impressive results (as noted above). Next, we will give more attention to relatively rigorous methods including Formal Grounded Theory (FGT), Reflexive Dimensional Analysis (RDA), and Integrative Propositional Analysis (IPA).

To demonstrate these methods, I will integrate two conceptual systems using each of the soft and rigorous methods. One conceptual system is from the study of service systems while the other conceptual system is from the study of natural systems. Finally, I will analyze and compare the resulting integrated conceptual system to determine the extent to which it might be considered an advancement of the field.

The goal of this paper is to provide a better understanding of how to create more effective conceptual systems by using rigorous methods. Additionally, this paper is expected to engender new learning and new insights into bridging the "theory-gap" between disparate disciplines (such as the separate studies of natural and service systems) through a deeper understanding of conceptual systems.

Data Set: Two Conceptual Systems (Theories)

We are assuming, for the purpose of this conversation, that these conceptual systems are based in some sort of rigorous empirical analysis. In this section, I simply present two sample conceptual systems. All of the subsequent analyses will refer to these two conceptual systems as the data source for their analyses. Each conceptual system contains a set of propositions as shown here:

A sample Service Systems Theory

A Complex Adaptive Systems Model of Organization Change from: (Dooley, 1997: 82; drawing on Thietart and Forgues, 1995)

- Organizations are potentially chaotic.
- Organizations move from one dynamic state to the other through a discrete bifurcation process (second-order change).
- Forecasting is impossible for organizations, especially at a global scale and in the long term (unpredictability).
- When in a chaotic state, organizations are attracted to an identifiable configuration (order out of randomness).
- When in a chaotic state, similar structure patterns are found at organizational, unit, group, and individual levels (fractal nature of chaotic attractors).
- Similar actions taken by organizations in a chaotic state will never lead to the same result.

A sample Natural Systems Theory

From: (Allison & Hobbs, 2004, drawing on Guderson *et al.*, 2002)

- The organization of regional resource systems emerges from the interaction of a few variables.
- Complex systems have multiple stable states. Complex systems can exhibit alternative stable organizations.
- Resilience derives from functional reinforcement across scales and functional overlap within scales.
- Vulnerability increases as sources of novelty are eliminated and as functional diversity and cross-scale functional replication are reduced.

These two conceptual systems (one of natural systems, the other of service systems) are both based in systems/complexity field. For scholars interested in creating more effective conceptual systems, the following section presents multiple methods for integrating these two conceptual systems.

Intuitive Methods of Integration

In this section I will present soft methods of integration including ad-hoc, cherry-picking, and intuitive methods. Then, in the following section, I will present the more rigorous methods of integration including Formal Grounded Theory (FGT), Reflexive Dimensional Analysis (RDA), and Integrative Propositional Analysis (IPA).

This is intended to provide a set of examples for these methods of integration. The goal is to highlight some strengths and weaknesses of each method. Because this paper is focused on the level of metatheory, we will look at the concepts within the conceptual systems, rather than on the application of the conceptual systems or the research from which the concepts were derived.

There are many soft methods available. For example, Mintzberg (2005: 361-371) suggests that personal characteristics are key to developing good conceptual systems. He suggests the benefit of creativity, intuition, and bravery. Ritzer (2009) suggests personal reflection while Hall (1999) suggests that social construction is a good path. It is difficult or impossible to test the extent to which these methods will yield useful conceptual systems. Primarily because there is no good way to test the bravery of a theorist or to say how much reflection is applied (or needed) to create or integrate conceptual systems.

Because things like bravery and reflection are difficult or impossible to measure, this paper will focus on the concepts and causal relationships within each conceptual system to see how they change as different methods of integration are applied.

Cherry Picking

Cherry picking is quite simply the process of choosing specific elements from two or more conceptual systems and combining them to create a new conceptual system. While it is expected that the scholar will use some form of reasoning to support the choice of concepts, it is understood that alternative reasoning would lead to alternative choices. For example, drawing on the above two conceptual systems, we might combine: "Resilience derives from functional reinforcement across scales and functional overlap within scales" with "Organizations are potentially chaotic." The derived conceptual system might be stated as: For chaotic organizations, resilience derives from functional reinforcement across scales and functional overlap within scales.

This may have created a conceptual system that is (to some extent) true and/or useful; however, it is important to note here that the derived conceptual system is smaller, less complete, than either of the two source conceptual systems. In short, cherry picking is reductionist. Additionally, in creating a cherry picked conceptual system, there has been some fragmentation of the field as a new theoretical focus has been created. Finally, there is the question of how we decide which part of a conceptual system to separate from the remainder of the conceptual system. Here, there are no rules in the academic world except that the choice should be supported by some rational argument. That, however, is a weak standard as anything may be rationalized. In short, it is problematic and inappropriate to use partial conceptual systems (Ritzer, 1990; Wallis, 2012a).

Ad-Hoc

Creativity and innovation are hallmarks of the ad-hoc process, which involves the combination of concepts from multiple conceptual systems. Because of the creative effect, additional concepts may be added that are not necessarily part of the original set of conceptual systems under investigation. Starting with the two conceptual systems presented above, an ad-hoc method might be narrated something like this:

> *From the CAS on organizational change, particularly proposition three, it is clear that there is very little opportunity for predicting the future of the organization. It can also be recognized that the lack of predictability also applies to natural systems. For example, where "complex systems can exhibit alternative stable organizations." Therefore, it appears that the important similarity between natural and service systems is a duality between predictability and chaos. Moving into other sources, an individual who is managing an organization (or,*

> *presumably, a natural system) should "expect to be wrong" (Richardson, 2009, 2009: 49). An idea that makes perfect sense if we have partial theories and chaotic situations. From this, a manager might conclude that there is no reason to study complexity, systems, or anything else for that matter—because that manager will always be in chaos and always be wrong.*

To some extent, this is a straw man argument—it is easy to argue against it and break it down. That is exactly the point. Indeed, most of our conceptual systems have a certain amount of ad-hoc logic within them. Thus, none stand for long. Each is replaced rapidly with some combination of other conceptual systems, concepts, and notions from a variety of sources until a "Frankentheory" is created. That conceptual system rampages across the pages of our publications until it is dismembered and recombined to create some new golem.

In this ad-hoc process, it may be seen that concepts were chosen through a reasonable process. Thus, there appears to be some validity. However, the same level of reasoning might have been used to choose different concepts. Thus, the ad-hoc process cannot be relied upon to integrate multiple conceptual systems with any useful level of rigor or repeatability.

The ad-hoc process does not require the scholar to engage the entire conceptual system—or any conceptual system. In short, like cherry picking, ad-hoc integration allows the scholar to look at conceptual systems in a non-systemic fashion.

Intuitive Integration

The intuitive process is, almost by definition, a process that cannot be governed by a rigorous process and explicit set of rules or guides. By way of brief explication, however, I will note that I have just read the conceptual systems closely... then put them aside and let my intuition emerge as I type...

> *Service and natural systems are different because natural systems become more brittle with decreasing complexity. Service systems in chaos, on the other hand exhibit similarity across levels of scale. It seems to me (intuitively) that such*

similarity would be quite the opposite of the complexity needed to maintain the natural systems. Thus, it seems that the two cannot be completely integrated. Another part of this (which may or may not have been mentioned in one of the theories) is that the social system is geared toward purposefully creating value for the other participants. Members of a natural system, on the other hand, seem geared toward the creation of value for themselves. Service systems create value for themselves by harvesting natural resources. Often without concern for the long-term sustainability of those resources. However, as the service systems evolve, they may learn to manage natural systems for long-term success.

Returning now to look at the conceptual systems, and reflecting on my intuitive effort, I note how I neglected to include a number of concepts. I may have misrepresented some concepts and created some conceptual linkages out of thin air. This example shows how intuition, like cherry picking and ad-hoc, is a soft approach to conceptual system integration.

Discussion on soft approaches to conceptual systems integration

My experience in investigating conceptual systems and their sources suggests that the most commonly applied methods are the soft ones. This is problematic for the field because intuition is not reliable (Meehl, 1992: 370). We are like gamblers—using intuition to make our wagers—only to go home empty handed.

An important key to the scientific validity of any study is the ability to replicate that study. If another scholar cannot replicate the study and arrive at similar conclusions, the validity of the study is thrown into doubt. The same criteria should apply to metatheoretical studies (Wallis, 2010b). If, for example, we ask ten graduate students to analyze the same ten conceptual systems using the same rigorous methodology they should arrive at similar conclusions. If not, the metatheoretical methodology must be called into question.

Because replication is a mainstay of science, it must be concluded that soft approaches cannot be considered useful from a science of metatheory perspective.

To summarize this subsection on soft methods of the integration of conceptual systems, scholars who use these methods are forced to rely on intuition and reduction. While it may be easy to use, intuition is clearly not to be trusted. And, similarly, reductionism leads us to address partial models that can sometimes lead to fragmented understanding and false assumptions.

Of course, there are benefits to a reductive approach. That direction may be useful and lead to more nuanced knowledge. However, it is also necessary that the deconstruction is followed by reconstruction for creating conceptual systems (Ritzer, 1990: 11); particularly if we want that knowledge to be systemic.

The lack of scientific replication of soft methods drives the final nail. The limited usefulness of soft methods is evidenced by the poor progress of the social sciences. Those methods, which have been used for so many decades, have resulted in low performing conceptual systems—clearly we need something better. In the next section, I will present and discuss more rigorous methods.

Rigorous Methods

In this section, I begin my briefly presenting three older methods of theory evaluation/integration. Finding that these provide limited levels of rigor, I go on to present three rigorous methods for integrating multiple conceptual systems. Formal Grounded Theory (FGT), Reflexive Dimensional Analysis (RDA), and Integrative Propositional Analysis (IPA). These methods follow a prescribed path, and are thought to provide a more repeatable approach to the scientific analysis and integration of conceptual systems. Next, each of those methods is used to analyze the subject conceptual systems.

Previous approaches have been suggested for the evaluation and analysis of theory. For example, Dubin (1978) suggests a measure of "efficiency" (pp. 109-111). These relate to:

1. The presence (or absence) of some unit of measure within a conceptual system
2. The casual directionality between two units of measure
3. The covariation of those units of measure.
4. The rate of change between them.

These are certainly good and useful things to have within any conceptual systems. And, all scholars should strive to include them. One limitation of this approach is that there is no concrete method to measure what a conceptual system contains within (or between) those levels. Dubin only provides a general guide.

Ritzer (2001: 53-55) suggests "architectonics" as an approach to compare and integrate conceptual systems of sociology. However, that approach is geared toward identifying fundamental similarities in human actions, rather than engaging in a highly rigorous study of the conceptual system, itself.

More recently, Shoemaker, Tankard Jr., & Lasora (2004: 170-178) suggest key steps to building a theory including:

1. Problem recognition;
2. Identification of key concepts;
3. Observation and creativity to identify and suggest the causes and effects of those key concepts;
4. Specify theoretical and operational definitions for all concepts;
5. Link some concepts to form hypotheses;
6. Specify rationale for hypotheses;
7. Attempt to think in terms of multiple hypotheses, and;
8. Attempt to place the hypotheses in an organized system.

They suggest that those steps will allow conceptual systems to be tested as to their testability, falsifiability, parsimony, explanatory power, predictive power, scope, and cumulative nature of the field, degree of formal development, heuristic value, and aesthetics. While their suggestions provide a good starting point, their steps are open to ad-hoc reasoning, cherry picking, intuition, and poorly defined measures.

While older kinds of methods offer some improvement over soft methods, the level of rigor does not seem sufficient to develop more effective conceptual systems because they still employ or allow for the application of soft methods. Therefore, for this section, I will focus on the more rigorous approaches that follow a specific methodology.

Formal Grounded Theory

Grounded Theory was developed by Glaser & Straus (1967) as a transparent process to create theory that is grounded in real word contexts (Glaser, 2002). In brief, experiences and insights are coded, categorized and related in a specific methodology to create a theory with a specific focus. Since then, others have used a Formal Grounded Theory (FGT) approach that uses extant theory as the data to create a new theory. For example, Apprey (2006) who suggests, FGT can be used to combine multiple theories and so gain more meaning and insight in an area of study. According to Charmaz (2006). The process includes:

1. Gathering data;
2. Coding the data and categorizing theoretical concepts;
3. Constant comparison between concepts;
4. Memo writing, and;
5. Creation of a theoretical construct.

The following subsections detail my efforts and results from following this approach.

Gathering data

This part is easily accomplished by choosing (as examples) the two conceptual systems presented above.

Coding the data

Initial Codes are:

1. Vulnerability;
2. Bifurcation process;
3. Sources of novelty;
4. Scale;
5. Forecasting possibility;
6. Functional diversity;
7. Actions lead to the same (or different) result;
8. Time;

9. Complex systems that have more states can exhibit more alternate stable organizations;
10. Emergence of organized regional resource systems;
11. Functional overlap within scales;
12. Functional reinforcement across scales;
13. Interactions of few variables;
14. Organization is attracted to identifiable configurations;
15. Resilience;
16. Similarity across levels of scale;
17. Organization is in chaos;
18. One dynamic state of organizations;
19. Organizations are potentially chaotic;
20. Second dynamic state of organization, and;
21. Cross functional replication.

Categorizing theoretical concepts

- Misc.: Time, Forecasting possibility.
- Emergence: Bifurcation process, Sources of novelty, Emergence of organized regional resource systems.
- Conditions of the system: Vulnerability, Organization is in chaos, Resilience.
- More Abstract Conditions of the systems: Complex systems that have more states can exhibit more alternate stable organizations, Functional diversity, One dynamic state of organizations, Organizations are potentially chaotic, Second dynamic state of organization.
- Scale: Scale, Functional overlap within scales, Functional reinforcement across scales, Similarity across levels of scale.
- Actions & Attractions: Interactions of few variables, Organization is attracted to identifiable configurations, Cross functional replication.

Constant comparison between concepts

I accomplished this part of the process by keeping the conceptual systems on a single page and referring back to them frequently.

Memo writing

Some memos include:

- Time is an interesting concept—many concepts include time implicitly, although only one mentions it explicitly.
- Some concepts do not seem to be well connected with others.
- Bifurcation may be understood as a process of creation and/or emergence
- I don't like ending up with a "miscellaneous" category—but I'm not sure where to place these two.
- Are resilience and vulnerability two sides of the same coin?
- One dynamic state and a second dynamic state are clearly in the same category—what might the opportunities be for defining what those states are?
- There seem to be few actions involved here.
- Scales seems to be a category—but there could be another one or two ways of looking at them. Within scales and between scales.
- Given the highly abstract category of conditions of the system, it seems that the more concrete concepts may not be needed as a part of the model. Chaos is not well defined (within the model) and vulnerability/resilience are represented more abstractly in "functional diversity"
- It is a common concept to consider the context or environment of the systems in question. Here, we are integrating natural systems and service systems. Where then is the context? Is it self-contextualizing? Is one the context of the other—if so, which one? Or, is the notion of context not relevant here… in this context?
- Where is "edge of chaos?"

Create theoretical construct

For FGT, the theoretical construct is based on a central question or focus. This introduces another point of ambiguity to the process. After all, if two scholars approach the conceptual data with different questions, they will likely create different constructs. Here, in addition to the subjectivity of creating themes, is another source of subjectivity for the FGT process. The combined theoretical construct I developed came out as:

> *Within and across levels of scale, there are overlaps and reinforcements. Organizations in chaos tend toward recognizable configurations and cross-functional replication. The larger and more complex system emerges from interactions of smaller systems. And, conversely, the larger system exhibits more stable systems and alternative stable states.*

Conclusion

In this process, some complex concepts were fragmented into multiple simpler concepts before being combined into categories. It is not clear if that extra step supports the creation of improved conceptual systems. Also, the categories are not rigorous—another scholar might legitimately undertake the same analysis and develop different categories. Thus, the process is not necessarily repeatable. Because this process is focused on concepts, rather than their relationships, it is too easy to find one's self with a conceptual system that is a collection of ideas, rather than a set of interrelated propositions. Thus, one may end up with a construct that is hardly a theory (or a system) at all. So, the usefulness of the resulting construct seems questionable.

Multiple concepts were categorized into fewer—suggesting that reductionism may be taking place. This might be countered in future versions by creating a new method of FGT that requires that each category represent an abstraction of the concepts. This opens some interesting possibilities. For a rather abstract example, if a conceptual system contained concepts of "square" and "rectangle" the abstract categories that are suggested might include "width" and height." The idea is not to force many ideas into fewer ideas. Rather, the goal is to seek highly abstract categories that can fully represent the concepts within the conceptual systems. As such, it is entirely possible that identifying all the abstractions might result in a conceptual system that is much larger and more complex than the subject conceptual system upon which it is based.

Reflexive Dimensional Analysis

Reflexive Dimensional Analysis (RDA) was derived in part from insights found in GT and has been used to integrate conceptual systems of Complex Adaptive Systems (Wallis, 2006) and Complexity Theory (Wallis, 2009). RDA differs from FGT because RDA specifically calls for the scholar to identify causal relationships at the sub-category level—and apply them to the category level. This provides an additional level of rigor above the FGT approach. RDA has six steps (Wallis, 2006: 7) :

1. Define a body of theory (conceptual systems).
2. Investigate the literature to identify the concepts that define it.
3. Code the concepts to identify relevant components.
4. Clump the components into mutually exclusive categories.
5. Define each category as a dimension.
6. Investigate those dimensions through the literature, looking for robust relationships.

Define a body of conceptual systems (theory).

The scope of the conceptual systems includes service systems and natural systems.

Investigate the literature to identify the concepts that define it.

This step has been accomplished by choosing the two conceptual systems presented above

Code the concepts to identify relevant components.

This step has already been accomplished in the FGT process above.

Clump the components into mutually exclusive categories.

This approach to categorization is more rigorous than other approaches to categorization. By calling for categories that are "mutually exclusive" there is more work to be done—and we end up with more categories. For ease of comparison, I will simply break out the category of "miscellaneous" into new categories of "time" and "forecasting possibility." I will also re-focus the category of "conditions of the system" to focus on vulnerability and resilience as that seems to be a state of the system that is of particular importance. Next, I will break out "scales" into overlap and reinforcement because they may be understood as exclusive.

This gives us the following categories:

1. Time;
2. Forecasting possibility;
3. Emergence;
4. Vulnerability;
5. Conditions of the system;
6. Overlap within scales;
7. Reinforcement across scales;
8. Similarity across scale, and;
9. Actions/Attractions.

Define each category as a dimension.

Here, we simply define each category as a scalar dimension. This gives it the ability to represent a wider variety of states. For example, "time" may be seen as "more time" (or, conversely, less time).

1. More Time;
2. More Forecasting possibility;
3. More Emergence;
4. More Vulnerability;
5. More (differing) Conditions of the system;
6. More Overlap within scales;
7. More Reinforcement across scales;
8. More Similarity across scale, and;
9. More Actions/Attractions.

Investigate those dimensions through the literature, looking for robust relationships.

Here is a difference between FGT and RDA. FGT simply asks the scholar to identify relationships between the categories. Thus, the scholar may intuitively assign relationships. This kind of approach is not so rigorous as it might be. RDA, in contrast, calls for those relationships to be defined by the data itself. Therefore, at this stage, we must go back to the propositions within each category to see if they contain linkages to other categories.

From the service systems model, More Time causes Less Forecasting ability—thus casually linking those two categories. Also from the service systems model, the states as one Condition of the system will lead to bifurcation found in the Emergence, which leads back to create more states of the organization in Conditions of the systems. Therefore, there are some linkages between those categories. I continued the process in this way—for each category, investigating the causal propositions of the concepts within that category. Those concept-level connections were then used to justify category-level connections. The result is a RDA model integrating the two conceptual systems that looks like this:

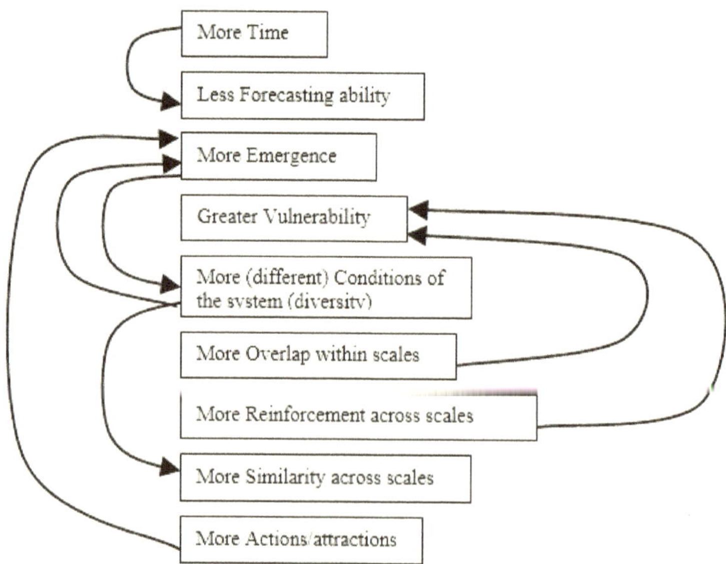

Figure 1 *RDA integrated model of Service Systems and Natural Systems*

To conclude this subsection, RDA is a more rigorous way to integrate conceptual systems from the service systems and natural systems. However, room remains for interpretation and intuition. This may be beneficial if one values creativity (which, I hope, we all do). However, that openness and flexibility becomes problematic when we are trying to make a more rigorous science.

To apply one test—a thought experiment—we might consider giving these two conceptual systems to ten scholars and ask that they all use the RDA method to create an integrated conceptual system. My hunch is that they would end up with ten different integrated conceptual systems. In short, where advances have been made in the area of rigor, there are accompanying difficulties in the convolutions that may work to reduce the effectiveness of the results. A more straightforward approach may be found in Integrative Propositional Analysis.

Integrative Propositional Analysis

Propositional analysis (PA) is used to determine the complexity and interrelatedness of a conceptual system or body of conceptual systems. This is a structural approach which extends and deepens the work of well-known authors who suggest a correlation between the structure of a conceptual systems and the effectiveness of that theory (Dubin, 1978; Kaplan, 1964; Stinchcombe, 1987) by providing reliable quantification. Where PA is generally understood to analyze the set of propositions within a single conceptual system, Integrative Propositional Analysis (IPA) explicitly accepts propositions from multiple conceptual systems as an input to the process and integrates them into a single conceptual system as an output to its process.

The process of PA includes the following six steps (Wallis, 2008):

1. Identify propositions within the conceptual system.
2. Compare with one another to identify overlaps, and drop redundant aspects.
3. Investigate propositions for conceptual relatedness.
4. Link causal aspects with resultant aspects.
5. Identify "Concatenated" aspects (those aspects that are explained by, or resultant from, two or more other concepts).

6. Divide the number of Concatenated aspects by the total number of aspects in the conceptual system (to provide a number between zero and one).

In short, PA starts by creating a diagram of the causal relationships found in the propositions of a conceptual system or body of conceptual systems. Below, I have diagramed each of the subject theories for clarity. Each concept is placed within a box (and numbered); an arrow represents each causal relationship.

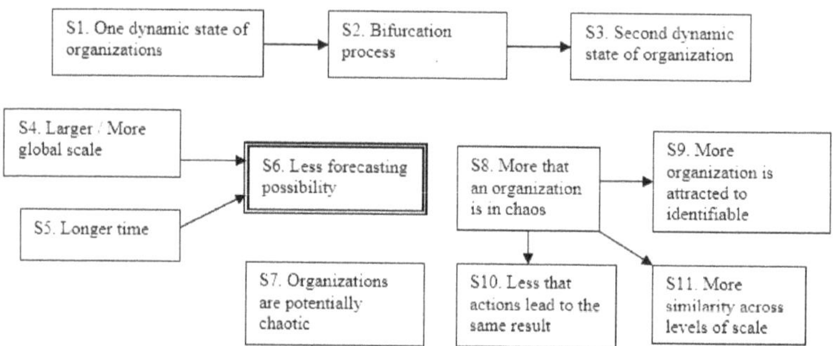

Figure 2 *Sample Service Systems conceptual system*

By counting the aspects within the conceptual system it is clear that the complexity of the conceptual system is C = 11. There is only one aspect that is the resultant of two or more causal concepts (see box #6). Therefore, the Robustness or interrelatedness of the system of conceptual system is R = 0.09 (the result of one divided by eleven). Performing the same analysis on the Natural systems conceptual system, we have:

Here it can be seen that there are nine different aspects, so the complexity of the conceptual system is C = 9. There are two aspects that are concatenated (#6 & #9) because they are the resultant of two or more other aspects. Therefore, the Robustness of this conceptual system is R = 0.22 (the result of two divided by nine). There is some small possibility for alternative interpretations. For example, it may be that resilience is the inverse of vulnerability. However, here we will stay with a direct representation of the author's text in order to maintain rigor.

Reflecting briefly on the two studies from a PA perspective, it seems that the Robustness for both conceptual systems is rather

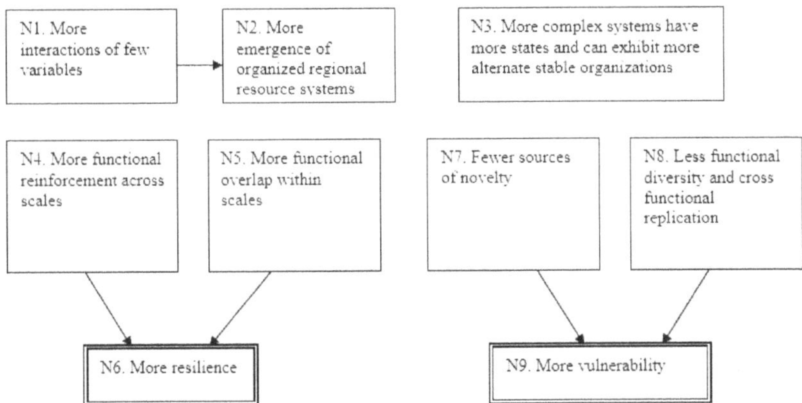

Figure 3 *Sample Natural Systems conceptual system Diagramed*

low. This is not unusual for conceptual systems of the social sciences. That low level essentially reflects how the aspects of each conceptual system are interconnected. That level of interconnection may be understood as systemicity (systemicity may be defined as the extent to which a set of things may be considered a system—in this case, a conceptual system).

The presented conceptual systems are not highly systemic. Thus, neither is likely to be highly useful in practical application. Each conceptual system may be improved through research that identifies causal linkages between the aspects within the conceptual system.

Seeking to integrate the two conceptual systems, a strict application of PA requires that we identify aspects within each conceptual system that are identical. Where identical aspects are identified, overlaps exist and the conceptual systems may be connected. While there are a number of similar aspects between the two conceptual systems, there do not seem to be any exact matches.

Allowing for some interpretive license (for the purpose of the present discussion), there are some possibilities for linking the two conceptual systems. First, we might interpret the conceptual systems to suggest that some of the aspects are really the same thing—only with different names. This kind of renaming is not uncommon in the social sciences! Another approach would be to seek a higher level of abstraction—and so link the two conceptual systems under a more abstract concept that adequately accounts for the more concrete phenomena. Third, we might infer casual linkages between aspects of the two models (although, in the name of rigor, this should not be done without empirical analysis).

In these kinds of integration it is much too easy to fall into the trap of intuition or ad-hoc thinking. For example, N4 and N5 include the concept of scales, as does S11. So, one may be tempted to integrate all of those aspects. However, N4 and N5 discuss Reinforcement and Overlap, while S11 is about Similarity. Therefore, it is not clear from the propositions that Similarity would be causal to Overlap and/or Reinforcement. This view suggests a fourth approach.

We are on more solid ground by addressing simpler aspects. For example, the linear logic represented in S1-S2-S3 might be abstracted to a derived proposition, SD1 "More bifurcation process creates More states" Similarly, N3 (which is represented as a single aspect because of the wording provided by the author) might be deconstructed to a derived proposition, ND1 "More states cause more complex systems that cause more alternatives to be exhibited. The derived propositions would be diagramed as in Figure 4.

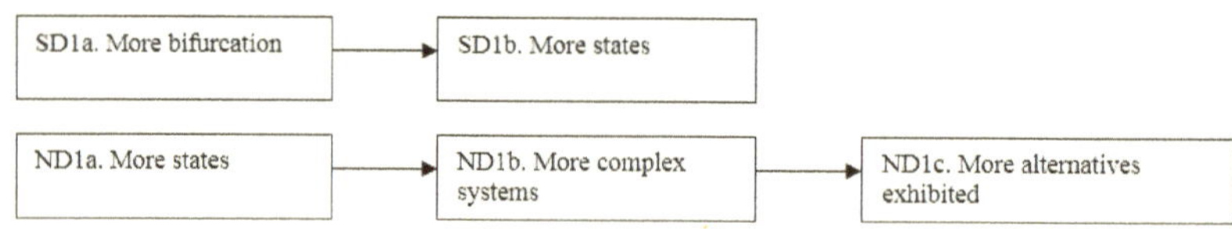

Figure 4 *derived propositions*

This opens the door for a match between SD1b and ND1a to create an integrated model as in Figure 5:

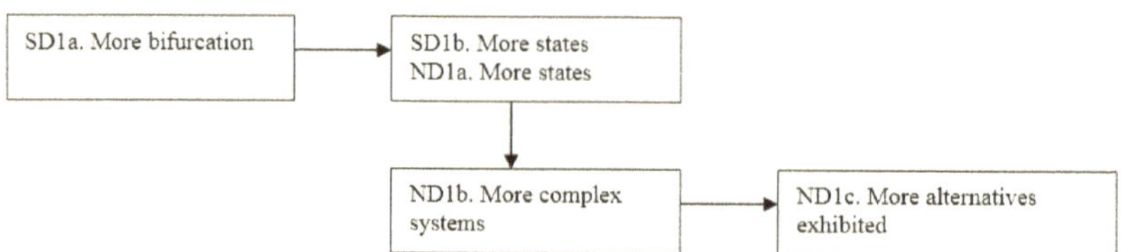

Figure 5 *Integrated derived propositions*

With their derived aspects legitimately integrated, the other aspects of the conceptual systems may be added to the structure in Figure 6:

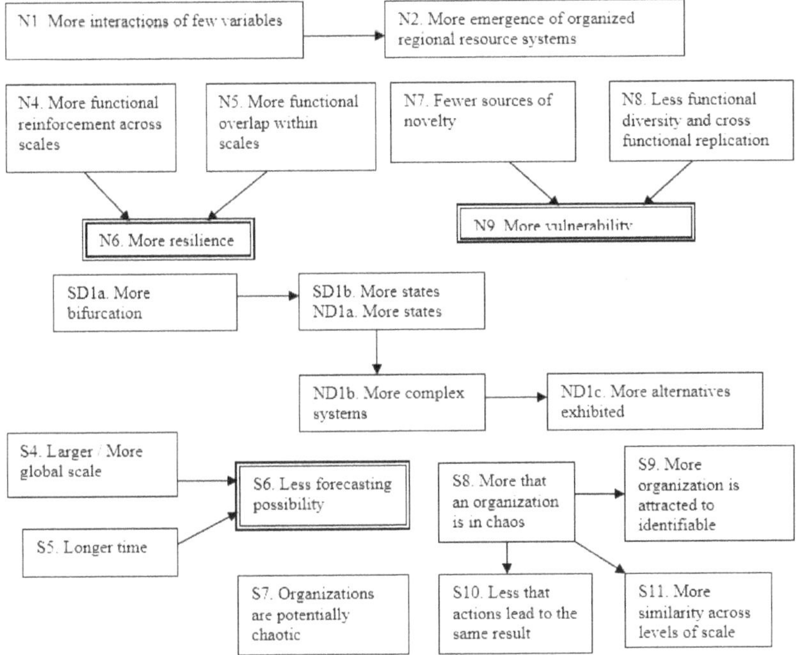

Figure 6 *Integrated conceptual systems*

Integrated Propositional Analysis (IPA) is a useful approach to integrating conceptual systems from within and between disciplines. The approach has more methodological rigor than other approaches and requires the use of whole conceptual systems and a systemic perspective. The integrated conceptual system in Figure 6 has 20 aspects, therefore it has a Complexity of C = 20. Three of those aspects are concatenated so the Robustness is R = 0.15 (the result of three divided by 20).

The complexity of the integrated conceptual system is much higher than either of the source conceptual systems. This may be seen as a step forward in the evolution of the conceptual system. However, level of interrelatedness between the aspects of the conceptual systems has not increased. Therefore, while the integrated conceptual system may be more effective than either of the source theories individually, it is not expected to be highly effective in practical application.

To increase the effectiveness of the derived conceptual system, there are two basic approaches. First, a scholar may add additional concepts to increase the Complexity of the conceptual system. Thus may be done by including additional theories and propositions. Recall here that the preference is for using whole theories to maintain the integrity of the conceptual systems. The second approach is to identify causal relationships between the concepts to increase the Robustness. This may be accomplished my additional research and/or by identifying casual relationships from the literature.

One potential weakness of this approach is the creation of highly complex conceptual systems. Some scholars (as well as editors and potential clients for consulting engagements) may look on complex conceptual systems with some suspicion because they are too large to print or too difficult to understand. For editors considering the inclusion of complex conceptual systems into their journals, I suggest creating links between print and online sources. This is one way that the key ideas may be communicated in print without losing the large-field view of the entire conceptual system.

In order to use highly complex conceptual systems to inform organizational interventions in service and/or natural systems, I suggest that practitioners adopt a team-based approach. Each practitioner might focus on a "chunk" of the larger conceptual system. Causal relationships between the chunks should be clearly defined and directly related to the coordination between practitioners.

For scholars, the same kind of approach also seems viable. That is, to involve a team of scholars to research a highly complex theory in a way that has each scholar addressing a specific "chunk" of the broader conceptual system. Importantly, each chunk should have clear causal connections with one or more other chunks.

Another important benefit of the IPA approach is that the integrated conceptual system combines multiple conceptual systems.

Thus, instead of fragmenting the field (as occurs with softer methods) this integrative approach serves to unify the field.

It should also be noted that the integration process creates new insights and new challenges for testing each conceptual system. For example if we integrate "more A causes more B," with "more A causes more C", we might ask if B & C are the same thing because they have the same causal relationship with A? Or, is there an abstraction that is relevant? Or, is there some dimension of similarity we might find between them? These are challenges that would not arise if we looked at one conceptual system or the other.

Similarly, juxtaposing the conceptual systems creates a challenge and opportunity for research that will clearly advance the coherence of the conceptual system. Using the above set of integrated conceptual systems, it is clear this process is only one step toward a higher level of systemic integration. Each set of disconnected boxes represents an opportunity for research to define the two as a causal relationship. And, defining those relationships will increase the internal coherence and usefulness of the conceptual system.

Conversations and Additional Concepts

In brief, the use of soft methods (cherry picking, ad-hoc, intuitive) for integrating conceptual systems gives the appearance of making sense, but do not seem particularly useful in the creation of more effective conceptual systems. First, they have been used through history, without highly effective results. Second, they are reductionist and non-systemic. Third, they support the fragmentation of the field without providing for re-integration.

Other approaches (FGT and RDA) follow a more rigorous methodology. And, they are more systemic because they seek to identify and connect causal relationships between concepts within conceptual systems. However, both of those methods allow the scholar to address partial conceptual systems. FGT, for example, would allow the theorist to cherry pick portions of the subject conceptual systems that seem most relevant to the analysis. In contrast, IPA requires the rigorous integration of whole theories—each as a "closed system" (Dubin, 1978: 116) unto itself.

Investigating theories as conceptual systems unto themselves appears to be a useful and effective path to improving theories of natural systems and theories of service systems. Research suggests that conceptual systems should evolve towards greater complexity first, and afterward evolve toward greater systemic interrelationship (Wallis, 2010a). Therefore, the IPA method presented above seems to be a good first step toward creating more complex conceptual systems—and pointing the way for additional research that will support the development of more systemic conceptual systems.

However, this approach seems to be counter the prevailing current of the social sciences. In preliminary studies of conceptual systems of psychology and sociology, it seems that conceptual systems of the social sciences have been declining in complexity (Wallis, 2012b). This may be due to scholars following the false call of parsimony that has been decried in theory (Dubin, 1978; Meehl, 2002) and research (Wallis, 2010a, 2011).

In the following section, I will discuss some additional insights and approaches that may be useful in developing theories that are more effective in practical application.

The Case Against Ease of Use

The case in favor of parsimonious conceptual systems is simple. "The simplest theory is the best" (Shoemaker *et al.*, 2004: 172). There is, however, no a priori reason or proof that a conceptual system "should" or "must" be easy to use. Indeed, given the astonishing complexity of our lived world, it is more reasonable to assume that our conceptual systems should be highly complex. Indeed, policies which are more complex tend to be more successful in practical application (Wallis, 2011). Also, students with conceptual systems that are more complex and more interrelated tend to have higher scores on papers (Curseu, Schalk & Schruijer, 2010).

It seems reasonable that we will be more effective as individuals and organizations if we use more complex conceptual systems. Given the limitations of the human mind, however, such an approach calls for greater collaboration between scholars. For example, in an organizational intervention, it may be better to have three consultants instead of one—with each consultant focusing on a different area of a shared conceptual system. This perspective also calls for more resources (including funding and administrative coordination) to support teams of scholars and teams of practitioners.

Conclusion

To briefly summarize and conclude, soft methods of integrating conceptual systems (intuitive, ad-hoc, and cherry picking) have been applied through the history of the social sciences without great benefit. Our present conceptual systems for understanding and engaging natural and service systems tend to be simple and have very low levels of internal coherence (systemicity). Indeed, this is part of the reason why we have the problems we do—because we do not have effective conceptual systems to understand and improve our situations.

Soft methods may support the fragmentation of the social sciences, as they lead toward the creation of conceptual systems that are atomistic and spurious. Further, the social sciences seems to be following the false promise of parsimony—and creating conceptual systems that are simpler instead of better.

What we have here is a kind of three-body problem. First, there is the world (which is irreducibly complex). Second, there is the scholar (who has abductive moments of inspiration). Third, there is the theory (the conceptual systems which have been historically difficult to evaluate). None of these provides a clear "fixed point" frame of reference. Instead, each influences the other in ways that have not been well defined. Ultimately, that pattern may be fractal and perhaps indefinable. It should come as no surprise that there is no easy way to understand the situation. Rigorous methodology such as IPA makes the conceptual systems quantifiable and opens the door for a science of conceptual systems.

A systemic view of conceptual systems suggests a number of alternative approaches to reverse the trend and create conceptual systems that we may use to more effectively address our social-ecological issues. First, using more rigorous methods (FGT, RDA, and particularly IPA) will serve to re-integrate the many fragmented conceptual systems. Second, the use of whole conceptual systems should be preferred to the use of partial conceptual systems.

I expect that this paper will provide new tools that scholars and practitioners might use to more effectively decide which conceptual systems will be more useful for research and practice. It will also provide a better understanding of how to more rigorously create more effective conceptual systems. The integrative effort and resulting conversation of this paper is expected to engender new challenges and new insights into bridging the theory-gap between disciplines including the study of natural and service systems through a deeper understanding of conceptual systems. These, in turn, will allow us to gain better understandings of our society and support more effective change.

References

Allison, H. E. and Hobbs, R. J. (2004). Resilience, adaptive capacity, and the "Lock-in Trap" of the Western Australian Agricultural Region. *Ecology and Society*, ISSN 9(1): 25. http://www.ecologyandsociety.org/vol9/iss1/art3

Appelbaum, R. P. (1970). *Theories of Social Change*, ISBN 8410-4019-2.

Apprey, M. (2006). A formal grounded theory on the ethics of transfer in conflict resolution. *Mind and Human Interaction*, ISSN 0533-3164, 14: 51-74.

Boudon, R. (1986). *Theories of Social Change* (J. C. Whitehouse, Trans.), ISBN 0-7456-0950-3.

Burrell, G. (1997). *Pandemonium: Towards a Retro-Organizational Theory*, ISBN 0 8039 7777-8 (pbk).

Charmaz, K. (2006). *Constructing Grounded Theory: A Practical Guide through Qualitative Analysis*, ISBN 9780761973539.

Curseu, P. L., Schalk, R. and Schruijer, S. (2010). The use of cognitive mapping in eliciting and evaluating group cognitions. *Journal of Applied Social Psychology*, ISSN 0021-9029, 40(5): 1258-1291.

Dooley, K. J. (1997). A complex adaptive systems model of

organization change. *Nonlinear Dynamics, Psychology, and Life Sciences*, ISSN 1(1): 69-97, link.

Dubin, R. (1978). *Theory building* (Revised ed.), ISBN 0-02-907620-X.

Dunning, D., Heath, C. and Suls, J. M. (2004). Flawed self-assessment: Implications for health, education, and the workplace. *Psychological Science in the Public Interest*, ISSN 5(3): 69-106.

Ehrlinger, J., Johnson, K., Banner, M., Dunning, D. and Kruger, J. (2008). Why the unskilled are unaware: Further explorations of (absent) self-insight among the incompetent. *Organizational Behavior and Human Decision Processes*, ISSN 105(1): 98-121.

Glaser, B. G. (2002). Conceptualization: On theory and theorizing using grounded theory. *International Journal of Qualitative Methods*, ISSN 1(2): 23-38, link.

Glaser, B. G. and Strauss, A. L. (1967). *The Discovery of Grounded Theory: Strategies for Qualitative Research*, ISBN 0202302601.

Hall, J. R. (1999). *Cultures of inquiry: From epistemology to discourse in sociohistorical research.*

Hitt, M. A. and Smith, K. G. (2005). Introduction: The process of developing management theory. In K. G. Smith and M. A. Hitt (Eds.), *Great minds in management: The process of theory development*: 1-6. ISBN 0-19-927681-1.

Kaplan, A. (1964). *The conduct of inquiry: Methodology for behavioral science*, ISBN 0765804484.

Kovera, M. B. and McAuliff, B. D. (2000). The effects of peer review and evidence quality on judge evaluations of psychological science: Are judges effective gatekeepers? *Journal of Applied Psychology*, ISSN 85(4): 547-586.

Meehl, P. E. (1992). Cliometric metatheory: The actuarial approach to empirical, history-based philosophy of science. *Psychological Reports*, ISSN 0033-2941, 71: 339-467.

Meehl, P. E. (2002). Cliometric metatheory: II. Criteria scientists use in theory appraisal and why it is rational to do so. *Psychological Reports*, ISSN 0033-2941, 91: 339-404.

Mintzberg, H. (2005). Developing theory about the development of theory. In K. G. Smith and M. A. Hitt (eds.), *Great minds in management: The process of theory development*: 355-372. ISBN 0-19-927681-1.

Richardson, K. A. (2009). Exploring the Implications of Complexity Thinking for the Management of Complex Organizations. In S. E. Wallis (Ed.), *Cybernetics and Systems Theory in Management: Tools, Views, and Advancements.*

Ritzer, G. (1990). Metatheorizing in sociology. *Sociological Forum*, ISSN 0884-8971, 5(1): 3-15.

Ritzer, G. (2001). *Explorations in social theory: From metatheorizing to rationalization*, ISBN 0-7619-6773-7.

Ritzer, G. (2009). Metatheory. In G. Ritzer (Ed.), *Blackwell Encyclopedia of Sociology Online*, Vol. 2010. Hoboken, New Jersey: Blackwell. http://www.sociologyencyclopedia.com

Shoemaker, P. J., Tankard Jr., J. W. and Lasorsa, D. L. (2004). *How to Build Social Science Theories*, ISBN 0-7619-2667-4.

Smolin, L. (2006). *The trouble with physics: The rise of string theory, the fall of a science, and what comes next*, ISBN 0618551050.

Spicer, M. W. (1998). Public administration, social science, and political association. *Administration and Society*, ISSN 0095-3997, 30(1): 35-53.

Stinchcombe, A. L. (1987). *Constructing social theories*, ISBN 0-226-77484-8.

Van de Ven, A. H. (2007). *Engaged scholarship: A guide for organizational and social research*, ISBN 978-0-19-922630-6.

Wallis, S. E. (2006). *A sideways look at systems: Identifying sub-systemic dimensions as a technique for avoiding an hierarchical perspective.* Paper presented at the International Society for the Systems Sciences, Rohnert Park, California. 0-9740735-7-1.

Wallis, S. E. (2008). From reductive to robust: Seeking the core of complex adaptive systems theory. In A. Yang and Y. Shan (eds.), *Intelligent complex adaptive systems*: 1-25. ISBN 978-1-59904-717-1.

Wallis, S. E. (2009). The complexity of complexity theory: An innovative analysis. *Emergence: Complexity and Organization*, ISSN 1521-3250, 11(4).

Wallis, S. E. (2010a). The structure of theory and the structure of scientific revolutions: What constitutes an advance in theory? In S. E. Wallis (Ed.), *Cybernetics and systems theory in management: Views, tools, and advancements*: 151-174. ISBN 978-1-61520-668-1.

Wallis, S. E. (2010b). Toward a science of metatheory. *Integral Review*, ISSN 1553-3069, 6(Special Issue: "Emerging Perspectives of Metatheory and Theory"), link.

Wallis, S. E. (2011). *Avoiding policy failure: A workable approach*, ISBN 978-0-9842165-0-5.

Wallis, S. E. (2012a). The right tool for the job: Philosophy's evolving role in advancing management theory. *Philosophy of Management - Special Issue (Guest Editors: Stephen Sheard, Mark Dibben*, ISSN 11(3): In press, publication anticipated 2013.

Wallis, S. E. (2012b). *Theories of Psychology: Evolving Towards Greater Effectiveness or Wandering, Lost in the Jungle, Without a Guide?* Paper presented at the 30th International Congress of Psychology: Psychology Serving Humanity, Cape Town, South Africa.

Wallis, S. E. (2013). *Propositional Analysis for Evaluating Explanations through their Conceptual Structures.* Paper presented at the International Society for Emergence and Complexity, Paris.

Chapter 9
Strategic Knowledge Mapping—The Co-Creation of Useful Knowledge

Best Paper award: Innovations and future directions in education track.

Wallis, S. E., & Wright, B. (2015, March 4-6). Strategic Knowledge Mapping: The Co-creation of Useful Knowledge. Paper presented at the Association for Business Simulation and Experiential Learning (ABSEL) 42nd annual conference, Las Vegas, CA.

Abstract: *Strategic planning typically involves conducting research and setting objectives. It is a difficult and expensive process with no guarantee of success. Recent research shows that managers with more "structured" knowledge will be more successful. Using Integrative Propositional Analysis (IPA) we can objectively determine the potential usefulness of a Strategic Knowledge Map (SKM). Creating an effective SKM is a precursor to more easily creating a more effective strategic plan. The present game is focused on players co-creating an SKM. Their play is scored in such a way that they will receive more points for creating a more structured map. The resulting map may be easily used in the "real world" to support dialog, decision making, and the creation of specific objectives for strategic plans.*

The game is unorthodox. It is not a simulation where play begins with a pre-set "world." Similarly, the game is not educational in the traditional sense where players attempt to acquire or test knowledge using an existing database. Instead, ASK MATT is a model-building game where knowledge is co-created within the game by the players. Further, the game goes beyond finding "insights"; instead, the results of the game may be directly applied as a guide to real world situations.

In the present paper, we explore the background, difficulties, and opportunities for improving strategic planning and policy planning using strategic knowledge mapping from a systemic perspective. We explain the play of the game, its scoring, anticipated outcomes, our experiences playtesting the game with small groups, plans for playtesting with larger groups, and opportunities for developing a version of the game that may be played online and/or as an APP.

Introduction

In this paper, we present ASK MATT, a rather unusual game with an odd history. The game began with research on the "theory of theory" or "metatheory." For our purposes, a theory is generally understood to be a set of interconnected concepts; metaphorically, it is a lens to understand and engage the world. In short, a theory may be understood as useful knowledge. A metatheory is a theory specifically directed to understanding how theory is created, structured, or used. Studies of metatheory, published in the academic literature, showed how we might better understand what we understand. In essence, the research shows how we may evaluate theories based on the internal structure of their logical propositions. That structure will show the potential usefulness of the theory—before testing the theory in practice or application as enacted knowledge. From those rather opaque investigations, it was recognized that there must be a simpler way to go about developing our understanding. After many attempts at simplification in the academic literature, popular literature, and conversations with a wide variety of people, one thing became clear. A better way to engage minds could be found by playing games.

In the present paper, we begin by presenting a brief background from the academic literature. Essentially showing a heterodox method for evaluating our "conceptual systems" (e.g., theories, policies, strategic plans). A key part of that presentation will be strategic planning and "Strategic Knowledge Mapping." This is important because if an organization is to be successful, it is crucial for that organization to have a reliable map. Yet, until recently, our ability to evaluate those maps has been severely limited. The present paper shows how maps may be evaluated.

Next, we present how the process of map creation and evaluation was gamified. That is followed by a description of game play, and then by our experiences with alpha testing the game. The paper concludes by discussing a few of the many potential directions for developing the game—and a call for your suggestions to help us move effectively forward in bringing this game to a place where it can be used to support more successful strategic knowledge mapping, strategic planning, and greater success for individuals and organizations.

The Necessity and Difficulty of Strategic Planning

Strategic planning is typically understood as clarifying strategic goals. And, importantly, this can be a problem if one does not have a good understanding of the organization and the business environment. To set goals without a map is foolhardy. What traveler standing on point A would say "we should go directly to point B" without knowing the intervening terrain? The leader may be unknowingly pointing toward a precipice!

Managers want and need useful insights (Czarniawska, 2001). Yet, our academic world seems unable to provide them with useful theories (Weick, 2003).

Strategic planning is a very important process for corporations because it helps to support organizational success through the efficient allocation of resources. Such an approach should provide a systemic view of the firm's resources and situation. Such an approach also supports managerial learning (for existing managers and new managers alike) to increase the leadership team's ability to make effective strategic decisions (Lorange, 1978). However, wrong decision may destroy a career. So the process is also scary. That fear leads managers to use relatively simple interpretations and traditional strategies (Martin, 2014).

It has been argued that top managers should spend up to 10% of their time engaging in strategic planning activities (Bryson, 2011). This is a considerable investment of time for questionable returns. While strategic planning is undoubtedly important and promotes organizational success, it is also argued that the results may be irrelevant, dysfunctional, and lead to excessive rigidity (Miller & Cardinal, 1994). Some studies have found strategic planning process such as SWOT analysis to be ineffective (Hill & Westbrook, 1997).

Importance of SKM (Strategic Knowledge Map)

The knowledge map is a close cousin of causal maps, strategy maps, concept maps, and cognitive maps. For the present paper, we will refer to these generally as "maps" or strategic knowledge maps (SKMs). A good map is one that is useful and effective for making organizational decisions, for improving organizational cohesion, communication, encouraging the development of new knowledge, and boosting economic returns (Wexler, 2001). As Wexler (2001) explains, a knowledge map is a useful tool for overcoming data smog or information overload. Maps are also used to make sense of existing knowledge and to identify where exploration might lead to new and useful knowledge—an important "strategic asset" (Zack, 2000) especially for entrepreneurs (Casson, 2005).

However, in the same way that we have poor theories of business and poor approaches to strategic planning, our ability to develop useful theories and practices of knowledge management is also in question. In a recent survey of 12 theories of organizational learning, Wallis (2009a) found that the concepts of organizational learning theory were not well understood, suggesting that the process of organizational learning is not well understood. This is a problematic situation if the success of a firm is tied to its people's development and use of knowledge because the process of cognition is also a resource to the firm (Alvarez & Busenitz, 2001).

Despite (or because of) its importance, "Knowledge mapping is a daunting task" (Wexler, 2001: 262). So much so that it may seem more efficient to have specialists who are engaged in the creation of maps. However, this creates a gap between those who make the maps and those who use them. In this age of transient knowledge, it is crucial that the map-makers be the same people who are the map users. Therefore, the task of mapping must be made accessible, something that any manager may learn and use. This gap between the importance of having maps and the difficulty of making maps prompted the present authors to develop a new approach. One that would leave open the content of the map to the greatest degree possible while maintaining a rigorous approach to how the presented knowledge should be structured.

But how do we know if a map is good when, for every firm, there is a need for unique knowledge? The lessons of one firm cannot be applied in whole to another. The map used by a CEO leading a Fortune 500 firm through an economic boom is unlikely to be useful to a mom-and-pop operation trying to find their way through a recession. Or, metaphorically, a map of Disneyland won't help me navigate through the Sahara desert. In short, we need new insights to create better SKMs.

Difficulty of Systems Thinking

Systems thinking (ST) and complexity theory (CT) have been suggested as ways to gain a better understanding of situations in the policy world (Dennard, Richardson, & Morçöl, 2008; Morçöl, 2010), as well as in a variety of business operations (Brown & Eisenhardt, 1998), management activities (Wheatley, 1992), as well as individual ways of thinking and interacting (Senge, 1990; Senge, Kleiner, Roberts, Ross, & Smith, 1994).

Generally, these systemic approaches are focused on looking at relationships. For example, the use of "feedback loops" and "balancing loops" may indicate unanticipated sources of interruption in development and production processes. Insights from systems thinking may be used to identify and resolve hidden problems.

However, systems thinking is difficult to learn (Nguyen, Graham, Ross, Maani, & Bosch, 2012). Or, more prosaically put, "Complexity thinking is hard" (Tait & Richardson, 2011: v). In short, complexity science has not been effective at creating tools for practitioners.

There is an interesting confluence between the difficulty of strategic knowledge mapping and the difficulty of systems thinking. Where either one would seem to require a very high level of education and effort, there seems to exist a point of leverage where each may be used to make the other more accessible. And, importantly for practical application, more useful for managers. By applying systems thinking to the creation and evaluation of SKMs, we aim to achieve this new level of insight and find a path to improve both of them.

Integrative Propositional Analysis (IPA)

In the middle of the 20th Century it was suggested that our mental models, our understandings of the world, might have some systemic "structure" (Kelly, 1955). Subsequently, researchers developed Integrative Complexity (IC) as a method to analyze the structure of mental models (Suedfeld, Tetlock, & Streufert, 1992). Briefly, IC was used to evaluate paragraphs of text (from writing samples, policies, government announcements, etc.) and rate them on a scale of one to seven (one being low level of structural interrelatedness between the ideas and seven indicating a high level of structural interrelatedness). A simple statement of universal truth would score low, while a complex statement (found in many philosophical and academic writings) would receive a high score. Studies using IC found that higher scores were significantly correlated with managerial effectiveness (Wong, Ormiston, & Tetlock, 2011), political success (Raphael, 1982), and higher student scores (Curseu, Schalk, & Schruijer, 2010).

The ideas of complexity and systemic interrelationship that are found in the underlying assumptions of IC are also found in systems thinking. IC is, therefore, in some sense, a relatively straightforward tool for applying insights from systems thinking to evaluate managerial thinking. However, IC was developed to analyze paragraphs, not the diagrams associated with strategic knowledge mapping. More recently, Integrative Propositional Analysis (IPA) was developed for the purpose of evaluating conceptual systems such as theories and policies (Wallis, 2014b). IPA has a track record of evaluating theories in a variety of fields (e.g., Wallis, 2009a; Wallis, 2009b, 2011b, 2012), including policy (Wallis, 2010c, 2011a, 2013), entrepreneurship (Wright & Wallis, Under submission), and even ethics (Wallis, 2010b). As with IC, IPA has shown that conceptual systems with more complexity and more systemic structure are more effective in practical application (Wallis, 2010a).

IPA rates a model based on its Complexity (the number of concepts contained) and its Systemicity (the percent of those concepts that are concatenated). A concatenated concept is one that is "resulting" from two or more "casual" concepts. Or, diagrammatically, a concatenated concept is a box with two or more arrows pointing towards it. A map with a higher Complexity score may be said to cover more ground (breadth of understanding). A map with a higher Systemicity score may be said to hold a greater understanding of the area that is covered (depth of understanding).

For example, Ohm's Law (E=IR) contains three concepts (so it has a Complexity of C = 3). Each of those concepts is concatenated from the other two (so it has a Systemicity score of S = 1.0). Thus, it is a highly effective map—within a very narrow area. In contrast, our research into theories of the management sciences tend to have Complexity scores of approximately 4-24 and Systemicity scores around 0.20. This low Systemicity reflects greater breadth and less depth than Ohm's Law. It also reflects the low level of effectiveness of these theories. One might also look at the equation and infer how changes in one will lead to changes in the next, and then the third. Thus, in some sense, an algebraic equation may be seen as a loop.

This emerging perspective is useful for evaluating and improving theories, models, and policies. However, the underlying thinking is heterodox and often difficult for people to grasp. Therefore, to bring this new method to a larger audience in a more accessible and fun way, the authors have worked to create the ASK MATT game.

ASK MATT is a simple acronym for the more complex title "Accessing Strategic (or Special) Knowledge Meta Analysis Think Tank." In brief, the collaborative input of information and specialized structuration of the play and resulting model are combined to turn an ordinary group of people into an extraordinary think tank.

Gamification

Understanding game design and testing the benefits is not without its challenges (Butler, Markulis, & Strang, 1985). From its humble origins, gamification has become a rapidly emerging phenomena (Faria, 2000). While there are still opportunities for improvement in the field as a whole (Gold, Markulis, & Strang, 2014), the process does appear to be useful for teaching about strategic management (Burch *et al.*, 2014) in the real world as well as in the classroom (Jakubowski, 2014).

The process of gamification includes the creation of gameful experiences using elements of games, but directed to more purposeful activities such as learning and skill-building (Deterding, Dixon, Khaled, & Nacke, 2011).

In the present game, the players do not need to understand anything of IC, IPA, complexity theory, or systems thinking as those difficult conceptual elements have been gamified. They need merely pay attention to the scoring system. And, importantly, the play of the game and the scoring system will lead players to create a map with a higher Complexity and higher Systemicity (or Mapicity). Thus, the game process will lead to the creation of a map that will be more likely to be effective when applied to strategic planning and strategic management activities.

While the gamification process has been useful for making the process more accessible, it is difficult to place the game into categories that are common to the world of gamification. First, as yet, we have found no direct precedents for this game. Therefore, like the underlying methodology, the game itself is unorthodox. Second, the game is not "problem based" (c.f. Bidgelow, 2004). Nor is it based on an existing knowledge base or data base that the players must learn (or test their knowledge against) as in "trivia" quiz type games. Neither is it a simulation where play begins with a preset "world." Instead, ASK MATT is a model-building game where knowledge is co-created within the game by the players. Further, the game goes beyond finding "insights;" instead, the results of the game may be directly applied as a guide to real world situations.

Game Process

ASK MATT is a game designed to help members of organizations and coalitions to better understand their business environment. As the game is played, bits of information held by individuals are linked into a more coherent map—a map that is more useful for guiding organizations and making key strategic decisions with greater confidence. And... it is fun! See Appendix C for simple sample map and Appendix E for an example of play.

Essentially, the individual players engage in a three-step process.

1. Think about what concept (or connection between concepts) might be best added to the map;

2. Write the concept on a card, and;
3. Place the card on the map.

Before the start of the game, the group decides on the focus of the Game based on the topic that the Group wishes to explore. That focus is written as the title for the MAP. For example, "XYZ Company's Map for Marketing," "Game Theory," or "Relationships in the Health Care Coalition of ABC County." Players' names are written on the score sheet and players are given a stack of pieces.

- One person is selected to be the Scorekeeper—he or she should have a calculator.
- A non-player may be chosen as a facilitator (to help conversations move forward smoothly).

The Strategic Knowledge Map (MAP) is created as players take turns adding pieces to the game's play surface. Each player may place one piece per turn. Turns are taken starting with the Scorekeeper and proceeding around the table in a clockwise direction. When a piece is played, the other players immediately vote to de-

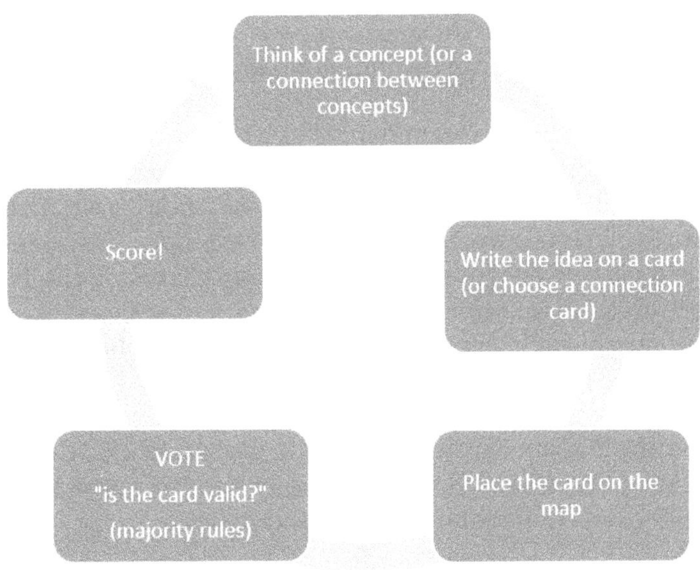

Exhibit 1 *General Overview of the Basic Game*

termine if the piece is to stay on the map—and how many points the player should receive.

There are two basic kinds of pieces that may be played (see Appendix A). The first piece is a "Point of interest" (POI)—these are the concepts that players think of and place on the map: the places on the map. Players take turns choosing and creating each Point of Interest and naming it by writing a title on the POI playing piece.

- One point is awarded for each POI that the player places on the map.
- One bonus point is awarded if the POI is measurable (determined by popular vote of the players).

Instead of POIs being "national parks," here, they can be anything that the player feels is important or interesting for the organization (within the focus of the game)—from "pencils in the stockroom," to "employee motivation," and even "interest rate." These might also be understood as "destinations" or "endpoints." However, it is important that each is more of a category than a specific number. For example, a POI should be titled "annual income" instead of "$5 million per year." Under the POI title, the player writes how the POI will be measured. For example, "income" might be measured in "dollars per month" and "customer satisfaction" might be measured by "phone survey."

There may be some argument as to whether a POI is "real" or otherwise acceptable. Here is a little bit of clarification. If a player places a POI, one point is automatically awarded—even if others think the POI is not real or important. For example, "invisible fairies" is a valid POI. When voting, the group decides if the POI is measurable. So, unless the player happens to have a way to identify the number of invisible fairies, the group will probably not vote to award the bonus point!

"Causeways" are like roads on a map (see Appendix A). During his or her turn, each player may choose to place a Causeway on the MAP to show the direction of cause-and-effect.

For example, one POI might say, ("Training") with a Causeway played to link the POI for "Training" with another POI "Customer

Satisfaction." Essentially, the causeway tells how changes in on POI results in changes in another POI. Here, wild ideas should be avoided. For example, Players should avoid thinking "I saw a show about how dogs can be taught how to drive, therefore, it is OK to say that we can use dogs to drive our delivery trucks." A good way to evaluate a Causeway is to state the relationship in a sentence. For example, "More of this POI will cause more of that POI."

When a Causeway is placed, the other players vote, to decide if it is a valid/reasonable play. For example, a Causeway showing POI (more dollars spent on advertising) will lead to (cause) POI (more sales) might be voted as valid. In contrast, a Causeway showing that POI (more dollars spent on advertising) will lead to POI (faster production speeds) might not be considered valid.

In addition to the two basic pieces (POI and Causeway) that are played during individual turns, there are three other pieces that may find their way onto the MAP.

"Fog" (see Appendix A) is a marker placed on Causeways when the group cannot come to a fast voting decision about whether the Causeway is valid. Fog indicates the need for additional research.

If there is no majority for a vote on the measurability of a POI or the validity of a Causeway, or if the argument gets too heated (we've all been on road trips like that) any player may suggest that a FOG piece be placed. Fog is placed by general consensus (or a majority vote). Discussion on the piece is halted for the time and play continues to the next person.

- A Causeway that is under FOG may remain on the map but does not count toward any player's point total.
- A POI under FOG may remain on the table but the player does not receive a bonus point for measurability.

The two final pieces are placed to help keep track of bonus points for key developments of the map (see Appendix A). A "Gold Star" is placed on a POI the first time that a second Causeway is placed pointing towards the POI. The player who places that Causeway is awarded two bonus points. This may be understood as "Merging."

With a real map, multiple roads provide players with multiple options for reaching their destinations by Merging onto and off of various routes. The Gold Star shows the importance of having multiple roads to reach POIs.

A "Blue Ribbon" (see Appendix A) is placed at the center of a loop when a player places a Causeway that completes a beltway (circle of Causeways). The player is awarded five points.

A beltway is a loop which includes four or more POIs, with all the Causeways between the POIs pointing in the "same" direction (so a vehicle could travel in a loop back to its starting point without violating any traffic laws). The Blue Ribbon shows the importance of recognizing positive and negative reinforcing loops (also called virtuous cycles and vicious cycles).

Simple Strategies

To gain a higher score, be sure that the POIs played are measurable. Some vaguely stated concepts ("morale", for example) are important, but not easily measurable, whereas "increased employee satisfaction" is easier to measure. Those may lead to arguments, delays, and loss of potential points. Players should strive for Gold Stars by identifying valid Causeways between multiple POIs. Similarly, strive to play Causeways that are reasonable, so, again, the group may easily award points. Finally, a Blue Ribbon provides the greatest opportunity to gain points. However, it must be built carefully from good POIs and Causeways (or else your Beltway may collapse). Remember, a Beltway can be very large, so you may be able to create loops by identifying connections between POIs at opposite ends of the map.

Basic Game Mechanics

The scorekeeper goes first; play follows around the table clockwise. When each player has taken a turn and placed a piece, it is the end of the round. The Group score is calculated at the end of the round.

1. Players take turns placing one game piece per player per round (Causeway or Point of Interest (POI)) on the MAP.
   a. If the played piece is a POI, the player must write what the POI represents (e.g., income or creativity) and how it

might be measured (e.g., dollars).

   i. The player receives one point for any POI played.

   ii. The group votes if the POI is measurable. If the group agrees it is, the player is awarded a second point.

b. If the played piece is a Causeway, it must be placed to indicate a link between two POIs and the player must mark the appropriate word on the Causeway to show one POI causes "more" or "less" of the other POI.

   i. The group votes if the Causeway indicates a reasonable or sensible connection. If a majority of players votes that it does, the player receives one point. If not, the Causeway is removed from the map.

   ii. f the vote is tied, a FOG piece is placed on the Causeway and no points are awarded.

   iii. If the Causeway is the second Causeway pointing to a POI, a Gold Star is placed on the POI and the player is awarded two bonus points.

   iv. If the Causeway completes a beltway (four or more Causeways in a loop) a Blue Ribbon is placed in the center and the person who played the last Causeway is awarded five bonus points.

2. Points are recorded on the Scorecard by the Scorekeeper immediately after voting occurs.

There are some additional rules to facilitate understanding and play.

- Scoring of the game may be "individual score," "group score," or both. The decision rests with the team and the culture of the organization.
- Pieces may be rearranged by group consensus—especially to make the map more readable and/or to make room for more pieces.
- Additional POIs and Causeways may be inserted by group consensus. For example, if POI "A" leads to POI "C," the group may want to insert POI "B" (and another causeway) between the two.

- Rules may be changed with the consensus of all players (and the consent of the facilitator if one is used). All changes to the rules should be reviewed monthly to see if they are helping or hindering improvement of the MAP and the navigation of the organization.

During play, it is useful for players to ask themselves (or, for the facilitator to ask the players to consider) such questions as:

- What concepts are important to me in this situation?
- What concepts might be added as POIs?
- What "blind spot" ("white space" or unexplored territory) exists on the map because a POI does not have a Gold Star?
- What new insights might be emerging that suggest new actions or changes in course?

When a FOG piece is placed on a piece to indicate where additional work is needed "outside the room"—this may indicate the need for additional research which is then brought back to the table at a later date. Then, FOG pieces may be removed if voted off the map by a majority of the group.

Post Game

It is certainly possible to put the game on "pause" and resume playing another day. It is also very useful to schedule some time at the end of the game (at least 30 minutes) to have a general conversation about the Game. Players might discuss:

- What did you learn from the Game?
- What are the implications in the organization for workflow, communication, tracking, and feedback?
- What might each player change in her or his part of the organization?
- What "leverage points" might exist for change—where one improvement will cascade into many (here, look especially at the loops)?
- Is the Mapicity (Systemicity) score greater than 0.50? If not, you should probably keep playing until it is.

- What concepts and connections might be missing from the MAP that might be considered for a future game?
- If you have a question about any facet of the organization—any operation, workflow, direction, or strategic challenge—now is the time to ASK MATT. If the group, guided by the map, cannot answer, it is time to schedule another game.

Road Trip Rules

The "Road Trip" begins with having a complete map from the basic game. Essentially, the Road Trip is about using the game map for navigating an organization through the real world. Before starting the Road Trip, we recommend having Gold Stars on more than half the POIs. The Road Trip focuses on tracking the progress of the organization by tracking the changes in each POI on a monthly basis.

For the Road Trip:

- A "Travelogue" piece (See APPENDIX A) should be placed on each POI A Travelogue is simply a 4X6 piece that lists each month, and has space for each month's data (for example, hours worked, or $ in bank).
- Each player volunteers to gather data for the Travelogue (or is assigned based on their roles within the organization).
- If this is to be part of a strategic plan, write the goals for each POI on the Travelogue.
- The whole group meets on a monthly basis. Each person brings their data.
- Between meetings, each player records their key decisions/actions on their personal "Turning Points" sheet. This is a sort of diary that is related back to the MAP. For example, see Appendix D.
- Each monthly meeting:
    - Record the data on the Travelogues (one point awarded when a data point is added).
    - Players take turns sharing their "Turning Points" (major decisions). The group votes as to whether

an action counts as a major decision (one point awarded for each).
    - Discuss results of actions and potential results of differing actions using the map as a guide. Discuss what additional research might make a better map.
    - Repeat each month.
- At the end of twelve months...
    - Crunch the numbers—do they add up? That is, did the predicted changes in POIs lead to the predicted changes in other POIs? Was the map followed as planned? Why, why not?
    - Replay the game—add additional POIs and Causeways to reflect the new knowledge.

During the Long Game, players have the opportunity to gain extra points with meta-level actions including:

- Look at a complex POI and play a New Game to understand it (the title of the original POI becomes the title of the new game) (five points for each player of the new game). In other words, create a detail map for an area within the original map.
- Play the Game with other departments, organizations, and coalitions (ten points for each person to convene a new game).
- Integrate multiple Games. Starting with two or more MAPs, look to see if there are any POIs that are the same for both MAPs. If there are, the two maps may be linked by their POIs. Or, if there are no overlaps between the MAPs, strive to identify new Causeways between POIs of differing MAPs. This will lead to greater understanding (twenty bonus points for the player or players who convene the integration effort and five bonus points for each participant).

Players

The game may be played with one person or more. For convenience, it may best be played with four to eight people. If players are not familiar with the game, or if there are many people playing, it is recommended that the organization should recruit a game coach or facilitator—someone who is not playing the game, but is familiar with the game process and responsible for making sure the game runs smoothly. If many players are involved, it may also be useful to play the game with multiple smaller groups—and then integrate the maps later on.

Equipment

This game includes five pieces (POIs, Causeways, FOG, Gold Stars, and Blue Ribbons). There are also score sheets (for tracking progress), marker pens (for writing information on pieces). If deemed necessary, a timer may be used for moderating conversations. The playing surface provides a place to place the pieces. The Game may also be played on a table-top or any other surface where the pieces may be set. Because players may want to save and/or move the MAP, it is often useful to play on some large sheets of chart-paper or butcher-paper. That way, the map may be rolled up your map and carried to meetings (the MAP may also be moved and connected with other MAPs). Finally, the Game includes a set of facilitator suggestions for stimulating conversation, improving the flow of the Game, and improving the end result as well as additional rules for the Advanced Game.

Scoring (Overview)

Individuals gain higher scores by placing Causeways and POIs. One point is awarded for each POI that is played. One bonus point is awarded if the POI is considered measurable by a vote of the group. One point is awarded and for each Causeway that is considered reasonable by a vote of the group (often, this should mean that they are supported by research). One Gold Star is placed on a POI the first time that a second Causeway is pointing towards it. The player who places the second Causeway is awarded two bonus points automatically if the group votes to keep the piece on the map. When a player places a Causeway that creates a loop (including four or more POIs with all Causeways pointing in the same direction) a Blue Ribbon is placed at the center of the loop and the player is immediately awarded five bonus points if the group votes to keep the Causeway as part of the map.

The Group gains higher scores by having a MAP that is more complex and more interconnected (which, for a map, makes it more practical and useful for navigation).

In this game, in a sense, each player is competing with other players to gain a higher score. That scoring sets a tone of friendly competition where each individual is striving to bring out the best knowledge in themselves and each other. The idea is not to reduce the score of others. Because, at the same time, all players are working together to gain a higher Group score—more like a running race, than a football game. That scoring reflects a quest for greater shared understanding of the world.

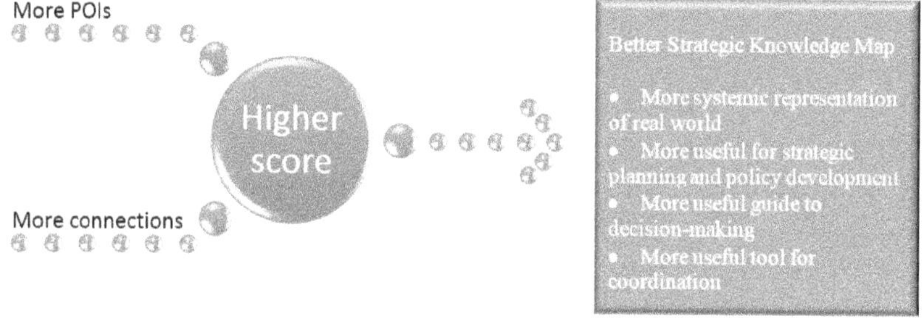

Exhibit 2 *General overview of scoring and usefulness*

To conclude this section, it is important that players engage the game, and one another, with a sense of respect, open inquiry, and playfulness.

Results of Alpha Playtesting

At the time of this writing, we've conducted six "alpha" tests of the game. These have been small-scale, often with people we know (or among the game developers, ourselves). In this section, we will report on some key themes that emerged. We have made changes to the game and developed beta testing plans based on ideas and comments that emerged in alpha testing.

Playtest groups ranged in size from one to five. Topics included marketing, business startup, operations, and politics in the Middle East. Game developers were always in attendance and provided the playtesters with a set of rules, pieces, and a brief overview of the game. Developers were also on hand to answer any questions as they arose.

As players read the rules and discussed the possibilities of play, they generally liked the cover on the box. Many found the rules confusing. Additionally, there was some confusion over terminology (because examples in the game were written for business players... but some players were interested in playing the game for other purposes, such as a research project or to solve a personal problem).

Many players liked the opportunity for team-scoring, recognizing that the group will always "win" whatever the individual scores may be. They also recognize that this is not an "ordinary" game because there were no set pieces or "right answers" of specific goals to reach. Indeed, that lack of clear or traditional objective was sometimes confusing.

Many testers gave great thought to the process of play; and, even before play began, were attempting to "strategize" their play to gain higher points. This led to additional "what if questions" and conversations.

Once play began, some confusion continued around the rules, actions to be taken during play, point allocation, and the objectives of the game.

Some recognized quickly that much confusion could be alleviated by having a facilitator or experienced player who could manage the flow of the game so that the players could concentrate on playing.

For some, there was the recognition that this game was something similar to their previous experience with flow-charting; others experienced confusion about the process and purpose of play.

It was found that there is often the need to be able to rearrange pieces.

One purpose of the game is to develop new insights around how people understand their organization and its environment. As play progresses, new insights and understandings emerge. Thus, it became a challenge for some people to move back and forth between "playing" the game and reflecting on the insights that were rapidly emerging from the map created during play.

The game is designed to develop a map. A higher score is achieved by creating a map with more Complexity (more POIs) and more Mapicity (more Causeway connections between POIs). From playtesting, the maps ranged in Complexity from 3-29. They ranged in Mapicity from 0 to 0.5.

Most players expressed interest in the game. Common comments included: "It was fun," "more fun than expected," and "intellectually stimulating." Even in those tests where players were not highly enthusiastic, they were clearly very engaged in the game.

Game play (and after play) included many good discussions, questions, and answers about the understandings being generated. Some discussions included using the map as a guide for organizational understanding and action.

There were many suggestions about how the game might be improved, simplified, marketed, and developed in new ways. For example, the game could be played by individuals or small groups who would each develop a small map; then, those maps could be connected to create a larger, more intricate map to gain still greater insights. Another important suggestion was that the game could be moved to an online environment. This could a much more positive user experience (Deterding, Sicart, Nacke, O'Hara, & Dixon, 2011).

Other suggestions include the idea that we might provide online instructions and videos as well as sample maps or samples of play. These would be very useful to help new players to see how the game should be played.

To summarize this section, the alpha testers provided an excellent range of suggestions—many of which have been incorporated into the revised rules. It is interesting to note that sometimes this interfered with the play of the game—as some testers were "redesigning" the game while others were busy playing it.

The underlying process and function of the game is generally considered to be successful because it stimulates a great deal of conversation between the players and insights by individuals. Game play supports the creative generation of new ideas and new ways to evaluate old ideas. Where one player might have thought some business practices were clear and simple, another might disagree. The game provides a forum for working out those differences. Importantly, the structure of the game allows and encourages participants to develop a shared understanding that is more complex and more complete than any of the individual understandings.

Generally, players were interested in playing the game again (which we took as a good sign). We are greatly in their debt for the wide range of suggestions and conversations which clarified many issues around play.

Competencies, Teaching, and Learning

As discussed, the focus of this game is on surfacing, crystalizing, and evaluating knowledge as formalized in a Strategic Knowledge Map (SKM). Thus, we expect that managers and other leaders who follow the process and scoring of the game will become more proficient in creating and evaluating SKMs.

Because the game involves surfacing real-world knowledge about existing business situations, the players may expect to gain a great deal of knowledge related to their specific situation.

Because the play requires a great deal of communication and collaboration within the process of play, we expect that the game will tacitly build competency in:

- Communication;
- Collaboration;
- Decision making;
- Systems thinking;
- Meta analysis;
- Finding unique insights that may be used to achieve goals more quickly, and;
- Coordination.

In some sense, the game creates a model—including visual representation, reduction, and a pragmatic feature (Karl, 2014). However, those design elements are chosen by the players rather than the designers. The game creates a model similar to a work-flow map. Therefore, the players may expect to gain knowledge around how work actually flows within their organization. This, in turn, may suggest insights for improving workflow and communication flow within the organization.

When the game is played by members of a community coalition, participants will also gain knowledge of related organizations, their effects on the community, and opportunities for increased collaboration.

It is interesting to note that the knowledge gained through play is "real world" rather than "hypothetical" or "business school" knowledge. Thus, the knowledge gained can be expected to be more useful in immediate situations and less transferable to alternative situations.

Insights Into the Future

Using the insights form alpha testing, we have made numerous changes to improve the clarity and playability of the game. These changes should be sufficient to create a more playable "beta" test version. Particularly if play is facilitated by an experienced individual or "mapping coach" (perhaps "cartography coach"?).

Further development in additional directions is being considered, including:

- For adventurous and creative groups, working with the group to decide their own "rules of the game" before the start of play.
- For those who want a pre-written set of POI, researching the related literature to find some concepts and relationships that might apply to their situation.
- Adding new and more complex elements to game play as players develop more skills.
- Developing a more effective explanation for play. The game itself is not terribly difficult. The underlying concept, however, is rather unusual and sometimes difficult to grasp.

- Testing and further developing a "long game" where the game is revisited on a monthly basis to bring in new data—and played out again in full on an annual basis. This would help participants to continually improve their understanding.
- Developing an online and/or APP version of game.
- Developing an easier game with POIs that are pre-made. That is, common business concepts are already written on them.

The game is based on notions of causality, measurability, complexity, and systems thinking. While our goal has been to simplify the process of understanding complexity, it is worth mentioning that there are also opportunities for creating an advanced game that will require a higher degree of experience and skill to play. Explorations in the literature suggest that more nuanced understandings await. These include abstraction (Wallis, 2014a), core/belt (Wallis, 2014c), orthogonal understandings of taxonomy, and others that await discovery on this new road to making better maps.

Conclusion

In this paper, we have presented and discussed a new game, ASK MATT. This game provides a novel approach to developing innovative understandings of organizational activities and environments. In this paper, we have presented the rules, process of play, and the results of our alpha testing.

As we have played the basic game with our alpha testers, most have exhibited a high degree of interest, engagement, and enthusiasm. Additionally, consultants we have spoken to also tend to show a high level of interest in the game. In part, we believe, because they recognize that gamification has a great future in the business world.

A high scoring game is expected to be highly effective for supporting Strategic Knowledge Mapping, strategic planning, policy development and analysis, and more. This game also offers significant potential benefits to individuals and organizations seeking to improve communication, collaboration, and their shared understanding of the complex world in which we work. Indeed, the online version of the game may well generate massive insights on organizational level. Consider, for example, what happens when individuals, teams, and experts are all placing their interconnected understandings into a large virtual-space version of the game. The resulting map will be far beyond the comprehension of any individual. Yet, having this map to serve as a guide, an individual may leverage his or her ability to make positive change.

Importantly, this game is different from existing practices for strategic planning (which, sadly, have rarely proved effective). Where typical strategic plans are often based on a single shared goal to be reached by a single path, the map developed in the ASK MATT game allows each player to have a different preferred destination or goal. Yet, the many goals are interwoven so that individual success is linked to shared success.

We look forward to developing this game in its table-top and online forms. And, importantly, we look forward to hearing your comments, suggestions, and opportunities for collaboration.

References

Alvarez, S. A., & Busenitz, L. W. (2001). The entrepreneurship of resource-based theory. *Journal of Management, 27*(6), 755-775.

Bidgelow, J. D. (2004). An online situation for problem-based learning in a junior-level management course. *Developments in Business Simulation and Experiential Learning, 31*, 120-124.

Brown, S. L., & Eisenhardt, K. M. (1998). *Competing on the Edge: Strategy as Structured Chaos*. Boston: Harvard Business School Press.

Bryson, J. M. (2011). *Strategic planning for public and nonprofit organizations: A guide to strengthening and sustaining organizational achievement* (4 ed.). San Francisco: Jossey-Bass.

Burch, G. F., Batchelor, J. H., Heller, N. A., Shaw, J., Kendall, W., & Turner, B. (2014). Experiential learning—What do we know?: A meta-analysis of 40 years of research. *Developments in Business Simulation and Experiential Learning,, 41*, 279-283.

Butler, R. J., Markulis, P. M., & Strang, D. R. (1985). Learning theory and research design: How has ABSEL fared? *Developments in Business Simulation and Experiential Learning, 12*, 86-90.

Casson, M. (2005). Entrepreneurship and the theory of the firm. *Journal of Economic Behavior & Organization, 58*(2), 327-348.

Curseu, P., Schalk, R., & Schruijer, S. (2010). The use of cognitive mapping in eliciting and evaluating group cognitions. *Journal of Applied Social Psychology, 40*(5), 1258-1291.

Czarniawska, B. (2001). Is it possible to be a constructionist consultant? *Management Learning, 32*(2), 353-266.

Morçöl G. (eds.) (2008). *Complexity and policy analysis: Tools and concepts for designing robust policies in a complex world.* Goodyear, Arizona: ISCE Publishing.

Deterding, S., Dixon, D., Khaled, R., & Nacke, L. (2011). *From game design elements to gamefulness: Defining "gamification".* Paper presented at the MindTrek'11, Tampere, Finland. http://85.214.46.140/niklas/bach/MindTrek_Gamification_PrinterReady_110806_SDE_accepted_LEN_changes_1.pdf

Deterding, S., Sicart, M., Nacke, L., O'Hara, K., & Dixon, d. (2011). *Gamification: Using game design elements in non-gaming contexts.* Paper presented at the CHI 2011, Vancouver, BC, Canada. http://gamification-research.org/wp-content/uploads/2011/04/01-Deterding-Sicart-Nacke-OHara-Dixon.pdf

Faria, A. J. (2000). The changing nature of simulation research: A brief ABSEL history. *Developments in Business Simulation and Experiential Learning, 27,* 84-90.

Gold, S., Markulis, P., & Strang, D. (2014). ABSEL research: A perspective on the quality of the research presented in the proceedings. *Developments in Business Simulation and Experiential Learning, 41,* 15-29.

Hill, T., & Westbrook, R. (1997). SWOT analysis: It's time for a product recall. *Long Range Planning, 30*(1), 46-52.

Jakubowski, M. (2014). Gamification in business and education: Project of gamified course for university students. *Developments in Business Simulation and Experiential Learning, 41,* 339-342.

Karl, C. K. (2014). Solving the simulation paradox: How educational games can support research efforts. *Developments in Business Simulation and Experiential Learning,, 41,* 132-139.

Kelly, G. A. (1955). *The psychology of personal constructs.* New York: Norton.

Lorange, P. (1978). Corporate planning: An executive viewpoint. from http://dspace.mit.edu/bitstream/handle/1721.1/47152/corporateplannin00lora.pdf?sequence=1

Martin, R. L. (2014). The Big Lie of Strategic Planning. *Harvard Business Review, January/February.*

Miller, C. C., & Cardinal, L. B. (1994). Strategic planning and firm performance: A synthesis of more than two decades of research. *Academy of Management Journal, 37*(6), 1649-1685.

Morçöl, G. (2010). Issues in reconceptualizing public policy from the perspective of complexity theory. *E:CO, 12*(1), 52-60.

Nguyen, N. C., Graham, D., Ross, H., Maani, K., & Bosch, O. (2012). Educating systems thinking for sustainability: Experience with a developing country. *Systems Research and Behavioral Science, 29*(1), 14-29.

Raphael, T. D. (1982). Integrative complexity theory and forecasting international crises: Berlin 1946-1962. *The Journal of Conflict Resolution, 26*(3).

Senge, P. (1990). *The Fifth Discipline: The Art and Practice of The Learning Organization.* New York: Currency Doubleday.

Senge, P., Kleiner, K., Roberts, S., Ross, R. B., & Smith, B. J. (1994). *The Fifth Discipline Fieldbook: Strategies and Tools for Building a Learning Organization*. New York: Currency Doubleday.

Suedfeld, P., Tetlock, P. E., & Streufert, S. (1992). Conceptual/integrative complexity. In C. P. Smith (Ed.), *Handbook of Thematic Content Analysis* (pp. 393-400). New York: Cambridge University Press.

Tait, A., & Richardson, K. A. (2011). Guest editorial: From theory to practice. *E:CO, 13*(1-2), v-vi.

Wallis, S. E. (2009a). Seeking the robust core of organizational learning theory. *International Journal of Collaborative Enterprise, 1*(2), 180-193.

Wallis, S. E. (2009b). Seeking the robust core of social entrepreneurship theory. In J. A. Goldstein, J. K. Hazy, & J. Silberstang (Eds.), *Social Entrepreneurship & Complexity*. Litchfield Park, AZ: ISCE Publishing.

Wallis, S. E. (2010a). The structure of theory and the structure of scientific revolutions: What constitutes an advance in theory? In S. E. Wallis (Ed.), *Cybernetics and systems theory in management: Views, tools, and advancements* (pp. 151-174). Hershey, PA: IGI Global.

Wallis, S. E. (2010b). Towards developing effective ethics for effective behavior. *Social Responsibility Journal, 6*(4), 536-550.

Wallis, S. E. (2010c). Towards the development of more robust policy models. *Integral Review, 6*(1), 153-160.

Wallis, S. E. (2011a). *Avoiding policy failure: A workable approach*. Litchfield Park, AZ: Emergent Publications.

Wallis, S. E. (2011b). The complexity of complexity theory: An innovative analysis. In P. M. Allen, K. A. Richardson, & J. A. Goldstein (eds.), *Emergence, Complexity and Organization: E:CO Annual* (Vol. 11, pp. 179-200). Litchfield Park, AZ: Emergent Publications.

Wallis, S. E. (2012, July 22-27, 2012). *Theories of psychology: Evolving towards greater effectiveness or wandering, lost in the jungle, without a guide?* Paper presented at the 30th International Congress of Psychology: Psychology Serving Humanity, Cape Town, South Africa.

Wallis, S. E. (2013). How to choose between policy proposals: A simple tool based on systems thinking and complexity theory. *Emergence: Complexity & Organization*, 15(3), 94-120.

Wallis, S. E. (2014a). Abstraction and insight: Building better conceptual systems to support more effective social change. *Foundations of Science, 19*(4), 353-362. doi: 10.1007/s10699-014-9359-x

Wallis, S. E. (2014b). Structures of logic in policy and theory: Identifying sub-systemic bricks for investigating, building, and understanding conceptual systems. *Foundations of Science, in press*.

Wallis, S. E. (2014c). A systems approach to understanding theory: Finding the core, identifying opportunities for improvement, and integrating fragmented fields. *Systems Research and Behavioral Science, 31*(1), 23-31.

Weick, K. (2003). Commentary on Czarniawska. *Management Learning, 34*(3), 379-382.

Wexler, M. N. (2001). The who, what and why of knowledge mapping. *Journal of Knowledge Management, 5*(3), 249.

Wheatley, M. J. (1992). *Leadership and the New Science*. San Francisco: Barrett-Koehler.

Wong, E. M., Ormiston, M. E., & Tetlock, P. E. (2011). The Effects of Top Management Team Integrative Complexity and Decentralized Decision Making on Corporate Social Performance. *Academy of Management Journal, 54*(6), 1207-1228.

Wright, B., & Wallis, S. E. (Under submission). A revolutionary method to advance entrepreneurship theories.

Zack, M. H. (2000). Competing on Knowledge. *Handbook of Business Strategy, 1*(1), 81-88.

APPENDIX A

Playing pieces

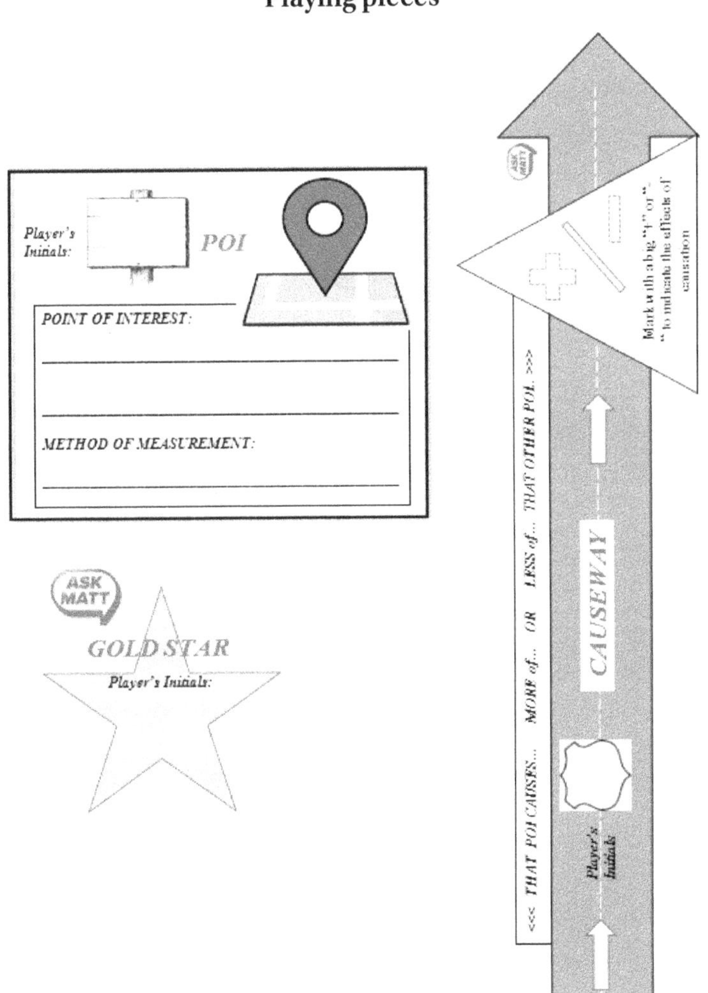

TRAVELOGUE:

Track data for POIs Here:

ASK MATT

Month — Data:

Jan:______________________________

Feb:______________________________

Mar:______________________________

Apr:______________________________

May:______________________________

June:______________________________

Jul:______________________________

Aug:______________________________

Sept:______________________________

Oct:______________________________

Nov:______________________________

Dec:______________________________

19

APPENDIX B

Score Sheet

TURNING POINTS
A PERSONAL LOG OF MAJOR DECISIONS

Name of Game: ____________________________
Name of Player: ____________________________

Date:	Decision/ Action	Anticipated Results (based on Causeways & POIs)

Title for this Game (focus of topic):______________________________

Players' Names: Turn:	1	2	3	4	5	6	7	8	9	10	11	12	
													Continue playing with a new scoresheet!
GROUP Strategic Knowledge Score													
Number of Blue Ribbons:													
Total Players' Points (for the turn)													
Total number of *POIs:* (MAP Complexity)													
Number of Gold Stars:													
Mapicity (Gold Stars divided by Complexity)													
TOTAL GROUP SCORE: (Total Players' Points X Mapicity) X (Number of Blue Ribbon)													

APPENDIX C

Sample Map

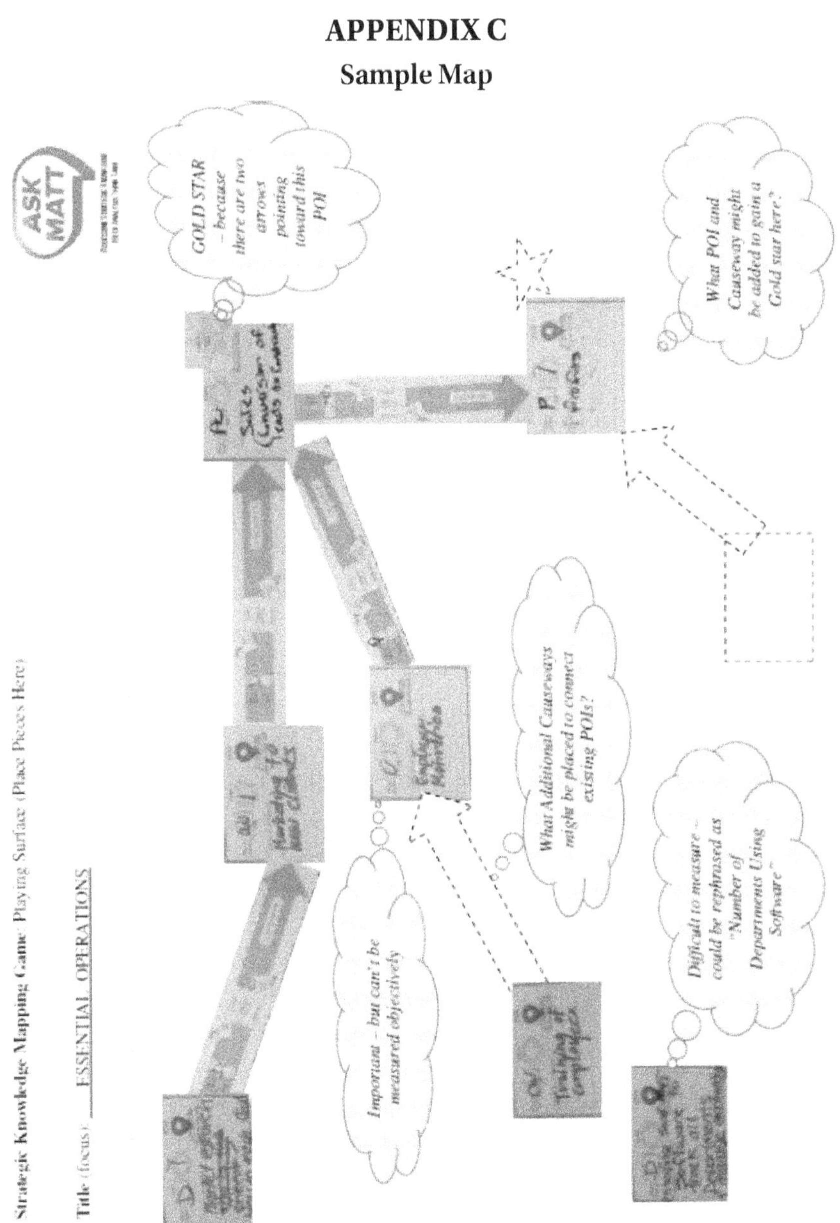

APPENDIX D

Sample entry on a "Turning Points" Tracking Sheet

.pril 5: Customer needs accelerated delivery of parts. Checked MAP, saw that *delivery me was Gold star - resulting from more *motivation, and more *raw materials.

alled Bob to check inventory – Bob says they can handle some extra production and wi :celerate our own order for new material. Also, noted that *motivation results (in part) om *bonus pay. Talked to production department and suggested bonus pay followed by fternoon BBQ for Saturday "work party."

nticipate that all will go well.. Also anticipate potential problems because reduced delivery time causes more *quality problems. So, I called Mary in QC to warn her to be xtra careful.

.pril 12: parts shipped – express. Customer happy, workers happy, my boss is very appy!

APPENDIX E

Example of Map Evolution Through the First Four Turns

Public Visits
LINKS

Chapter 10
The Science of Conceptual Systems—A Progress Report

Wallis, S. E. (2016). The science of conceptual systems: A progress report. Foundations of Science, 21(4), 579–602.

I wish to express my appreciation for the reviewers' close reading and detailed suggestions which have resulted in a more readable and effective paper. All remaining mistakes and weaknesses belong to me.

Abstract: *In this paper I provide a brief history of the emerging science of conceptual systems, explain some methodologies, their sources of data, and the understandings that they have generated. I will also provide suggestions for extending the science-based research in a variety of directions. Essentially, I am opening a conversation that asks how this line of research might be extended to gain new insights—and eventually develop more useful and generally accepted methods for creating and evaluating theory. This effort will support our ability to generate theory that is more effective in practical application as well as accelerating the development of theory to support advances in other sciences.*

Introduction

This paper reports on the "science of conceptual systems." Metaphorically, each conceptual system provides a lens or framework of understanding which enables the user to understand and engage the world. Generally, a conceptual system is any form of theory, model, mental model, policy, etc. Here, because we are addressing a wide range of disciplines—and each discipline has its own language—we will use those terms interchangeably.

The purpose or direction of this science is to gain a better understanding of conceptual systems so that we can create better ones. Better conceptual systems are those that allow us to more effectively understand and engage the world—to reach our goals more easily. Here, conceptual systems are our primary source of "data." And, our methodologies are focused on the relationships between those data. Additionally, we compare the usefulness of the conceptual system with their effectiveness when applied in the world.

As we all know, the world faces grave problems—including poverty, war, injustice, crime, and economic instability. To understand and resolve these kinds of problems, we turn to science. Because these are generally social problems, we turn more specifically, to the social/behavioral sciences.

A science may be defined as "the pursuit of knowledge and understanding of the natural and social world following a systematic methodology based on evidence" (Science Council, 2010). Science includes measurement, data, experimentation, observation, critical analysis, and other closely related requirements. A science of conceptual systems may then be understood as the pursuit of knowledge and understanding of conceptual systems using rigorous methodologies.

Since the scientific revolution, science has been generally successful in the process of finding facts. Using inductive, deductive, and abductive approaches, science has developed new ideas and insights, used data to build theories, and tested theories to develop more data. Within the natural sciences, these kinds of efforts have led to the development of very useful theories and laws. Within the social/behavioral sciences, however, such efforts have not been so successful in creating more effective theories.

While most of the social/behavioral sciences have been focused on empirical studies, there seems to be an underlying assumption that once we have enough data (or the right kind of data, or data from well-funded research) that useful theories will somehow emerge. This akin to believing that if we have enough bricks, a house will somehow build itself. In short, the "normal" view of theory building has not proved effective in advancing the social/behavioral sciences. While there have been inroads, those paths seem to have led mainly to cul-de-sacs of science.

One direction of thinking suggests that each theory is a lens or frame that we use to understand and/or engage the world around us (Bolman & Deal, 1991. For example, one may look at a business as a supportive family or a political battlefield—and take very different actions, accordingly! It is undoubtedly useful to be aware of one's frames of reference, particularly in a business setting (Senge, Kleiner, Roberts, Ross, & Smith, 1994). However, those kinds of approaches have generally been fuzzy. They have not quantified their findings. Thus, the ability to create more effective theories has been elusive. Without rigorous methodologies, it is difficult to think of such efforts as fully scientific. Indeed, in trying to describe a good test for business strategies, Parnell (2008) could find no reliable advice. Instead, he suggested that organizational leaders rely on "artistry." Similarly, others rely on intuition to evaluate theories (Shotter & Tsoukas, 2007). Like artistry, however, intuition is not a reliable tool (Meehl, 1992: 370).

Another direction suggests that it is not possible (or even desirable) to create formal theories (Burrell, 1997; Shotter, 1994). There, it seems that the authors are trying to "prove a negative"—a very difficult task. And, in that process, they implicitly create their own theories. Therefore, I am inclined to reject their claims as tautological.

Part of the classically accepted path is that theory may be evaluated by testing its generalizability, sensemaking ability, testability, plausibility, and usability (Sussman & Sussman, 2001: 90-92). Kaplan (1964: 311-322) suggests tests based on norms of correspondence (does the theory fit observed facts), coherence (does the new theory fit the existing theory), and pragmatism (does the new theory work when it is applied). Correspondence to "facts," however, is problematic because all observations are theory-laden. So what counts as a fact depends very much on the theory. Also, what is pragmatic for one individual may be abhorrent to another. Finally, if a new theory coheres (matches) perfectly to the existing theory, they would be identical—and there would be no new theory.

Within the social sciences, it is often difficult to evaluate the application of a theory. For example, in the sphere of economic policy, we cannot simultaneously compare two policies if one says to raise interest rates and the other says to lower them. In brief, our experiments are not repeatable. You can't dip your toe in the same river twice. Thus, developing theories and policies to solve the many problems of the world has proven impossible—so far.

A new approach to understanding theory is needed because we have an overly large number of problems in this world, and a great shortage of useful theories that we might apply to solve them. Such an approach should be rigorously scientific and should also pursue a different path from those that have already been followed. The study of conceptual systems may also be considered as an important direction because it supports the development of other sciences—especially the social/behavioral sciences—which so far have not proved highly useful.

Recently, there is increasing interest in the topic of theory creation. Indeed, there have been a number of useful and insightful texts discussing important aspects to building theory (M. Edwards, 2010; M. Edwards, G., 2009; Shoemaker, Tankard Jr., & Lasorsa, 2004; Van de Ven, 2007; Wallis, 2010b).

While we do have plenty of data, part of the problem (and a new direction), may be what we plan to do with that data. Shaw and Allen (2012) note that theory construction becomes problematic when the conceptual components are chosen in a way that is not systematic. In general, those who develop theory, "rarely build their theories in textbook fashion from the bottom-up. Real theory building is less organized and a good deal more fun. You start at the top, with an intuition for how some systems, some structure of things or concepts works. From there you feel your way, by intuition trial and error luck and logic, to what looks like the right answer" (Friedman, 1997: 40). This approach is far from scientific or systematic. It should come as no surprise, then, that our theories are of rather limited usefulness in understanding our systemic world. In short, a non-systemic approach to building theory has not led to the creation of useful theories. So, we may benefit from a more systemic perspective.

While concepts are already considered important within the science of cognitive systems, conceptual systems are too often thought of in terms of language and its analogical relationship to the world (e.g., Gentner, 1983; Johnson-Laird, 1980). Again, an approach that is not highly rigorous.

At the same time, within the systems sciences, interdisciplinary thinkers have worked to understand social systems, biological systems, chemical systems, and so on. A gap in understanding exists because comparatively little has been written about our systems of concepts (which may be understood as a set of interrelated ideas, concepts, or propositions). Such conceptual systems are in contrast to (for example) biological systems (consisting of interconnected cells) or a chemical system (consisting of interconnected atoms). Examples of conceptual systems include theories, models, mental models, policies, and others.

A New Path

Although the social/behavioral sciences have worked to systemically understand our world, we have not made the same rigorous effort to understand our conceptual systems. This is strange because it is those conceptual systems that we use to understand and engage the world.

Therefore, in taking a more purposefully scientific and systemic approach, we will begin by defining a conceptual system more specifically as a set of interrelated propositions. Examples of those propositions may be found in theories, models, schema, conceptual maps, strategic plans, and policies. Any place where a concept is described in relation to another concept. Mental models may also be useful data for analysis, particularly when surfaced through writing and/or speeches. On the edges of the science, valid data may also include partial systems such as individual propositions and hypotheses. Such partial systems are useful in striving to understand the "building blocks" of larger systems.

The process of this science is to quantify aspects of conceptual systems; and, importantly, to find links between the measures of those systems and their usefulness in the world.

In the classical approach to science, we live in a "world of data." So, theories are evaluated on the strength of their data. That is to say, the *correspondence* between concepts and reality. In contrast, this emerging science works under the assumption that we live in a "world of systems." And, such a world will be better understood by theories which are more systemic. So, theories are evaluated for their internal *coherence.*

Coherentist perspectives have been applied to evaluate scientific revolutions. And, have arguably provided better explanations (Šešelja & Straßer, 2014). Naturally, we need both coherence (interconnecting logics within a theory) and correspondence (empirical evidence or facts relating to a theory) in order to have a useful theory.

In this paper I will provide a brief history of this emerging science of conceptual systems, explain some methodologies, their sources of data, and the understandings that they have generated. I will also provide suggestions for extending the science-based research in a variety of directions. Essentially, I am seeking to open a conversation that asks how this line of research might be extended to gain new insights—and eventually develop more useful and generally accepted methods for creating and evaluating theory to generate theory that is more effective in practical application to understand and resolve the problems of the world.

A Brief History

The science of conceptual systems is set within the broader science of cognition begun in the 1950s. Cognitive science is focused around the investigation of mental representations, mental processing of reasoning and inference, and the meaning of words and phrases (Johnson-Laird, 1980). However, where cognitive science investigates the more complex issues of thinking, the present science is focused on the structures that result from, and/or guide, the thinking.

From Aristotle's "pictures in the mind," we have understood that our minds contain some kind of image or understanding of the world. Because there are some things that cannot be pictured (especially abstract notions such as freedom) Kenneth Craik (1943) began talking about a "mental model" instead of a picture. This was a subtle, through significant shift because having a model implies having some structure. Which, in turn, suggests that it may be amenable to analysis.

This was not to be an easy path of investigation. Lane (1992, drawing on Lakoff and others) notes that, "there is overwhelming evidence that we humans do in fact maintain overlapping and inconsistent conceptual systems" (p. 27). It also appears that we don't do much better when transferring those conceptual systems to paper. Even our ideas of philosophy, "are often themselves a contradictory and confusing patchwork of fragments of various philosophers' representations of the world" (Ledoux, 2012: 14). In short, in our minds and in our writing, we have a great deal of potential data—but it is rather jumbled. We need rigorous analysis to make sense of it all.

The path continued when Kelly (1955) suggested that mental models must have a kind of structure to them. Thus bringing us a step closer towards scientifically rigorous analysis.

Over the years, a number of tools have been used to focus on the "internal" aspects of texts. These include scientometrics and bibliometric approaches to quantify the content of theory (c.f. Calas & Smircich, 1999; Fiske & Shweder, 1986; Flower & Mellon, 1989; Jean-Pierre & Edward, 2000; Knorr-Cetina, 1981; Kostoff, del Rio, Humenik, Ramírez, & García, 2001; Ritzer & Smart, 2001; Zimmer, 2006). Some related work may also be found in formalizing structures of sentences (Gentner, 1983) such as formal language or "natural language." More attention should be paid to using and developing the tools from these areas. However, these approaches have not purposefully addressed the systemic structure of theories.

An important step forward occurred when, building on Kelly's insights, scholars developed Integrative Complexity (IC) as a method to analyze the structure of mental models (Suedfeld, Tetlock, & Streufert, 1992) using the paragraph completion test.

Briefly, IC was used to evaluate paragraphs of text (from writing samples, speeches, government announcements, etc.) and rate them on a scale of one to seven. A score of one indicates a low level of interrelatedness between the ideas of the text and a score of seven indicates a high level of interrelatedness. For example, a simple statement of universal truth would receive a very low score.

In contrast, a complex statement (as may be found in many academic and philosophical writings) would receive a high score.

Studies using IC found that higher scores were significantly correlated with managerial effectiveness (Wong, Ormiston, & Tetlock, 2011), political success (Raphael, 1982), and higher student scores (Curseu, Schalk, & Schruijer, 2010). This ongoing path of research has shown that we can indeed evaluate the structure of our conceptual systems in a way that is rigorous and scientific. And, importantly, those conceptual systems with higher levels of structure are more useful and support success. Therefore, we may be guided into developing conceptual structures with higher levels of structure so that we may learn to more effectively address the problems of the world.

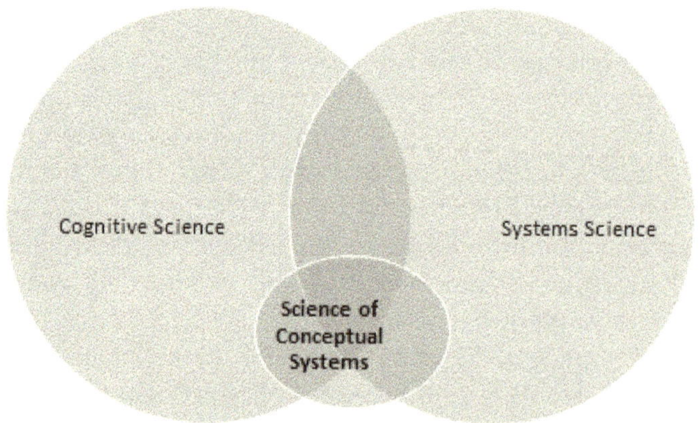

Figure 1 *Relationships between some sciences.*

Because that approach includes rigorous methodology and measurement (IC), data (evidence in the form of text), I suggest that IC represents the start of a science of conceptual systems. Further, that this new science exists at the intersection of cognitive science and systems science (Figure 1).

Integrative Propositional Analysis

Before long, it was generally agreed that theories must also have some kind of structure (Dubin, 1978; Kaplan, 1964; Stinchcombe, 1987). Studies into the structure of theories seem to have great promise because we are studying the internal coherence—comparing concepts to concepts instead of comparing concepts to non-conceptual objects (Wallis, 2014a). That is to say, to learn more about apples, we should compare them with other apples. We can't learn much about apples by comparing them to concepts (e.g., freedom).

However, little was done to quantify the structure of those interrelationships. In response to this gap, Integrative Propositional Analysis (IPA) was developed specifically to provide rigorous and quantitative analyses of conceptual systems such as academic theories and policy models.

By way of background, IPA is derived from Reflexive Dimensional Analysis (RDA) which was founded in Grounded Theory and dimensional analysis (Wallis, 2008). IPA also includes aspects of scientometrics and bibliometrics (Wallis, 2009a), as well as structuralist perspective (Wallis, 2009b), integral thinking (Wallis, 2010d), systems thinking (Wallis, 2010c), complexity theory (Wallis, 2011), and narrative analysis (Wallis, 2013b).

IPA begins by identifying specific conceptual systems for analysis. Within each (theory, policy, model, mental model, etc.), causal propositions are then identified. Each proposition includes one or more "things" and the relationships between them. Things are expressed as concepts which may be as concrete as (dare I say) "concrete" or abstract as "abstraction." Between things, causality is the best way to describe specific relationships.

Johnson-Laird (1980) neatly explains the importance of causality for improved understanding (p. 107). And, it has been shown that maps based on causal relationships have been used effectively in business settings (Axelrod, 1976). In short, causality is important because it is the best path to scientific understanding (Pearl, 2000).

While it is generally held that theories, and their propositions, are best when derived from observational data, creative insights may also be useful—perhaps even necessary (c.f. Ambrose, 1996; Stacey, 1996; Thagard & Stewart, 2011; Uzzi & Spiro, 2005).

The structure of those propositions may take many forms (e.g., atomistic, linear, circular, branching). While there are varying de-

Figure 2 *Simple concatenated logic structure.*

grees of usefulness to those structures, the one that seems most useful to date is the "concatenated" logic structure (Wallis, 2014c).

In Figure 2, A, B, and C represent concepts and the arrows represent causal connections between them. Specific changes in A and/or C (e.g., increase or decrease), will cause changes in B.

IPA is a six step process:

1. **Identify propositions** within one or more conceptual systems (models, etc.);
2. **Diagram** those propositions with one box for each concept and arrows indicating directions of *causal* effects;
3. **Find linkages** between causal concepts and resultant concepts between all propositions;
4. Identify the **total number** of concepts (to find the Complexity);
5. Identify **concatenated** concepts, and;
6. **Divide** the number of concatenated concepts by the total number of concepts in the model (to find the Systemicity).

Using Figure 2 as a very simple example of an IPA analysis, one may identify three concepts (A, B, and C). Therefore, the Complexity of the theory is C=3. Among these, there is one concatenated concept (B). Dividing the concatenated concept by the total concepts gives us a Systemicity of 0.33.

One might say that the Complexity of the conceptual system is a measure of its "breadth" while Systemicity is a measure of its "depth."

We can use IPA to analyze text where causal relationships are described. Useful sources of this kind of data include academic journals, books, policy statements, and similar material. The results of such studies may be presented in a variety of ways to generate new insights into the subject matter. In the following section, we will look at the results of some recent research. In each of the following figures, each dot represents the analysis of one theoretical model drawn from the literature.

Examples of Research Results

One type of study is to use IPA to evaluate changes in Systemicity of a theory over time. Here, the goal is to identify macro trends, rather than attempt to make fine-grained inferences about the efforts of individual theorists. Although, of course, the fine-grained approach represents an interesting direction for additional research.

Figure 3 shows the evolution of a theory from ancient times (low Systemicity and low level of usefulness) through the scientific revolution (rising Systemicity and rising level of usefulness), through its development into a “law” (perfect Systemicity and high level of usefulness).

In using Complexity and Systemicity to choose one theory from among many, the first choice would be to use the theory with the highest level of Systemicity (greater usefulness and greater depth of understanding). The second choice would be to use a theory with higher Complexity (more concepts and greater breadth). By using a theory with higher level of Complexity, one may expect some increase in usefulness (including explanatory and predictive power). There is a cost, however. Engaging a theory with a large number of concepts means that one should track data from all those many concepts—often a difficult task. Realistically, when choosing from a range of theories, a scholar might be forced to compromise between a theory with higher Complexity and one with higher Systemicity. The important thing is that such a choice should be made consciously. It (almost) goes without saying that those theories are best when supported with empirical studies and include concepts related to the topic of interest.

Understanding the evolving level of Systemicity is useful because it provides an alternative view of the scientific revolution. Where traditional views suggest that the revolution was due to

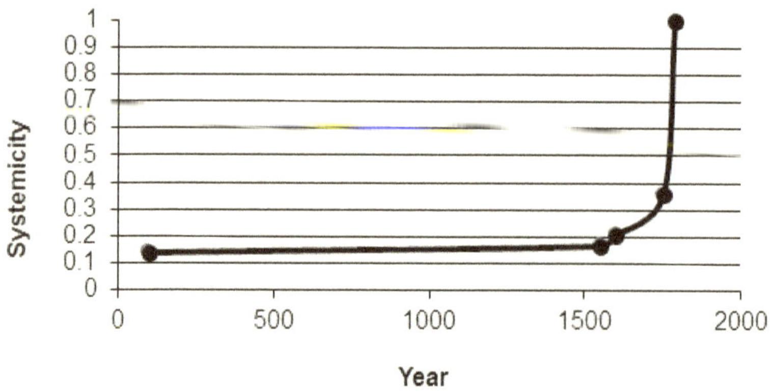

Figure 3 *Systemicity of electrostatic attraction theory (adapted from Wallis, 2010a).*

changing metaphors of electricity and increased reliance on empirical data, this new perspective suggests that the success of the revolution was also due to a rapidly increasing understanding of how the empirical data fit together with specific structures of logic.

This is an important shift of perspective because it provides a more objective evaluation of the structure of a theory—along with a more reasonable goal of creating better theory by developing a measurably higher structure. In contrast, the idea that a "change of metaphor" might be causal to a scientific revolution may lead some to believe that any change of metaphor is tantamount to a revolution in science. Indeed, many publications claim that their work represents a paradigmatic shift—when the actual changes to the science are not so impressive (Clegg, Cunha, & Cunha, 2002).

Another result of IPA highlights the Complexity of theory. Figure 4 shows the same theory over the course of its evolution. This time, however, we are looking at the Complexity (the number of concepts) within the theory. Interestingly, the theory became more complex during the revolution. This seems to have occurred as scientists created larger theories to explain new and different ideas. While it may prove a difficult mountain to climb, ascending a Complexity peak of this sort may provide a positive course to advancing theory in any science.

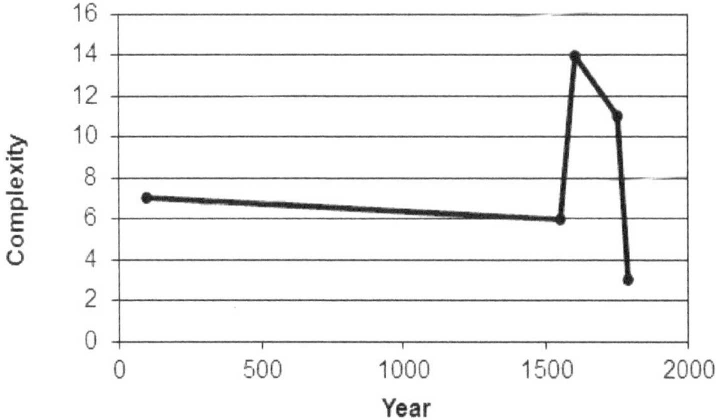

Figure 4 *Complexity of electrostatic attraction theory (adapted from, Wallis, 2010a.*

Figure 3 and Figure 4 both show very little change over the centuries before the scientific revolution. This suggests that the longest-lived theories are not necessarily the best.

More comparative studies may be conducted with other theories from the natural sciences. The results are expected to be similar. This is because those theories all began with a low level of usefulness and ended with a high level of usefulness. And, similarly, they ended up being amenable to algebraic manipulation—which is much the same thing as having a Systemicity of 1.0.

Within the social/behavioral sciences, it is difficult to compare such theories as those sciences generally present us with large numbers of theories—with little confirmable usefulness. For example, Figure 5 presents a set of well-known theories of conflict from sociology (drawn from Turner, 1986). Here, we see that the level of Complexity of those theories appears to be in slow decline. Similarly, in Figure 6, the Systemicity of those theories is also in slow decline. There is no evidence of a revolutionary increase in the Systemicity of these theories. And, outside the ivory tower, we continue to face the tragedy of war. So, it seems that our ability to understand and prevent war is also not improving.

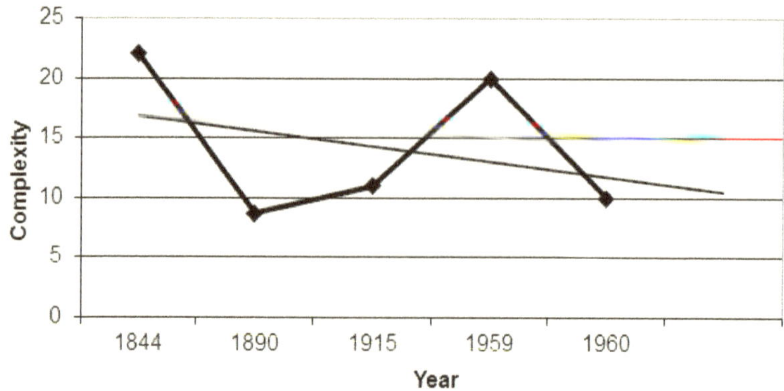

Figure 5 *Complexity of theories of conflict from sociology (adapted from Wallis, 2015, Under submission).*

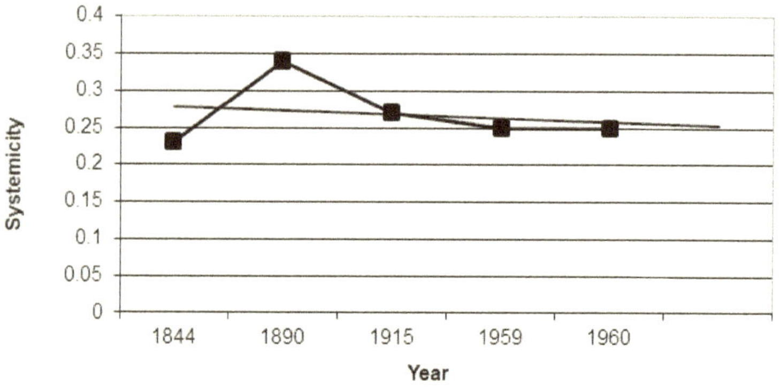

Figure 6 *Systemicity of theories of conflict from sociology Wallis, 2015.*

I will resist speculating on the events of history, or the personal lives of the theorists, that may have impelled changes in these paths of evolution. However, such directions of thinking do indicate new directions for research that is more fine-grained than is presented here. It is beyond the scope of the present coarse-grained analyses to suggest (for example) that the outbreak of war impelled a theorist to new insights. Or to revise a theory to higher levels of Complexity or Systemicity. Similarly, additional analyses might investigate changes in theory as developed by a single scholar—regardless of world events. For example, does the theory become less Complex and more Systemic as the scholar develops a deeper

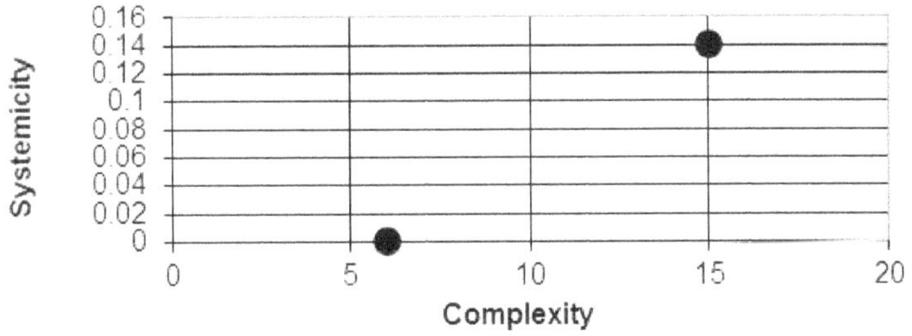

Figure 9 *Comparing two economic policies as predictor of success (adapted from Wallis, 2013a).*

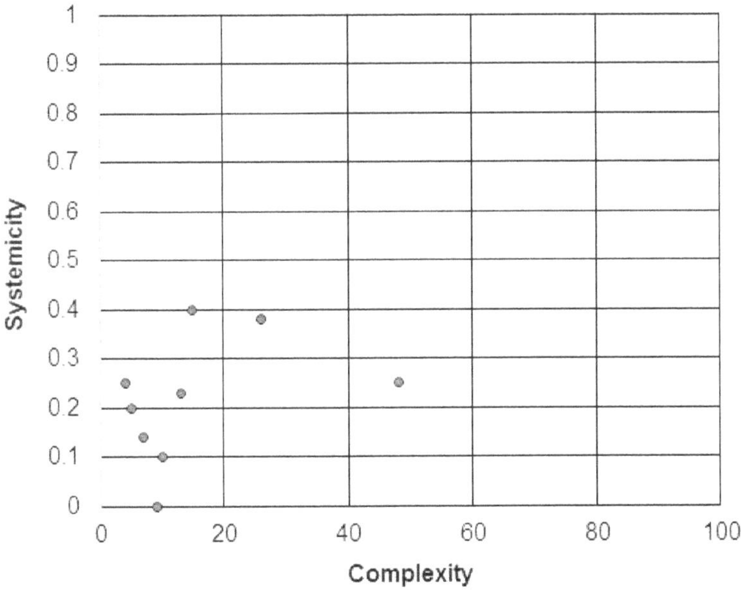

Figure 10 *Theories of entrepreneurship (from data in Wright & Wallis, Under submission).*

understanding (which may be on a "personal" level and might not be completely represented within the pages of the literature)?

In Figure 7, a similar study (this time with our data consisting of theories drawn from the highest cited paper of each decade) shows that theories of motivation are increasing in Systemicity. However, that increase is unsteady and does not show promise of reaching high levels of Systemicity any time soon.

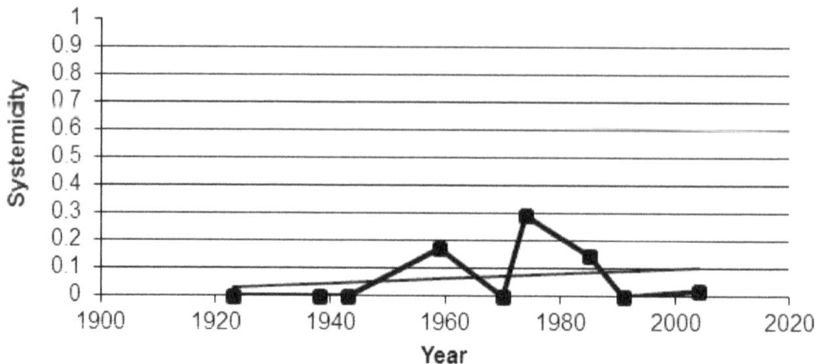

Figure 7 *Systemicity of theories of motivation from psychology (adapted from Wallis, 2012).*

As may be seen in Figure 8, the Complexity of those theories is likewise unstable. However, here the Complexity is declining.

Here, again, it is beyond the scope of the present paper to explain or speculate usefully on the events (historical and/or personal) which may have influenced these changes. There are simply too many variables and too little understanding. Such explorations should be encouraged among interested scholars.

These studies provide a new view into the evolution of theories. By charting the history of a body of theory on this large-grained scale, we may gain a sense of where it has been, where it is going, and how long it may take to get there. However, as may be inferred from the relatively small sample size, they are only early explorations. Future studies should be conducted using a statistically relevant number of theories as data.

The main point to be conveyed here is that "something is happening" that is deserving of more study. And, importantly, this new perspective suggests a direction for gaining new insights into how those world events support the improvement of theory.

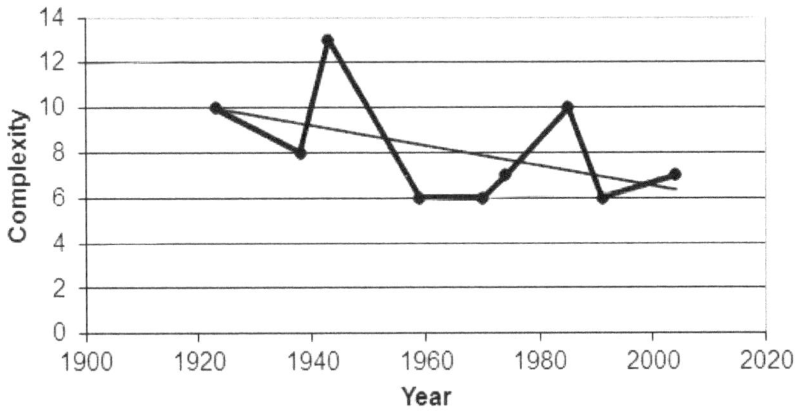

Figure 8 *Complexity of theories of motivation from psychology (adapted from Wallis, 2012).*

One area where we have been able to compare differing conceptual systems of the social sciences is in the study of policy models. By conducting a case comparison of policy models we have found that those with higher Complexity and Systemicity tend to be more effective in practical application. That is to say, they are more likely to be useful in helping nations to reach their goals (Wallis, 2011). Or, to achieve higher goals at lower cost.

Figure 9 shows the Systemicity and Complexity of two policies. The policies were drawn from the websites of two political parties before a recent election. IPA provides a new perspective for comparing these policies. Where individuals may be swayed by partisan rhetoric, this analysis suggests a strong argument in favor of adopting the policy with the higher Complexity and Systemicity because that policy is more likely to be useful and successful.

Another form of analysis is presented in Figure 10. This data set is of nine theories of entrepreneurship (Wright & Wallis, Under submission). This scatterplot diagram may be understood as taking a "snapshot" of a set of theories within a field or sub-field. To obtain this sample, we surveyed Google Scholar ™ to find a set of theories that were highly cited (more than 200). Those six theories were supplemented by three more (a sample of convenience). We were surprised to find that theories with fewer citations (under 100) were also those with the lowest scores of Complexity and Systemicity (found in the lowest left-hand corner of the diagram).

This finding hints at a new path for improving one's citation rate in the academic literature.

For the Complexity, note how most theories are at C=10 and below. Like Figure 5 and Figure 8, the Complexity of theories seems constrained. This may be due to a number of influences—all of which are deserving of additional investigation if we are to advance our theories and our sciences more rapidly. One possibility is the simple limitation of space in journals. Another is the difficulty faced by scholars in collecting and presenting such a large number of concepts. In some sense, those may be considered perfectly reasonable explanations. However, if our goal is to create more effective theories and advance the social/behavioral sciences into something that is more useful, then we have a problem with the way we apply our social/behavioral science analyses.

For the Systemicity, in Figure 10, please note how no theory exists above the 0.4 level—and most are much lower. Like the theories presented in Figure 6 and Figure 7, the social sciences do not seem to be producing theories with a high level of Systemicity. Even though that seems to be required to develop theories that are highly useful in practical application—based on past philosophical perspectives (Dubin, 1978; Kaplan, 1964) and more recent analyses (reflected in the present paper).

In short, in Figure 10, theories are clustered in the lower left-hand quadrant. This suggests a valuable opportunity to identify (and/or create) theories in the other three quadrants. Perhaps theories inhabiting other quadrants may prove more useful in solving the problems of the world.

As you may well infer from the relatively small sample sizes presented in the above figures, this science is still in an early stage of exploration. Naturally, we cannot readily extend these results to suggest that all theories of the social sciences are in such poor condition. That kind of conclusion must wait upon more research. There is a vast opportunity to map the "theory genome" or create a "periodic table of theory" and gain new views into our world of theory across the social/behavioral sciences—and beyond.

A Few Key Insights

These results lead us to a few key ideas about the structure of theory within the social/behavioral sciences. First, is that "parsimony" (although popular) is not (by itself) a useful or viable goal in theory construction. Dubin (1978: 316), for example, notes that the test of parsimony is often applied and misapplied. Meehl (2002) argues brilliantly against the claim of parsimony. Yet, others claim that "the simplest theory is the best" (Shoemaker *et al.*, 2004: 172) and that there is a tradeoff between theories that are "general," "accurate," and "simple" (Van de Ven, 2007).

The comparison in the present paper suggests that the relationship is not so simple. See especially Figure 3 and #4 which show that a theory with low Complexity and low Systemicity is not useful. While, in contrast, a theory with a high level of Systemicity coupled with a low level of Complexity is useful. There also exists a middle ground where a theory with a high level of Complexity and a medium level of Systemicity will also have some level of usefulness. In short, at a minimum, scholars should avoid claiming validity for their theories based only on parsimony because it appears that Systemicity is a more accurate indicator for usefulness.

Certainly, a move toward larger theories will cause some difficulties—such as the limits of human understanding. However, the benefit of creating more effective theories means that these issues should be addressed, not avoided. One way to address the difficulty of engaging more complex theories is to encourage more collaboration between scholars. If each scholar were to focus on a set of concepts within a theory, their efforts could be coordinated along the lines of the causal connections between the concepts. This kind of effort would also support linking scholars for interdisciplinary efforts.

Another difficulty is the space available for publishing in journals. This might be alleviated by providing links to online content. Naturally, all of these efforts will require more coordination—and that suggests the need for more funding.

An additional idea is that the social/behavioral sciences should be striving to identify more causal connections between concepts within the model. Metaphorically, this approach provides a useful "compass" to guide researchers towards developing better theory through research. It also provides a new and more useful way to focus the efforts of researchers. Where Lakatos (1970) suggested that theories have a core of unchanging ideas, another researcher has thrown that into question—and suggested that the core of a theory is the set of concepts with a higher density of causal connections (Wallis, 2014b). Thus, researchers may coordinate themselves by identifying whether their work involves adding new concepts to the belt of the theory or adding casual connections to strengthen the core.

Generally, and metaphorically, this paper suggests that we should choose to create and use roadmaps with many dots connected by many lines; not maps with many disconnected dots.

Using the tools and perspectives of a science of conceptual systems, there are many opportunities to accelerate the advancement of science. Future studies may be conducted in at least three general directions. The first is to deepen the existing studies. Where the above studies were conducted with only a few theories as data, they should be replicated with hundreds of theories. In such an endeavor, we have no lack of data because each publication essentially presents a new theory. Although many scholars use the same title to refer to their theories, I can't recall finding two identical theories—although many scholars will use the same title for theories built of differing concepts.

Theories chosen for these kinds of study should not be limited to those which are most cited or most popular. A more useful approach is to choose them according to more objective criteria—such as the strength of their underlying empirical research. Another approach is to draw upon thousands of theories, chosen at random and compare those theories along multiple dimensions over time (Faust & Meehl, 2002; Meehl, 1992, 2002, 2004).

Between the present studies (drawing on a few theories) and future studies (involving thousands) researchers drawn to this emerging field might consider analyzing theories found in existing meta-studies, benchmarks, or reviews of theories—such as the one used above for theories of conflict (Turner, 1986).

Second, the research stream could be broadened to include new areas. For example, comparing theories from all of the sciences. Here, of course, all sciences may provide useful data in the form of theories and their constituent propositions. As a more concrete example within the field of business, an effective study could analyze a wide range of strategic plans and/or mission statements. Naturally, such a study would need to control for many confounding variables. Was the business environment as stable as expected? Where the plans implemented properly? How much research went in to the creation of the plans? How many changes occurred before a new formal plan was deemed necessary? Studies of this sort might begin using a case-comparative research method in tandem with IPA (e.g., Wallis, 2011).

A third area for advancement is to experiment with conceptual systems in practical application. We need to take our theories and put them to use. This is an excellent opportunity for the growing number of emeriti professors. They might, for example, provide advice to their local city councils and county boards based on theories with a high level of Systemicity.

The science of conceptual systems (in general) and IPA (specifically) have the potential to advance science and practice in a number of ways. Potential projects may include:

1. Evaluating theories—and encouraging schools to teach theories that are of a higher level of Systemicity. This could lead to significant changes in our textbooks.
2. Teams of researchers may strive to integrate multiple theories for collaboration in research and practice.
3. Use this structural perspective to highlight high-leverage opportunities for research and allocate funding based on those leverage points.
4. Evaluating programs and policies before implementation to predict their success (or failure).
5. Creating a "Periodic Table" of theories
6. Exploring the "empty quadrants" (above, in discussion of Figure 10). Where are the theories that are highly complex and/or highly systemic?

7. Identifying "core" (more structured) & "belt" (less structure—more in need of research) for bodies of theory to facilitate and coordinate research efforts.
8. Looking at Theories of Everything (TOEs)s. How do they compare in terms of Complexity and Systemicity?
9. Identifying a theory genome or "web" of interconnected theories within and between disciplines.
10. Evaluating submissions to journals. Naturally, the more rigorous journals should prefer submissions that are more systemic.
11. Reviewing research proposals. Naturally, more funding should be provided for projects which hold greater promise for advancing theories to higher levels of Systemicity.
12. Doctoral candidates developing their literature review might be encouraged to compare existing theories and develop or propose theories with higher levels of Complexity and Systemicity. Their research, then, should lead to the creation of theories with still higher levels.

With more effective ways to create and evaluate theories, we may be able to create a social-political system that is as effective and useful as our sciences. At least one effort, currently under development, aims to evaluate bills before congress using IPA and other methods: http://scipolicy.org/ Another group is striving to integrate theories across multiple branches of science using IPA and other approaches (McNamara & Troncale, 2012). And, a dissertation used IPA to gain new insights into US policy (Shackelford, 2014) noting the importance of complexity theory and this new methodology, "are vitally important to the field of public policy" because we need to methods to adapt to an increasingly complex and nonlinear world. Indeed, "Change makers must be able to change the way they play the game" (p. 146).

Advancing the Science of Conceptual Systems

The history of science clearly shows that theories come and theories go (Kuhn, 1970). Indeed, in the social sciences, it has been suggested that 90% of the theories rise very rapidly, only to disappear with equal speed (Oberschall, 2000). So, it is reasonable to expect that IC and IPA will also evolve, change, and (hopefully) be replaced by something better. In this section, I will briefly suggest a few directions where progress and evolution might be encouraged.

First, Paul Meehl suggested an ambitious program of evaluating many theories over time according to many variables (Faust, 2005; Faust & Meehl, 2002; Meehl, 1992, 2002, 2004). The results of such studies can be expected to suggest additional insights into the creation of more useful theories.

Second, would be to explore the idea that theories might be "nested" within one another. Such an approach might open new directions for thinking about theories from physics to sociology. For an explanation, by way of comparison, let me note that IPA, so far, has looked to see if theories might be "connected" based on overlapping concepts. However, there may be no direct analogy between two theories and so no direct connection. There may, however, be "emergent properties" (Robertson, 2014). Emergent properties have been studied between physical, biological, and social systems. However, they have not been studied in conceptual systems. Here, it may be speculated that emergent properties on one level may provide links to theories on another level. For example, theories of emotion do not seem to directly relate to theories of evolution. However, emotions may guide actions which in some way influence evolution. Alternatively, theories built of more concrete concepts might be nested within theories built of more abstract concepts. These are rich areas for study starting on the philosophical level.

Third, investigations might study and quantify "causal loops" as structures within theories. Loops have already be used to business organizations (Senge *et al.*, 1994). Indeed, it is possible that loops might even be better than concatenated structures as markers to quantify the usefulness of theories. An idea that should make cyberneticists happy (c.f. Dent & Umpleby, 1998; Yolles, 2006).

Finally, as discussed above, IPA requires causal relationships between concepts. However, IPA is neutral as to what form those causal connections might take. Alternative understandings of causal connections should be studied. For example, the deep and thoughtful work of Kent Palmer (2014) investigates the importance of causality from multiple perspectives. He notes how Aristotle describes four kinds of causality (formal, material, efficient, and final) which seem to relate closely to Kant's categorical and hypothetical imperatives. It seems rather important to identify causes that are "orthogonal" to one another as a way to better understand the structure of our theories.

For a concrete example, it is easy to say that planting more seeds will lead to more trees. It is more interesting to identify that the "seeds" and the "planting" are very different things—perhaps orthogonal to one another so that a better theory of tree growth would be developed by understanding "planting" and understanding "seeds" both separately and in conjunction. If we can create new rules for evaluating theories based on the orthogonality of causal inputs (and outputs), we will have made a very large step forward in understanding how to create more useful theories.

On a slightly speculative note, it is interesting that the above studies find so many theories with a Systemicity of about 0.25. This is curious because there are a number of studies suggesting that our organizational change methodologies tend to be effective around 20-25% of the time (e.g., Dekkers, 2008; MacIntosh & MacLean, 1999; Smith, 2003). Is this a coincidence? Or, is it possible that we might be able to double the effectiveness of our theories, practices, and policies by doubling their Systemicity?

Conclusion

The emerging science of conceptual systems is located at the overlap between cognitive science and the systems sciences. This paper has shown that the science of conceptual systems exists, not only as a philosophical curiosity, but rather as a tool to support other sciences so that they may advance more rapidly. Thus, the science of conceptual systems is not merely an "education multiplier" (e.g., teaching better ways to teach) it is a "science multiplier" (where advances in one science will impel advances in other sciences).

While it has been recently suggested that the advancement of the social/behavioral sciences is curtailed because we do not have an effective science of metatheory (Wallis, 2010b), the more recent work within the science of conceptual systems suggests that we now have the potential to make more rapid advances.

The approaches presented here do not provide a "map" for advancing the sciences. Indeed, none is available because we are advancing across that new terrain. What it does provide is a compass suggesting a new and potentially exciting direction to travel. Historically/traditionally, scholars have travelled about the countryside of the social/behavioral land making useful observations and gathering interesting data. They have often staked claims regarding the importance of their terrain. They could not advance more purposefully because they were guided only by intuition. So, they did not know what direction was "forward" although it seemed to have something to do with empirical analysis. Today, with the creation of a new science, we have a new direction

Another benefit of this approach is in counteracting the forces of fragmentation. Where increasing specialization has led to an explosion of new fields and sub-fields and all their associated theories, we are overwhelmed by their sheer number and variety. The science of conceptual systems offers paths for rigorously integrating those theories and connecting our fragmented fields (Wallis, 2014b). In short, where we see theories as data, our weakness becomes our strength.

However, IPA does not have an extensive track record. More studies are needed before we can have great certainty about its efficacy. Additionally, there are clear and interesting paths for developing alternative measures of conceptual systems. These should be investigated philosophically, qualitatively, and quantitatively.

History shows that theories emerge, evolve, and are replaced. It is very likely, therefore, that the same path of development will occur with IPA. In order to develop an improved version of IPA, it may be useful to use IPA in a wide variety of situations for research and evaluating the effectiveness of conceptual systems. That way, we can "push it" to find what its natural limits of usefulness might be. A world of scientific exploration awaits.

References

Ambrose, D. (1996). Unifying theories of creativity: Metaphorical thought and the unification process. *New Ideas in Psychology, 14*(3), 257-267.

Axelrod, R. (1976). *Structure of decision : The cognitive maps of political elites*. Princeton: Princeton University Press.

Bolman, L. G., & Deal, T. E. (1991). *Reframing Organizations: Artistry, Choice, and Leadership*. San Francisco: Jossey-Bass.

Burrell, G. (1997). *Pandemonium: Towards a Retro-Organizational Theory*. Thousand Oaks, California: Sage.

Calas, M. B., & Smircich, L. (1999). Past postmodernism? Reflections and tentative directions. *Academy of Management. The Academy of Management Review, 24*(4), 649.

Charmaz, K. (2006). *Constructing Grounded Theory: A Practical Guide through Qualitative Analysis*. Thousand Oaks, CA: Sage.

Clegg, S. R., Cunha, J. V., & Cunha, M. P. (2002). Management paradoxes: A relational view. *Human Relations, 55*(5), 483-503.

Council, S. (2010). Defining science, link.

Craik, K. (1943). *The Nature of Explanation*. New York: Cambridge University Press.

Curseu, P., Schalk, R., & Schruijer, S. (2010). The use of cognitive mapping in eliciting and evaluating group cognitions. *Journal of Applied Social Psychology, 40*(5), 1258-1291.

Dekkers. (2008). Adapting organizations: The instance of Business process re-engineering. *Systems Research and Behavioral Science, 25*(1), 45-66.

Dent, E. B., & Umpleby, S. A. (1998). Underlying assumptions of several traditions in systems theory and cybernetics. In R. Trappl (Ed.), *Cybernetic and Systems '98* (pp. 513-518). Vienna, Austria: Austrian Society for Cybernetic Studies.

Dubin, R. (1978). *Theory building* (Revised ed.). New York: The Free Press.

Edwards, M. (2010). Organizational transformation for sustainability: An integral metatheory*. New York: Routledge.

Edwards, M., G. (2009). A general research method for metatheory building.

Eisenhardt, K. M., & Graebner, M. E. (2007). Theory building from cases: Opportunities and challenges. *Academy of Management Journal, 50*(1), 25-32.

Faust, D. (2005). Why Paul Meehl will revolutionize the philosophy of science and why it should matter to psychologists. *Journal of Clinical Psychology, 61*(10), 1355-1366.

Faust, D., & Meehl, P. E. (2002). Using meta-scientific studies to clarify or resolve questions in the philosophy and history of science. *Philosophy of Science, 69*(3), 185-196.

Fiske, D. W., & Shweder, R. A. (1986). *Metatheory in Social Science: Pluralisms and Subjectivities*. Chicago: University of Chicago Press.

Flower, L., & Mellon, C. (1989). Cognition, Context, and Theory Building.

Friedman, D. (1997). *Hidden Order: The Economics of Everyday Life*. New York: Harper Business.

Gardner Jr, E. S. (2004). Dimensional Analysis of airline quality. *Interfaces, 34*(4), 272-279.

Gentner, D. (1983). Structure-mapping: A theoretical framework for analogy. *Cognitive Science, 7*, 155-170.

Glaser, B. G., & Strauss, A. L. (1967). *The Discovery of Grounded Theory: Strategies for Qualitative Research.* New Brunswick, NJ: Aldine Transaction.

Jacobson, N. (2001). Experiencing recovery: A dimensional analysis of recovery narratives. *Psychiatric Rehabilitation Journal, 24*(3), 248-256.

Jean-Pierre, V. M. H., & Edward, A. G. (2000). Meta-disciplinarity, belles lettres, and Andre Malraux: A bibliometric exploration of knowledge formation. *The Serials Librarian, 37*(4), 51.

Johnson-Laird, P. (1980). Mental models in cognitive science. *Cognitive Science, 4*, 71-115.

Kaplan, A. (1964). *The conduct of inquiry: Methodology for behavioral science.* San Francisco: Chandler Publishing Company.

Kelly, G. A. (1955). *The psychology of personal constructs.* New York: Norton.

Knorr-Cetina, K. (1981). *The Manufacture of Knowledge: An Essay on the Constructivist and Contextual nature of Science*: Pergamon Press.

Kostoff, R. N., del Rio, J. A., Humenik, J. A., Ramírez, A. M., & García, E. O. (2001). Citation mining: Integrating text mining and bibliometrics for research user profiling. *Journal of the American Society for Information Science and Technology, 52*(13), 1148-1156.

Kuhn, T. (1970). *The Structure of Scientific Revolutions* (2 ed.). Chicago: The University of Chicago Press.

Lakatos, I. (1970). Falsification and the methodology of scientific research. In I. Lakatos & A. Musgrave (eds.), *Criticism and the Growth of Knowledge* (pp. 91-195). New York: Cambridge University Press.

Lane, D. A. (1992). *Artificial worlds and economies.* Working Paper for the Santa Fe Research Program. Santa Fe, New Mexico, USA.

Ledoux, L. (2012). Philosophy: Today's manager's best friend? *Philosophy of Management - Special Issue (Guest Editors: Stephen Sheard, Mark Dibben, 11*(3), 11-26.

MacIntosh, R., & MacLean, D. (1999). Conditioned emergence: A dissipative structures approach to transformation. *Strategic Management Journal, 20*(4), 297-316.

McNamara, C., & Troncale, L. (2012). *SPT II.: How to find and map linkage propositions for a GTS from the natural sciences literature.* Paper presented at the 56th Annual Conference of the International Society for the Systems Sciences (ISSS), San Jose, CA.

Meehl, P. E. (1992). Cliometric metatheory: The actuarial approach to empirical, history-based philosophy of science. *Psychological Reports, 71*(2), 339-467.

Meehl, P. E. (2002). Cliometric metatheory: II. Criteria scientists use in theory appraisal and why it is rational to do so. *Psychological Reports, 91*(2), 339-404.

Meehl, P. E. (2004). Cliometric metatheory III: Peircean consensus, verisimilitude and asymptotic method. *The British Journal for the Philosophy of Science, 55*(4), 615-643.

Müller, K. H. a. T., N. . (2012). New cognitive environments for survey research in the age of science 2. *Društvena Istraživanja [Social Research - Journal for General Social Issues], 21*(2), 315-339.

Oberschall, A. (2000). Oberschall reviews "Theory and Progress in Social Science" by James B. Rule. *Social Forces, 78*(3), 1188-1191.

Palmer, K. D. (2014). *Setting off to Nowhere: Introduction - Search for a Deeper Theory of Everything.*

Parnell, J. A. (2008). Assessing theory and practice in competitive strategy: Challenges and future directions. *Journal of CENTRUM Cathedra, 1*(12), 12-27.

Pearl, J. (2000). *Causality: Models, reasoning, and inference.* New York: Cambridge University Press.

Raphael, T. D. (1982). Integrative complexity theory and forecasting international crises: Berlin 1946-1962. *The Journal of Conflict Resolution, 26*(3).

Ritzer, G., & Smart, B. (2001). Introduction: Theorists, Theories and Theorizing *Handbook of Social Theory* (pp. 1-9).

Robertson, P. P. (2014). Why top executives derail: A performative-extended mind and a law of optimal emergence *Journal of Organizational Transformation & Social Change, 11*(1), 25-49.

Senge, P., Kleiner, K., Roberts, S., Ross, R. B., & Smith, B. J. (1994). *The Fifth Discipline Fieldbook: Strategies and Tools for Building a Learning Organization.* New York: Currency Doubleday.

Šešelja, D., & Straßer, C. (2014). Epistemic justification in the context of pursuit: A coherentist approach. *Synthese, In Press*, 1-31.

Shackelford, C. (2014). *Propositional Analysis, Policy Creation, and Complex Environments in the United States' 2009 Afghanistan-Pakistan Policy.* (Doctoral Dissertation), Walden, Minneapolis, MN.

Shaw, D. R., & Allen, T. F. H. (2012). A systematic consideration of observational design decisions in the theory construction process. *Systems Research and Behavioral Science, 29*(5), 484-498.

Shoemaker, P. J., Tankard Jr., J. W., & Lasorsa, D. L. (2004). *How to Build Social Science Theories.* Thousand Oaks, California: SAGE.

Shotter, J. (1994). Conversational Realities: From Within Persons to Within Relationships, link.

Shotter, J., & Tsoukas, H. (2007, 7-9 June 2007). *Theory as therapy: Towards reflective theorizing in organizational studies.* Paper presented at the Third Organizational Studies Summer Workshop: 'Organization Studies as Applied Science: The Generation and Use of Academic Knowledge about Organizations', Crete, Greece.

Smith, M. E. (2003). Changing an organization's culture: Correlates of success and failure. *Leadership & Organization Development Journal, 24*(5), 249-261.

Sosa, E. (2003). In search of coherentism. In E. Sosa (Ed.), *Epistemic Justification: Internalism vs. Externalism, Foundations vs. Virtues* (Vol. 7, pp. 240). Malden, Massachusetts: Blackwell.

Stacey, R. D. (1996). *Complexity and Creativity in Organizations*. San Francisco: Berrett-Koehler Publishers, Inc.

Stinchcombe, A. L. (1987). *Constructing social theories*. Chicago: University of Chicago Press.

Suedfeld, P., Tetlock, P. E., & Streufert, S. (1992). Conceptual/integrative complexity. In C. P. Smith (Ed.), *Handbook of Thematic Content Analysis* (pp. 393-400). New York: Cambridge University Press.

Sussman, S., & Sussman, A. (2001). Praxis in health behavior program development. In S. Sussman (Ed.), *Handbook of program development for health behavior research and practice* (pp. 79-97). Thousand Oaks, CA: Sage.

Thagard, P., & Stewart, T. C. (2011). The AHA! experience: Creativity through emergent binding in neural networks. *Cognitive Science, 35*(1), 1-33.

Turner, J. H. (1986). *The structure of sociological theory* (4 ed.). Chicago: The Dorsey Press.

Umpleby, S. (2010). *From complexity to reflexivity: The next step in the systems sciences*. Paper presented at the Cybernetics and Systems 2010, Vienna, Austria, link.

Uzzi, B., & Spiro, J. (2005). Collaboration and creativity: The small world problem. *American Journal of Sociology, 111*(2), 447-504.

Van de Ven, A. H. (2007). *Engaged scholarship: A guide for organizational and social research*. New York: Oxford University Press.

Wallis, S. E. (2008). From reductive to robust: Seeking the core of complex adaptive systems theory. In A. Yang & Y. Shan (Eds.), *Intelligent Complex Adaptive Systems* (pp. 1-25). Hershey, PA: IGI Publishing.

Wallis, S. E. (2009a). The complexity of complexity theory: An innovative analysis. *Emergence: Complexity and Organization, 11*(4), 26-38.

Wallis, S. E. (2009b). Seeking the robust core of social entrepreneurship theory. In J. A. Goldstein, J. K. Hazy, & J. Silberstang (Eds.), *Social Entrepreneurship & Complexity*. Litchfield Park, AZ: ISCE Publishing.

Wallis, S. E. (2010a). The structure of theory and the structure of scientific revolutions: What constitutes an advance in theory? In S. E. Wallis (Ed.), *Cybernetics and systems theory in management: Views, tools, and advancements* (pp. 151-174). Hershey, PA: IGI Global.

Wallis, S. E. (2010b). Toward a science of metatheory. *Integral Review, 6*(Special Issue: "Emerging Perspectives of Metatheory and Theory").

Wallis, S. E. (2010c). Towards developing effective ethics for effective behavior. *Social Responsibility Journal, 6*(4), 536-550.

Wallis, S. E. (2010d). Towards the development of more robust policy models. *Integral Review, 6*(1), 153-160.

Wallis, S. E. (2011). *Avoiding policy failure: A workable approach*. Litchfield Park, AZ: Emergent Publications.

Wallis, S. E. (2012, July 22-27, 2012). *Theories of psychology: Evolving towards greater effectiveness or wandering, lost in the jungle, without a guide?* Paper presented at the 30th International Congress of Psychology: Psychology Serving Humanity, Cape Town, South Africa.

Wallis, S. E. (2013a). How to choose between policy proposals: A simple tool based on systems thinking and complexity theory. *Emergence: Complexity & Organization, 15*(3), 94-120.

Wallis, S. E. (2013b). A systems approach to understanding theory: Finding the core, identifying opportunities for improvement, and integrating fragmented fields. *Systems Research and Behavioral Science Journal (in press).*

Wallis, S. E. (2014a). Abstraction and insight: Building better conceptual systems to support more effective social change. *Foundations of Science, 19*(4), 353-362. doi: 10.1007/s10699-014-9359-x

Wallis, S. E. (2014b). Existing and emerging methods for integrating theories within and between disciplines. *Organizational Transformation and Social Change, 11*(1), 3-24.

Wallis, S. E. (2014c). Structures of logic in policy and theory: Identifying sub-systemic bricks for investigating, building, and understanding conceptual systems. *Foundations of Science, in press.*

Wallis, S. E. (2015). Are theories of conflict improving? Using propositional analysis to determine the structure of conflict theories over the course of a century, link.

Weick, K. E. (1989). Theory construction as disciplined imagination. *Academy of Management Review, 14*(4), 516-531.

Wong, E. M., Ormiston, M. E., & Tetlock, P. E. (2011). The Effects of Top Management Team Integrative Complexity and Decentralized Decision Making on Corporate Social Performance. *Academy of Management Journal, 54*(6), 1207-1228.

Wright, B., & Wallis, S. E. (Under submission). A revolutionary method to advance entrepreneurship theories.

Yolles, M. (2006). Knowledge cybernetics: A new metaphor for social collectives. *Organizational Transformation and Social Change, 3*(1), 19-49.

Zimmer, L. (2006). Qualitative meta-synthesis: a question of dialoguing with texts. *Journal of Advanced Nursing, 53*(3), 311-318.

Glossary

Adapted (and extended) from (Wallis, 2011: 99-124)

◊ **Aspect** See "Concept."

◊ **Atomistic Logic** A kind of logical structure found within a proposition that is reductionist such as "A is valid" or "A is true." Or, more concretely, "Apples are important." These would score very low in a study of Integrative Complexity.

◊ **Bibliometrics** Methods such as content analysis and citation analysis, typically used in library sciences, for the quantification of text from academic literature.

◊ **Branching Logic** A logical structure found within causal propositions including three or more concepts where a change in one concept causes change in two or more other concepts. For example, a Branching proposition might say that changes in A will cause changes in B and C. For a more concrete example, "More teamwork will lead to more cohesion, and more results, and more frustration.

◊ **Case Comparative Study** Investigating and comparing two or more cases along multiple dimensions to draw insights and inferences. A useful start to building new theory (Eisenhardt & Graebner, 2007).

◊ **Causal Relationship (causality)** Where two or more concepts are related so that a change in one causes a change in one or more others. A causal relationship is often expressed as a proposition, hypothesis, or a diagram. A causal relationship such as, "More A causes more B." may also be used as a general term in place of other more specific terms. Instead of saying "more" other indicators might be "better" or "less" (for example). Similarly, instead of "causes" more specific indicators might include such terms as "creates," or "engenders." In any case, the description must be specific to be valid. It is not useful to state (for example) that "A and B are interrelated" or "More A may cause more B" because the nature of the relationship is not causally defined. Logical structures often describe causal relationships (e.g., Linear, Branching, Concatenated).

◊ **Circular Logic** A logic structure where one cause leads back to itself such as Changes in A cause changes in B cause changes in C cause changes in A. Circular logics may be seen as feedback loops which are held to be very useful in understanding systems. However, they can also be misleading if researchers rely on too few loops to adequately represent the system. Then they may appear as tautologies (e.g., more A causes more A).

◊ **Cognitive Science** Interdisciplinary investigation of the human, social, and artificial processes around perception, information, reasoning, and decision making.

◊ **Coherentist Perspective** Coherence is focused on the relationships between beliefs. A statement is true/meaningful/useful based on its relationship with other statements (Šešelja & Straßer, 2014; Sosa, 2003). This is in contrast to a perspective of correspondence where a statement is held to be true based on its relationship to empirical data. The two perspectives are orthogonal and so most useful when used in combination.

◊ **Complexity** A measure representing the number of concepts within a conceptual system. This may also be understood as the diversity of ideas within a document. For an abstract example, consider a conceptual system containing the propositions: A is true; More B causes more C; More B causes more D. In such a model, there are four concepts (A, B, C, D). Therefore, the Complexity of the conceptual system is C = 4.

◊ **Concatenated Logic** A logical structure found within a causal proposition including three or more concepts where changes in two or more concepts cause change in another concept. For an abstract example, a Concatenated proposition might state that changes in concept A and concept B will cause changes in concept C. In that example, C is the Concatenated concept, while A and B exist within a Concatenated relationship but are themselves not Concatenated. For a more concrete example, "More collaboration and more shared goals will result in more teamwork." Here, "teamwork" is the Concatenated (and better understood) concept.

◊ **Concept** The part of conceptual system that represents a concept, idea, or notion. The concept may be as concrete as in "apple" or as abstract as in "truth." Concepts may be simple as in "numbers" or complex as in "left handed monkeys with undiagnosed trauma." A concept is typically detectable, that is to say empirically measurable, but that is not an absolute standard. Concepts are part of propositions.

◊ **Conceptual System** A set of interrelated concepts. Examples include theories, policy models, mental models, schemata, etc. Metaphorically, they serve as lenses to aid in understanding and effective engagement of the world (Wallis, 2014a, 2014c).

◊ **Correspondence Perspective** Commonly associated with "normal" science or "Science One" (Müller, 2012; Umpleby, 2010) and the use of Toulminian logics, correspondence is focused on the relationship between a statement and some empirically verifiable existence.

◊ **Dimensional Analysis** The analysis of relationships by focusing on their dimensions and units of measure on a fundamental level (c.f. Gardner Jr, 2004; Jacobson, 2001).

- ◊ **Grounded Theory** An approach to theory development that is non-positivist, yet is solidly based on the use of data—although the data are generally qualitative. Data, often from interviews, are coded and categorized. Then, relationships are identified between the categories to identify a useful theory (Charmaz, 2006; Glaser & Strauss, 1967).
- ◊ **Integral Thinking** Understanding (or attempting to understand) the world from a transdisciplinary perspective where those many perspectives are interrelated.
- ◊ **Integrative Complexity (IC)** "Integrative complexity is a measure of the intellectual style used by individuals or groups in processing information, problem solving, and decision making. Complexity looks at the structure of one's thoughts, while ignoring the contents. It is scorable from almost any verbal materials: books, articles, fiction, letters, speeches and speech transcripts, video and audio tapes, and interviews." link
- ◊ **Integrative Propositional Analysis (IPA)** Combined processes of qualitative and quantitative analysis involving rigorous hermeneutic deconstruction of propositions found in formal texts including the rigorous re-integration of propositions from those texts following a structured methodology. Also a process of meta-analysis for investigating conceptual systems to determine the Complexity of conceptual systems (diversity of concepts) and the Systemicity of the conceptual system (connectedness between concepts).
- ◊ **Linear Logic** A logical structure found within a proposition describing simple causal relationship between two concepts. Such as, "More A causes more B." Both A and B exist in Linear relationship to one another. Here, A is the causal concept and B is the resultant concept). Linear structures can be of any length (e.g., More A causes more B which causes more C which causes more D… which causes more Z). For a more concrete example, "Having more shared goals leads to more teamwork which, in turn, leads to more productivity." Within an explanation, this may also be phrased as, "A is true because of B and B is true because of C… because of Z."

◊ **Logic Model** A set of interrelated logic statements such as a theory or a Policy Model describing causal relationships between the elements of the model.

◊ **Mental Model** A representation within one's mind about how the world works. Useful for understanding and engaging the world and for making predictions.

◊ **Metapolicy** Generally, the study of policy—how the policy is created, applied, and evaluated. This may be rigorous as in the use of Integrative Propositional Analysis or fuzzier as in the use of historical narrative. May also be used to describe a policy on how to make policy.

◊ **Metatheory** Primarily the study of theory, including the development of overarching combinations of theory, as well as the development and application of theorems for analyses that reveal underlying assumptions about theory and theorizing.

◊ **Model** "A model is a simplified representation of a system at some particular point in time or space intended to promote understanding of the real system. As an abstraction of a system, it offers insight about one or more of the system's aspects, such as its function, structure, properties, performance, behavior, or cost." link

◊ **Narrative Analysis** Qualitative approach for understanding how people make sense of their world based on spoken or written accounts of their experiences and how those experiences are interpreted.

◊ **Parsimony** Generally, the understanding that a theory is better when it is smaller. Or, as small as possible while including ideas that are necessary or useful. Ockham's Razor is a common example.

◊ **Policy Model** A cognitive structure (like a theory) representing how a community or organization understands the world, thus enabling them to take specific actions to achieve their goals. As a sense-making structure, the policy model does not (strictly speaking) include goals or actions. Those are part of the broader policy.

◊ **Proposition** "A proposition is a declarative sentence expressing a relationship among some terms." (Van de Ven, 2007: 117). For example, "More travel leads to more discovery." (See Causal Relationship). Those terms are often understood to be "concepts."

◊ **Reflexive Dimensional Analysis (RDA)** A process for creating a unified conceptual system from multiple conceptual systems through a process of categorization, abstraction, dimensionalization, and the identification of causal connections.

◊ **Robustness** An older term for "Systemicity"

◊ **Schema** Most commonly used in computer science and logics in referring to a formula and/or set of data used to represent a system.

◊ **Scientometrics** Approaches of Bibliometrics applied to study the development of science within and between fields.

◊ **Social/Behavioral Sciences** The purposeful study and advancement of understanding in all fields of human interaction including (but not limited to) psychology, sociology, policy, ethics, business, management, human development, organizational development, economics, and social anthropology.

◊ **Systemicity** A ratio describing the interrelatedness between concepts of a conceptual system on a scale of zero to one. Systemicity is calculated by dividing the number of Concatenated concepts by the total number of concepts in a policy (see Integrative Propositional Analysis). Systemicity is a measure of how well integrated the propositions of a conceptual system are, the degree to which they are understood as existing in a systemic relationship, and the level of causality between the concepts. Systemicity is also related to the effectiveness of the conceptual system in practical application.

◊ **Systems Thinking** Understanding that the world is made of systems, this interdisciplinary way of thinking explores how those systems effect one another. Systems thinkers strive to obtain a deeper understanding by looking at the connections of many things, rather than seeking to understand things in isolation.

◊ **Theory** An ordered set of assertions. Weick (1989: 517; drawing on Southerland).

Chapter 11 Abstraction and Insight—Building Better Conceptual Systems to Support More Effective Social Change

Wallis, S. E. (2020). Evaluating and improving theory using conceptual loops: A science of conceptual systems (SOCS) approach. Cybernetics and Human Knowing, in press.

This paper is based on a presentation titled "Existing and Emerging Methods for Integrating Theories Within and Between Disciplines" at the 56th annual meeting of the International Society for Systems Sciences (ISSS). July 15-22, 2012, at San Jose State University, California

I appreciate the excellent insights of three anonymous reviewers whose suggestions have led to an improved paper. All remaining mistakes I claim as my own.

Abstract: *When creating theory to understand or implement change at the social and/or organizational level, it is generally accepted that part of the theory building process includes a process of abstraction. While the process of abstraction is well understood, it is not so well understood how abstractions "fit" together to enable the creation of better theory. Starting with a few simple ideas, this paper explores one way we work with abstractions. This exploration challenges the traditionally held importance of abstracting concepts from experience. That traditional focus has been one-sided—pushing science toward the discovery of data without the balancing process that occurs with the integration of the data. Without such balance, the sciences have been pushed toward fragmentation. Instead, in the present paper, new emphasis is placed on the relationship between abstract concepts. Specifically, this paper suggests that a better theory is one that is constructed of concepts that exist on a* *similar level of abstraction. Suggestions are made for quantifying this claim and using the insights to enable scholars and practitioners to create more effective theory.*

Introduction

I begin with the view that the social sciences have not yet developed theories that are inarguably highly effective in practical application. One reason for this lack of progress is that we have not yet developed an adequate understanding of conceptual systems such as theories, models, and policies. Lacking that understanding, the social sciences has instead heeded calls for standards such as parsimony, creativity, compelling authorship, and bravery. These pursuits, while useful to some extent, are clearly insufficient to our task of creating theories that are inarguably highly useful in practical application.

This is not to say that creativity is not useful. Indeed, it is necessary to the process of science (c.f. Ambrose, 1996; Stacey, 1996; Thagard & Stewart, 2011; Uzzi & Spiro, 2005). While the creation of innovative concepts is needed, it is not always clear how those innovative concepts may best be interconnected to form theory that might be more useful in practical application and/or research.

The science has also expended considerable effort in the pursuit of empirical evidence; again, without the creation of effective theories. Indeed, there may be a negative effect—because the creation of so much data may have impelled researchers toward ever more narrow sub-specializations (Phillips & Johnstone, 2007); thus accelerating the fragmentation of the sciences. Therefore, it seems that we need new understandings of how to create and evaluate theory so that we may fulfil the dream of the social sciences and develop theory for supporting and transforming our social systems.

While the present paper is focused on the social sciences, these same insights may be generalized to all sciences. This is because all forms of science and investigation use concepts, and all those concepts have some level of abstraction.

Here, I use the term "theory" in a broad sense to refer to a conceptual system that is similar to a model, schema, policy, or other set of interrelated concepts that may be used to understand and/or engage the world. Such concepts and conceptual systems may be understood as existing in a conceptual world that is related to, but distinct from the real world of physical objects. Here, I am addressing concepts and conceptual systems themselves—not delving into existing "conceptual systems theory" that might be applied for understanding learning and other behaviors (Fedigan, 1973). That is to say, I am primarily looking at the concept-to-concept relationship rather than the concept-to-reality relationship.

A concept may be understood as a named abstraction from a pattern recognition experience; and, a set of concepts may be understood as existing in some systemic relationship—much as reality is systemic (James, 1909). The way that concepts fit together allow our minds to grasp related ideas, understand things, and may even be necessary for thought, itself (Fodor, 1998). Because our reality is systemic, developing an improved understanding of how our concepts are systemically interrelated provides us with a better understanding of our world. Indeed, there is a general consensus that theories which are more systemic will be more useful (Dubin, 1978; Friedman, 2003).

Indeed, if we return to the concept of Plato's cave, we must recognize that we will never be able to understand "reality"—that the best we can do is to understand the systemic interrelationship between the shadows. Thus, understanding concepts in systemic relationship to one another seems to provide a new way to define (or at least understand) those concepts as they may be perceived or understood (perhaps creatively) without resorting to claims of what might constitute some absolute yet unreachable reality.

In this paper we will be looking at the similarities and differences between objects within and between those worlds because this is an area that requires deeper understanding if we are to make more progress in our science (Wallis, 2008). The present paper addresses ideas that are important to studying theories as conceptual systems in ways that are similar to the way we understand physical systems, biological systems, and social systems.

Increasingly, our understanding of the structure and complexity of conceptual systems is providing us with more insight into how we can develop more effective conceptual systems for more effective practice. While this kind of understanding has been suggested in the past (e.g., Dubin, 1978; Kaplan, 1964) it was not approached with scientific rigor or quantitative analysis. Building on insights from complexity theory, systems thinking and Axelrod's work into concept mapping, recent advances in this direction include insights into the systemic structure and complexity of theory (Wallis, 2010a), policy (Wallis, 2011) and ethics (Wallis, 2010c). Others engaging in similar studies have quantified the complexity and systemicity of conceptual systems for learning (Curseu *et al.*, 2010), studies of integrative complexity to support cognitive processes and management strategy (Tetlock, 1985), narrative analysis (Greenhalgh *et al.*, 2005), program evaluation (Rogers, 2008), and casual mapping for organizational change (Ackermann & Eden, 2004).

One kind of conceptual system, social theory, has not been shown to be highly effective. Broadly, social change theory seems to have failed (Appelbaum, 1970; Boudon, 1986). In the field of organizational change, high failure rates are associated with the application of Total Quality Management (MacIntosh & MacLean, 1999), culture change (Smith, 2003), and Business Process Reengineering (Dekkers, 2008). Also, there is a failure of organizational theory (Burrell, 1997) and theories of bureaucracy (Bernier & Hafsi, 2007). Despite its current popularity, management theory might be causing new problems as it tries to solve others—such as the collapse of Enron (Ghoshal, 2005). Even the vaunted discipline of economics has not proven very successful (Dubin, 1978, citing Rapoport). We are not sure if it is even possible to make theory (Kessler, 2001) and some have incited us to ignore theory (e.g., Shotter, 2005).

In the present paper, I focus on the idea of abstraction. What is an abstraction, where does it originate, how might it be manipulated, and how might abstractions be best used to create more effective theories. By understanding relationships between abstract concepts in a new way, we may learn to understand conceptual systems in a new and more useful way. Indeed, this approach may be understood as supporting broader efforts at understanding and integrating conceptual systems to create more effective theory for more effective application (Wallis, 2008, 2010b).

Foundation of Abstraction

Starting with the basics, Quine (1980) draws on Russell's theory of types to discuss abstraction in terms of "individuals" on one level, "classes of individuals" on another level and "classes of classes" on a still higher level of abstraction from concrete to abstract. Leading him to ask, "How much of our science is merely contributed by language and how much is a genuine reflection of reality?" (Quine, 1980: 78). Drawing on Korzybski's work (which was developed in more depth by Hayakawa, who describes the process as moving between levels on a 'ladder of abstraction'), it is noted that "Moving up the ladder entails looking at similarities and ignoring differences" (Seabury, 1991). Indeed, on each level, some details may be gained or lost thus changing one's view of (what may be considered as) 'reality'.

It may be that the notion of reality is not as useful as some believe. Fodor (1998) argues that each concept is atomistic—without structure. Although, of course, each concept is an abstraction of a whole or part of an object in the real world. Already, we have an interesting distinction between the conceptual world and the real world. In the real world, an object such as an apple is systemically interrelated with other things (e.g., tree, bird, farmer, water).

Yet, in the "isolation chamber" of a person's mind, it is possible to consider the concept of apple separately from those other things. Indeed, we may also consider (in our minds) that concept of an apple as existing in closer proximity with other concepts—such as the Andromeda galaxy. In contrast, distant galaxies arguably have negligible effect on real world apples. So, the systems of interrelationships we create in our minds do not necessarily reflect the systems in the real world. This is much the same as we are able to imagine a perfect circle—where none exists in the real world.

This is of great importance because systems scientists have spent considerable time and effort studying real world physical systems (e.g., the interaction of atoms within the apple), biological systems (e.g., the growth of the apple), and social systems (e.g., farming communities and recipes for apple pie). In contrast, scientists have spend very little effort studying conceptual systems. We've been stuck looking at relationships between things and thoughts; we've not been considering the relationships between thoughts and thoughts.

It is generally accepted that the process of description and abstraction is one way to build a theory (Morgeson & Hofmann, 1999; Ostroff & Bowen, 2000). It is also interesting to note that theories may be developed through the use of data that is derived from differing levels of abstraction. For example, Newton used observation to gain empirical data to build his theories of motion. Einstein, on the other hand, used data in the form of existing theories to create a more effective theory (Dubin, 1978).

A highly interesting thought here is that the empirical observations by Newton led to a useful abstract theory. In contrast, the integration of theories led to Einstein's theory that was more useful than Newton's. This is not to say that their work was solely on one side of the scale or the other. Only that one emphasized theory and one emphasized data more than the other. On a side note, one can, of course, see theory as data! However, this difference (even a difference of scale) suggests an important opportunity to use existing abstract theories in a process of rigorous integration to create theories that are more effective than we ever imagined possible. In short, the present conversation suggests that the social sciences should follow in the footsteps of Einstein rather than Newton.

Let us return now to the idea that nothing exists in isolation. This is analogous to the idea that "everything in the world is interrelated/interconnected/connected" that everywhere we look there are structures with "interaction patterns" (Axelrod & Cohen, 2000: 6). I suggest that this rule, which applies to objects, humans, and societies, also applies to concepts within our minds. Such an approach is foundational to the systems sciences where a system is, "A structure of interrelated elements" (Dancke, 1999: 230). Also, where our understanding takes, "into account the interrelationships and interdependence between the parts of a system" (Hammond, 2003: 11). Indeed, "a theoretical term has a *systemic meaning*" (Kaplan, 1964: 57—emphasis in original). Therefore, it follows that nothing can be usefully understood except in relation to other things.

As we delve into this level of nuance the questions emerges, what *kinds* of things should be placed in relationship to things in order to improve the usefulness of our understandings? My assertion, and the focus of the present paper, is that we generate more understanding and so things are better understood when they are seen in relation to things that have greater similarity to one another. This is the key—that greater similarity opens the opportunity for greater understanding. It does not seem to be so important that this similarity occur at any particular level.

It may be, however, that a greater similarity of levels results in greater insight for application while a greater difference in levels (as in highly imaginative or speculative connections) may result in more artistic insights. What traditional science might call "misinterpretation" we might call "alternative interpretation." That alternative allows for creativity, imagination, and insight—critically important features of life, research, and science. Clearly, there is the opportunity for more investigation on this topic. For the time being, the present paper, however, is focused on abstraction.

In contrast, things that are less similar do not lend themselves quite so well to a process of improved understanding (although those differences may open the door for greater metaphors—but that would be another investigation). This is important because we use our theories to generate understanding. With more understanding, we are more able to enact change in our organizations and communities.

The Physical World

For our discussion, let us start with an example from the physical world. Here, I will use examples that are easily accessible to readers from a wide variety of disciplines.

If we compare apples and oranges, we are talking about a class of things called apples and a class of things called oranges. Let us assume, for the purpose of this presentation, that we have little or no experience with these fruit. We may look at an apple and say, "It has color." We may improve our understanding of the apple by placing it on a table near an orange. In comparing the two fruits, the apple is seen to be more red and the orange to be more recognizably orange. By comparing the two, we can more easily identify differences in flavor, color, texture, etc. Indeed, such comparisons may well arise in the reader's mind at this moment.

However, such a broad difference indicates that we can judge/compare/understand apples only broadly. When I write "apple and orange" I suspect that most people will imagine one apple and one orange—a representative sample, if you will, or perhaps an idealized expression. Few minds will immediately conjure all of the many varieties of both fruits! Yet, there are many variations in flavors among all those different apples.

To improve our understanding of apples, we may remove the orange from the table and place the apple in proximity to another apple. For example, we might compare a golden delicious (that is more yellow in color) with a pippin (that is more green in color). Such a distinction among apples is not possible when comparing apples with oranges. This one brief thought experiment can be repeated successfully with many objects. Within the classroom, for example, comparing chairs to walls yields superficial differences. Comparing chairs with other chairs, however, impels students to identify more details.

Moving up some scale of nuance in our comparisons, we could compare two apples of the same kind. For example, two pippin apples. To an individual with a high level of discernment, it would be possible to detect slight differences between two pippin apples (variations in color, flavor, etc.). Those subtle differences are not so evident when comparing pippin apples (or apples generally) with oranges. Zooming / graining back out to compare objects of greater difference instead of less, apples are less well understood in relation to (for example) automobiles. A car might have an "apple red" color. Which provides one point of comparison. However, there is not a good way to consider the difference in structure and function between apples and cars. There are few points of comparison. What are car seeds? No such thing.

How about comparing apples with stellar nebula? The greater the difference, the fewer opportunities exist for comparison. Moving further away, it would be more difficult still to compare an apple with a highly abstract concept such as freedom. Freedom, does not have color, or flavor in the same way that an apple does.

Here, I am not saying one cannot find any comparison. As a side-note, another possible use of identifying similarities is to create abstraction. If we place apples and galaxies in the same category, we would be impelled to identify some similarity between them. That similarity, the name of the category, becomes a new abstraction. For example, apples and galaxies are both made of matter. My point is that there are fewer relationships and so fewer opportunities to identify nuances of difference. Thus, fewer opportunities to generate insight (including creative insight) and understanding.

Those differences, if one is impelled to find a connection, may generate more metaphors. And, indeed, we might span that gulf by the use of metaphorical relations and other poetic endeavors. While those too may be interesting, they are not useful for practical application. Indeed, they are likely to lead to misleading speculative endeavors. For example, one might "reach out and pluck an apple" but reaching out and plucking a galaxy is not likely to be a successful action or result in a tasty snack.

This is not to say that metaphors are without merit. Indeed, they can be a great spur to imagination and insight. Those benefits have been explored by others (e.g., Baake, 2003; Daneke, 1997; Fuller & Moran, 2000; Hatch & Yanow, 2008; Hung, 2002; Kuipers, 1982; Lakoff & Johnson, 1980; Letiche & van Uden, 1998; Morgan, 1980; Robbins, 2000; Speicher, 1997) and experienced by all of us. Here, however, I seek to identify useful approaches from a different direction. In short, while metaphor is certainly useful, "we are after bigger game than metaphor" (Wallis, 2009b: 86).

The usefulness generated by similarity extends in other directions as well. If one learns the appropriate technique for picking one type of apple from a tree, one may easily learn to pick other types of apples. Different techniques are required for picking oranges, grapes, peanuts, and so on. The further one drifts from the original, the less useful the insights are in practical application.

The Conceptual World

When comparing concepts, the same approaches and insights apply. That is, we are able to gain a more nuanced understanding of a concept when we compare concepts that are more similar. We gain more knowledge when we compare concepts that are closer together. Here, I would like to focus on the idea that those differences are particularly evident when the conceptual objects exist at different levels of abstraction.

For example, it is not very illuminating to compare the more concrete concept of apple with the more abstract concept of fruit. This may be (at least in part) because apples are a sub-set of fruit. Therefore, all concepts that are part of the apple are already included in the more abstract concept of fruit. Yellow apples, green apples—all are fruit.

For an applied example, if a child is learning basic math functions, the idea of multiplication is easier to learn if the child has already learned addition because the two concepts are similar (four times three is the same as four plus four plus four—or four plus four, *three times*). Learning the concept of multiplication would be much more difficult if the student were to begin with the concept of glassblowing because the concept of glassblowing is not as well connected to multiplication as the concept of addition.

For another example, one might compare the concept of "movement" with the concept of "size." There is not much overlap there—they seem quite distinct. However, if we compare the concept of "upward movement" with the concept of "lateral movement" clear similarities and differences begin to emerge. We are more easily able to acquire useful understandings.

In academic circles, in creating theories for the social sciences, we have attempted through the centuries to define and clarify the apparent connections between things (objects in the physical world) and concepts (objects within our mental world). This empirical (or, perhaps, positivist) approach has not met with great success. There is no evidence that the theory-based practices of the social sciences are more effective now than they were 50 years ago. Indeed, practitioners call loudly for useful theories (when they bother to look our way at all) and studies across a range of fields show that our theories are of very limited value (as noted above).

To summarize, more useful understanding may be gained by comparing things that are more similar. That is to say, if we want to make effective theories, it seems that they should be constructed of concepts that exist at the same level on some scale of abstraction. That approach optimizes the opportunity for insights to emerge that will add to our knowledge and the usefulness of the theories. Or, to phrase it a bit differently, a conceptual system such as a theory includes interrelated concepts. The closer those concepts are along some scale of abstraction, the more potential the theory has as an insight-generation device. The traditional approach of focusing on the relationship between objects and concepts (in empirical research) may be necessary; however, by itself it is clearly insufficient to the task of developing more effective theories. We must use both if we are to build effective theories.

As a point of proof about the comparability of concepts, I must risk a recursion that is potentially amusing, thought-provoking, and informative. Very briefly, in this paper, I have presented a number of concepts for comparison. Because you, the reader, have compared them in your mind, you have already conducted the appropriate experiment.

It is worth mentioning, that each concept may be understood as an attribute variable. And, that each person has different models. Thus, by better understanding the interrelationships between those concepts, we gain new insights into how to support multidisciplinary integration research. Indeed, this kind of approach opens a new door for integrating theories within and between disciplines to impel the sciences toward an interdisciplinary direction. Following this approach, a hypothetical group of scholars should first seek to shift their individual theories toward the same level of abstraction before striving to link them. The resulting, more effectively integrated theories, would pave the way for more effective interdisciplinary collaboration and greater effectiveness in practical application.

Comparisons

An example may be found in comparing two conceptual systems from the physical sciences; one that is effective in practical application and one that is not. About 1600, Gilbert developed a theory of electrostatic attraction that included a diverse set of abstractions. These included very concrete concepts (e.g., "magnets"), more abstract concepts (e.g., "form"), and some (e.g., "attraction") that may be somewhere in-between (Roller & Roller, 1954). That theory was not nearly as useful as the Coulomb's conceptual developed in the following century which included "distance," "force," and "charge;" arguably existing closer to one another on some scale of abstraction. What we now know as Coulomb's Law is far more useful than Gilbert's conceptual system. Coulomb's Law supports the design of electronic devices including cell phones and computers—far outstripping the usefulness of Gilbert's version.

In the social sciences, it is difficult to compare ineffective conceptual systems with effective ones. There are many of the former and vanishingly few of the later. Organizational learning theory as an example has not been shown to be highly successful (Wallis, 2009a). There, the abstraction of concepts range from the very abstract (e.g., "formalization of knowledge") to the rather concrete (e.g., "the creation of new firms).

A rare exception may be found in the comparison of two economic policies where one failed and one succeeded (Wallis, 2011). The failed policy (Prices and Income Accord—implemented in Australia in early 1980s) consists of a range of concepts from "centralized wage control" which may be seen as rather abstract (especially because it is contrasted in the policy with piecemeal wage control) to "spending" which is arguably more concrete. The more successful policy (Wassenaar Accord—implemented in the Netherlands about the same time) contained relatively few concepts (including economic recovery, competitiveness, price levels, and others). The Netherland's version may be seen as existing at a similar level of abstraction compared to the Australian version. However, more studies are necessary to clarify and quantify the relationship between similarities of abstraction and success.

Summary, Action and Conclusion

To summarize, this paper adopts a science of conceptual systems perspective and suggests that it is useful for developing a new understanding of theory. This applies to social and behavioral sciences such as business, psychology, sociology, economics, and policy. It may also apply to other fields and sciences. This paper first argued that a well-constructed theory should contain concepts that are of the same level of abstraction. For example, a theory should not contain one concept with a low level of abstraction (e.g., pippin apple), and another concept is more abstract (e.g., fruit). This implies that the internal validity of the theory will be enhanced to the extent that all the concepts are at the same level of abstraction. I know of no studies that have examined this relationship. So, it seems the gates are open for a new stream of research in this emerging science of conceptual systems.

Another opportunity for action involves the creation of a method to analyze and measure the levels of abstraction with some degree of objectivity. And, thereby, provide a new method for scholars and editors to evaluate the internal validity of a theory under construction or under submission to a journal. Such a methodology might reflect, for example, the percentage of all concepts within the theory that are at the same level of scale. That percentage, in turn, might be correlated with the scale of abstraction of the theory as a whole—as it relates to the world at large (unit level to grand level). This relationship would indicate the extent to which a theory remains true to its stated role. This kind of measurement might also be useful in creating a "periodic table of theories" as originally suggested by Feuer (1995). Here, I deepen and extend his idea by suggesting that such a table could include scale of abstraction along with other relevant features of the theory such as complexity and systemicity (Wallis, 2012).

This paper leaves open an important question about how we might build theories that work between levels of abstraction. The main suggestion I have in that regard is that we should develop a new understanding of theory—where theories at one level of scale (e.g., apples) are somehow nested within theories of another scale (e.g., fruit). The connection between levels, then, might be understood as an emergent property.

We might also seek to understand potential benefits of linking concepts at the very highest levels of abstraction, or the very lowest. And, we may seek to understand the benefits of creating conceptual systems with a greater number of concepts, fewer concepts, and those concepts which are linked through causal relationships (e.g., Wallis, 2013) to identify which may be more useful in practical application.

The above conversation around the scale of abstraction seems useful (and, potentially, important) because it provides scholars with a relatively objective approach to create more effective theory and evaluate our theory more effectively. Such theories are expected to be more useful and effective in practical application. By developing more rigorous approaches to understanding conceptual systems, we can learn to go beyond the empirical/positivist approach of Newton and leap forward to the poly-conceptual approach of Einstein.

References

Ackermann, F., & Eden, C. (2004). Using causal mapping - individual and group, traditional and new. In M. Pidd (ed.), *Systems Modeling: Theory and Practice* (pp. 127-143). Chichester, UK: john Wiley & Sons.

Ambrose, D. (1996). Unifying theories of creativity: Metaphorical thought and the unification process. *New Ideas in Psychology, 14*(3), 257-267.

Appelbaum, R. P. (1970). *Theories of Social Change*. Chicago: Markham.

Axelrod, R., & Cohen, M. D. (2000). *Harnessing Complexity: Organizational Implications of a Scientific Frontier*. New York: Basic Books.

Baake, K. (2003). *Metaphor and Knowledge: The Challenges of Writing Science* (Studies in Scientific and Technical Communication): State University of New York Press.

Bernier, L., & Hafsi, T. (2007). The changing nature of public entrepreneurship. *Public Administration Review, 67*(3), 488-503.

Boudon, R. (1986). *Theories of Social Change* (J. C. Whitehouse, Trans.). Cambridge, UK: Polity Press.

Burrell, G. (1997). *Pandemonium: Towards a Retro-Organizational Theory*. Thousand Oaks, California: Sage.

Curseu, P., Schalk, R., & Schruijer, S. (2010). The use of cognitive mapping in eliciting and evaluating group cognitions. *Journal of Applied Social Psychology, 40*(5), 1258-1291.

Daneke, G., A. (1999). *Systemic Choices: Nonlinear Dynamics and Practical Management*. Ann Arbor: The University of Michigan Press.

Daneke, G. A. (1997). From metaphor to method: Nonlinear science and practical management. *International Journal of Organizational Analysis, 5*(3), 249.

Dekkers (2008). Adapting organizations: The instance of Business process reengineering. *Systems Research and Behavioral Science, 25*(1).

Dubin, R. (1978). *Theory building* (Revised ed.). New York: The Free Press.

Fedigan, L. (1973). Conceptual systems theory and teaching. *Educational Leadership, 30*(8), 765-968.

Feuer, L. S. (1995). *Varieties of scientific experience: Emotive aims in scientific hypotheses*. Piscataway, NJ: Transaction.

Fodor, J. A. (1998). *Concepts: Where Cognitive Science Went Wrong* (Oxford Cognitive Science). new York: Oxford University Press.

Friedman, K. (2003). Theory construction in design research: Criteria: Approaches, and methods. *Design Studies, 24*(6), 507-522.

Fuller, T., & Moran, P. (2000). Moving beyond metaphor. *Emergence, 2*(1), 50-71.

Ghoshal, S. (2005). Bad management theories are destroying good management practices. *Academy of Management Learning & Education, 4*(1), 75-91.

Greenhalgh, T., Robert, g., Macfarlane, F., Bate, P., Kyriakidou, O., & Peacock, R. (2005). Storylines of research in diffusion of innovation: A meta-narrative approach to systematic review. *Social Science and Medicine, 61*, 417-430.

Hammond, D. (2003). *The Science of Synthesis: Exploring the Social Implications of General Systems Theory*. Boulder, Colorado: University Press.

Hatch, M. J., & Yanow, D. (2008). Methodology by Metaphor: Ways of Seeing in Painting and Research. *Organization Studies, 29*(1), 23-44, doi:10.1177/0170840607086635.

Hung, D. W. L. (2002). Metaphorical ideas as mediating artifacts for the social construction of knowledge: Implications from the writings of Dewey and Vygotsky. *International Journal of Instructional Media, 29*(2), 197.

James, W. (1909). *A Pluralistic Universe* (Hibbert Lectures at Manchester College on the Present Situation in Philosophy). Manchester, UK.

Kaplan, A. (1964). *The conduct of inquiry: Methodology for behavioral science* (Chandler Publications in Anthropology and Sociology). San Francisco: Chandler Publishing Company.

Kessler, E. H. (2001). The idols of organizational theory from Francis Bacon to the Dilbert Principle. [Essay]. *Journal of Management Inquiry, 10*(4), 285-297.

Kuipers, B. (1982). The "map in the head" metaphor. *Environment and Behavior, 14*(2), 202-220.

Lakoff, G., & Johnson, M. (1980). *Metaphors We Live By*: University of Chicago Press.

Letiche, H., & van Uden, J. (1998). Answers to a Discussion Note: On the 'Metaphor of the Metaphor.' *Organization Studies, 19*(6), 1029-1033, doi:10.1177/017084069801900606.

MacIntosh, R., & MacLean, D. (1999). Conditioned emergence: A dissipative structures approach to transformation. *Strategic Management Journal, 20*(4), 297-316.

Morgan, G. (1980). Paradigms, Metaphors, and Puzzle Solving in Organization Theory. *Administrative Science Quarterly, 25*(4), 605-622.

Morgeson, F. P., & Hofmann, D. A. (1999). The structure and function of collective constructs: Implications for multilevel research and theory development. *Academy of Management Review, 24*(2), 249-265.

Ostroff, C., & Bowen, D. E. (2000). Moving HR to a higher level: HR practices and organizational effectiveness. In K. J. Klein, & S. W. J. Kozlowski (Eds.), *Multilevel theory, research, and methods in organizations: Foundations, extensions, and new directions* (pp. 211-266). San Francisco: Jossey-Bass.

Phillips, B., & Johnstone, L. (2007). *The Invisible Crisis of Modern Sociology: Reconstructing Sociology's Fundamental. Boulder, Colorado: Paradigm.

Quine, W. V. O. (1980). *From a logical point of view* (2, Revised ed.). Cambridge, MA: Harvard University Press.

Robbins, D. (2000). *Vygotsky's Psychology-Philosophy: A Metaphor for Language Theory and Learning*. New York: Kluwer.

Rogers, P. J. (2008). Using programme theory to evaluate complicated and complex aspects of interventions. *Evaluation, 14*(1), 29.

Roller, D., & Roller, D. H. D. (1954). *The Development of the Concept of Electric Charge: Electricity from the Greeks to Coulomb* (Vol. 8, Harvard Case Histories in Experimental Science). Cambridge, Mass: Harvard University Press.

Seabury, M. B. (1991). Critical thinking via the abstraction ladder. *English Journal, 80*(2), 44-49.

Shotter, J. (2005). Inside the moment of managing: Wittgenstein and the everyday dynamics of our expressive-responsive activities. *Organization Studies, 26*(1), 113-135.

Smith, M. E. (2003). Changing an organization's culture: Correlates of success and failure. *Leadership & Organization Development Journal, 24*(5), 249-261.

Speicher, M., Marga (1997). Theory, metatheory, metaphor--introduction. *Clinical social work journal, 25*(1), 7-9.

Stacey, R. D. (1996). *Complexity and Creativity in Organizations*. San Francisco: Berrett-Koehler Publishers, Inc.

Tetlock, P. E. (1985). Integrative complexity of American and Soviet foreign policy rhetoric: A time-series analysis. *Journal of Personality and Social Psychology, 49*(6), 1565-1585.

Thagard, P., & Stewart, T. C. (2011). The AHA! experience: Creativity through emergent binding in neural networks. *Cognitive Science, 35*(1), 1-33.

Uzzi, B., & Spiro, J. (2005). Collaboration and creativity: The small world problem. *American Journal of Sociology, 111*(2), 447-504.

Wallis, S. E. (2008). Validation of theory: Exploring and reframing Popper's worlds. *Integral Review, 4*(2), 71-91.

Wallis, S. E. (2009a). Seeking the robust core of organizational learning theory. *International Journal of Collaborative Enterprise, 1*(2), 180-193.

Wallis, S. E. (2009b). Seeking the robust core of social entrepreneurship theory. In J. A. Goldstein, J. K. Hazy, & J. Silberstang (eds.), *Social Entrepreneurship & Complexity*. Litchfield Park, AZ: ISCE Publishing.

Wallis, S. E. (2010a). The structure of theory and the structure of scientific revolutions: What constitutes an advance in theory? In S. E. Wallis (ed.), *Cybernetics and systems theory in management: Views, tools, and advancements* (pp. 151-174). Hershey, PA: IGI Global.

Wallis, S. E. (2010b). Toward a science of metatheory. *Integral Review, 6*(Special Issue: "Emerging Perspectives of Metatheory and Theory").

Wallis, S. E. (2010c). Towards developing effective ethics for effective behavior. *Social Responsibility Journal, 6*(4), 536-550.

Wallis, S. E. (2011). *Avoiding policy failure: A workable approach*. Litchfield Park, AZ: Emergent Publications.

Wallis, S. E. (2012). A Systems Approach to Understanding Theory: Finding the Core, Identifying Opportunities for Improvement, and Integrating Fragmented Fields. *Systems Research and Behavioral Science*, 31(1): 23-31.

Wallis, S. E. (2013). How to choose between policy proposals: A simple tool based on systems thinking and complexity theory. *Emergence: Complexity & Organization, 15*(3), 94-120.

Chapter 12 Orthogonality— Developing a Structural/Perspectival Approach for Improving Theoretical Models

Wallis, S. E. (2020). Orthogonality: Developing a structural/ perspectival approach for improving theoretical models. Systems Research and Behavioral Science, 37(2), 345-359.

My thanks to the two anonymous reviewers whose comments and suggestions led to a much improved paper. Any and all remaining problems remain my own.

Abstract: *In order to create more useful theories we must attempt to choose better concepts (variables). Although we may strive to identify "independent" variables, that effort is problematic because we live in a world where everything is connected; also, attempting to obtain even nominally independent variables may require sophisticated analytical methodologies available only to well-funded researchers and practitioners. To help resolve those problems, and so provide an improved path for developing more useful theories, we explore the notion of orthogonality as a tool for rethinking concepts in a theory, from a structural perspective, in a way that clarifies research opportunities (from a data perspective) and clarifies situational/theoretical perspectives (from a relevance perspective).*

Introduction

In science, understood as a process (Hull, 1988), key tasks include gathering data (through research and practice), building theory, and communicating the results to others. Within the social/behavioral sciences there are effective/rigorous guides for conducting research (cf. Argyris, Putnam, & Smith, 1985; Bentz & Shapiro, 1998; Charmaz, 2006; Checkland, 2000; Yin, 2011), engaging in ethical practice (cf. Hammersley, 2003), and communicating the results in peer-reviewed journals (cf. Armstrong, 1997).

For building theories (as a form of knowledge) that are more useful in practical application for reaching goals, it is helpful to understand theories as systems of concepts (cf. Umpleby, 1997), having a kind of causal logic structure. Recent research focused on that structural perspective identifies new paths for rigorously evaluating and improving theories (Wallis, 2016a). Approaches to evaluating structure include measures of Systematicity (cf. Gentner & Toupin, 1986), Integrative Complexity (cf. Suedfeld, Tetlock, & Streufert, 1992), Integrative Propositional Analysis (cf. Wallis, 2016a; Wallis & Wright, 2019), loops (Wallis, 2019), abstractions (Wallis, 2014), and sub-structures (Wallis, 2016b).

Theories are more useful for enabling successful decisions and actions when they have more supporting data, greater meaning or relevance to the stakeholders and their situation, and higher levels of causal logic structure (Wright & Wallis, 2019). While appreciating the importance of data and relevance, in this paper we focus on that structural perspective to explore and develop an approach to evaluating and improving theories based on the concept of "orthogonality." Briefly, as this will be explored in greater depth below, orthogonality is an expression of difference between two concepts specifically with respect to their causal relationships representing some transformation (not merely that they are different).

For a simple example, Figure 1 abstractly represents two concepts (A and B) that, together, cause change in a third concept (C).

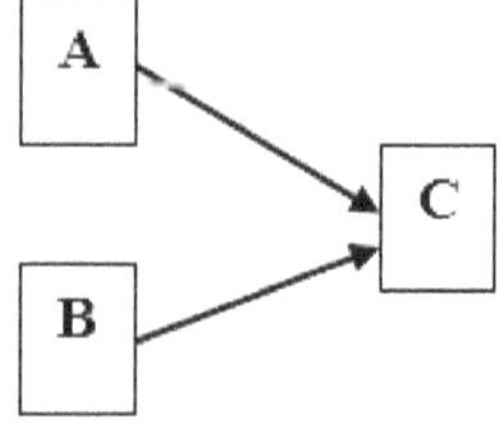

Figure 1 *Simple, abstract, causal theory (boxes represent concepts/variables, arrows represent causal relationships)*

The extent to which A and B are uncorrelated reflects, to some extent, our understanding of how change occurs in C. Both (and perhaps more) are required for change in C.

Alternatively, if there is some correlation between A and B (represented in Figure 2) our understanding is not so clear. Graphically, that correlation may be represented by overlapping boxes (Figure 2) reflecting some commonality between them.

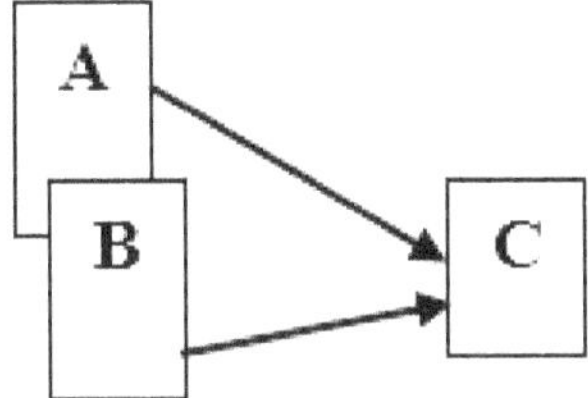

Figure 2 *Overlap between concepts reducing clarity of understanding*

Individual and teams are highly effective at identifying things, applying labels, and reaching conclusions. However, this kind of process often results in a pretense of knowledge with accompanying confusion, poor policy choices, and social turmoil (Hayek, 1974) rather than generating more useful patterns of understandings (Lissack, 2019). Even when this is done in the context of academic research, this process has led to fragmentation and confusion rather than clear advances in science and practice (cf. Newell, 2007).

Wallis (2016a) has surfaced this concern recently, suggesting that "If we can create new rules for evaluating theories based on the orthogonality of causal inputs (and outputs), we will have made a very large step forward in understanding how to create more useful theories" (p. 594). However, it seems that we have no great understanding of orthogonality—no way to say with any level of objectivity or rigorous process that pairs of concepts are "more orthogonal" than another so there is no "compass" to guide the theory builder.

This is important because orthogonality is one way we understand things as being different, yet connected. As such, it is a critical part of our understanding our world of things that are "differentiable yet inseparable" (Zude & Yolles, 2006).

If we can find better paths to evaluating and building theories, we may be able to more easily to determine the most useful theories for teaching, publication, and application. Finding such gaps in our theories to identify opportunities for improved research and practice, this structural approach to orthogonality in theory building, may serve as a science accelerator with attendant benefits for humanity.

Definitions

Before proceeding to greater depths, some relevant definitions are in order.

Theory

Generally, we may think of a theory as a system because:

> *...a system is a complex object whose parts or components are held together by bonds of some kind. These bonds are logical in the case of a conceptual system, such as a theory; and they are material in the case of a concrete system, such as an atom, cell, immune system, family, or hospital. The collection of all such relations among a system's constituents is its structure (or organization, or architecture).* (Bunge, 2004: 188).

Therefore, for this paper, a theory may be understood as a system of concepts (variables, categories, or abstractions of the real world) linked by causal connections. Other conceptual systems are models, theoretical models, mental models, policy models, metatheories, theories of the middle range, policy theories, sets of hypotheses, or any other similar thing which may be used as a framework of knowledge for understanding and engaging the world (Wallis, 2016a). For our purposes, a better theory is one that is more useful in practical application for reaching specified goals. Through primary and secondary research, such theories are created by building on higher levels of supporting data, greater relevance to users, and more systemic structure (Wright & Wallis, 2019). One approach to understanding and improving structure is orthogonality.

Orthogonality

For a brief overview (we will go into greater depth later) the term orthogonal is used in a rather imprecise, casual, and often incorrect way to present something that is "different." For example, (Peters, 2019) describes organizations (e.g., military) as having a dominant culture (e.g., work ethic) while also having sub-cultures (e.g., slacker) that are subversive to the dominant culture. He sees the difference between those cultures as orthogonal. On the surface, that use of the term seems to imply that the two cultures are "very different." Such is not a highly useful distinction. After all, a star in the sky is very different from a set of accounting principles used in a business; yet, the two would not (or should not) be considered orthogonal unless someone were able to show that the two are causal and necessary for some resulting phenomenon.

If Peters were to see the different cultures as combining to create something new (e.g., the conflict between them) then the cultures might be considered (to some extent) orthogonal. Instead, his discussion was on the differences between them as a source of different perspectives and the desired state of having cultural commonality or uniformity within an institution (Peters, 2019). Instead, because culture may include more things than perspectives (e.g., people, artefacts, and activities) there may be considerable overlap between the many aspects of the two cultures thus limiting our understanding of the situation. A racist may perceive substantial differences between individuals of different skin color while a psychologist may perceive few or none.

Here, we want to avoid having "orthogonality" become synonymous with "different" and instead develop a perspective of orthogonality that seeks to understand two or more concepts that are different and causally interrelated in some way.

That dual perspective (à la Bateson, 1979) of difference and relationship gives us the opportunity to see our conceptual systems with greater clarity and so develop more useful theories.

In the academic world, a common way to understand differences between concepts in a theory is to identify them as independent variables. The importance of having variables be independent from one another is generally accepted in theory building (cf.

Ahn, 1997). However, choosing independent variables or concepts is often difficult for scholars and practitioners, requiring statistical techniques such as the R-squared test which, although it is widely used, is often misused (Kvålseth, 1985).

So, even a comprehensive research project of vast scope (and cost) may not provide a highly useful reflection of reality. Also, such empirical approaches do not lend themselves to qualitative research or to the activities of politicians, managers, and others who must make decisions on a daily basis without having access to sufficient empirical data, statistical methodologies, or expert scholars. The difficulty is accentuated when conducting qualitative research and still greater when developing speculative theories because of the difficulty of intuitively choosing things that seem to be completely different in a world where everything seems to be connected.

Metaphorically, we might imagine designing a skyscraper where the plans describe clear differences between structural components but only vague descriptions of how they are to be assembled. Rather than placing girders at specific angles required to provide structural integrity, they end up instead being connected haphazardly; and, unable to support their own weight, collapse.

In short, to develop more useful theories (with better structure), we should better understand the orthogonality between concepts in each theory. And, importantly, the methods developed from that understanding should be accessible to a wide range of scholars and practitioners.

Steps to resolving the problems

The primary goal of this paper is to explore the meaning of orthogonality as it may be applied to understanding and improving conceptual systems such as theories. With that improved understanding, another goal of the present paper is to make it easier for scholars and practitioners (including emerging experts and citizen scientists) to advance or improve theory for practical application.

This is important for choosing concepts for theories that relate to real world phenomena because that process is often accompanied by a high level of ambiguity. "By ambiguity, I mean the co-existence of multiple interpretations of a phenomenon among reasonable people while there is not necessarily an easy way to choose between interpretations or eliminate some of them" (Dahler-Larsen, 2018: 7).

In the following section, we begin by briefly considering existing thinking on the closely related areas of causality and abstraction (understood as a kind of categorization) for building theory, then considering in some depth existing thinking on orthogonality from perspectives found in multiple fields. Next, we explore the implications of those perspectives on orthogonality as they may be applied to building theory. After that, we present a set of instructions for using orthogonality as a tool for theory building, and offer a range of questions and opportunities for continuing this stream of research.

Existing Thinking

Metaphorically, when building a theory, we are striving to assemble a panorama from a multitude of photographs. While it is possible to gain *some* kind of view from any combination of photographs, the best view is one where we are aware of (and deal appropriately with) gaps and overlaps among the individual photos.

It is not enough to 'simply' create a better panorama by including more photos (ending up with some sloppy bricolage); nor is it enough to create a better theory by including concepts. "Indeed, the art of dealing with the complicated and complex real world lies in knowing when to simplify and when, and how, to complicate" (Rogers, 2008: 30). In this section, to better understand that art, with an eye toward creating more of a science, we briefly review some existing thinking on concepts (or categories) and causality before touching on the idea of relevance and then taking a deeper dive into understanding the existing thinking around orthogonality from multiple fields.

Concepts and Categories

For our purposes, a concept (sometimes referred to as a category) is anything that can be named (e.g., atoms, cars, hats, planets, along with various activities and so on). The thing may be a concrete object in the real world or something more abstract such as "democracy." Concepts are, by nature, abstractions.

Many terms have been used to identify such things including concept, category, aspect, vertex, variable, point, object, node, and so on. More philosophically, they may be referred to as "normal types" or "ideal types" (Fink, 2014). For the present paper we may use those terms interchangeably but will generally rely on the term "concept."

Additionally, and importantly for our present discussion of orthogonality, each concept *may* have multiple sub-concepts in much the same way that a physical object has parts. For example, the concept of "productive worker" may be said to include sub-concepts of productivity and person.

While each concept is an abstraction, it is worth noting that each is at some *level of abstraction*. For example, apple trees, fruit trees, and trees represent three levels of abstraction where they (also understood as categories) may hold the same "things." A theory will be more useful when its concepts are at a similar level of abstraction (Wallis, 2014) whether they are chosen from the start of a study or changed to be more similar as a result of developing the theory.

A theory that contains concepts of both "fruit" with "apples" may lead to confusion. For example, imagine someone asks you how many apples farmer Jane has available for sale that day. You call the farm and are told that there is one basket labeled "apples" containing ten items and another basket labeled "fruit" containing 10 items. There may be ten apples, there may be twenty... or somewhere in between.

Moving too far up some scale of abstraction may not be useful. A farmer, for example, might want to work at a level of abstraction where she categorizes "male cattle" and "female cattle" for the purposes of breeding, milking, and so on. If she were to categorize on a higher level, such as "animal," the corral might be filled with all kinds of creatures—leading to pandemonium instead of a functional and productive farm.

While such concrete things are easy to differentiate in daily life, theories may contain concepts that are less clear, such as democracy or freedom. So, an important point to consider is that the concepts in a theory should be "mutually exclusive" in the sense that there should be no overlap between them. This reduces the chance of misunderstanding and confusion on the farm (this will be discussed in greater detail, and with fewer metaphors, below).

Another important point is that such abstractions of concepts are not known in advance. Instead, in qualitative research in the social/behavioral sciences, it is common to create themes or categories when coding and "clumping" qualitative data as those themes seem to emerge from the data as perceived by the researcher (cf. Goulding, 2002; Yin, 1984). In contrast, through history, some have attempted to create "perfect" or "ultimate" lists of concepts. However, they have not gained wide acceptance.

Just to briefly mention some examples from philosophy, Aristotle lists the following as the ten highest categories of things: "Substance (e.g., man, horse), Quantity (e.g., four-foot, five-foot), Quality (e.g., white, grammatical), Relation (e.g., double, half), Place (e.g., in the Lyceum, in the market-place), Date (e.g., yesterday, last year), Posture (e.g., is lying, is sitting), State (e.g., has shoes on, has armor on), Action (e.g., cutting, burning), Passion (e.g., being cut, being burned)" (link).

In contrast, Grossman identifies eight highest categories: "Individuals, Properties, Relations, Classes, Structures, Quantifiers, Facts, Negation". A Hindu perspective suggests seven: "Individuals, Properties, Relations, Classes, Structures, Quantifiers, Facts, Negation". A Hindu perspective suggests seven: "Ganeri: substance (dravya), quality (guṇa), motion or action (karma), universal (sāmānya), particularity or differentiator (viśeṣa), inherence (samavāya), and absence (abhāva)".

Einstein might have said that there were two categories, matter and energy. And, those are essentially the same thing! Seven, eight, ten, two, one. We do not seem to have a definitive answer as to what that the appropriate number of categories actually is (or what those categories should be).

More pragmatically, for organizational improvement, it is common to use pre-determined categories for planning purposes. For example, the SWOT model (Dyson, 2004) suggests that there are four categories (Strengths, Weaknesses, Opportunities, Threats). The "business model canvas" (Joyce & Paquin, 2016) includes nine categories (Partners, Activities, Resources, Value propositions, Channels, Customer relationship, Customer segments, Costs, Revenues) while the "balanced scorecard" has eight (Kaplan & Norton, 1996). Four, eight, nine... which is the appropriate number, if any (and what should they be)?

As noted above, having overlaps between concepts/categories is more likely to engender confusion rather than understanding. So, considering categories is relevant for an orthogonal perspective because a problem arises when one thing shows up in two or more categories (concepts). For an example from the business world, consider the four categories of SWOT analysis. Imagine how something listed as a strength might have considerable overlap with something listed as an opportunity (e.g., a strength is our ability to find new customers while an opportunity is finding new customers). One might even list something like "intense focus on a task" as both a strength (ability to get work done) and a weakness (inability to shift focus when necessary). That overlap seems to provide at least a partial explanation, from a structural perspective, as to why SWOT has not proven highly useful for supporting successful business decisions (Hill & Westbrook, 1997). There is confusion on the farm due to insufficient differentiation between concepts.

Relevance / Meaning

While some perfect list of concepts may be someday found (a kind of theory of everything, or TOE), few scholars today are actually seeking such a thing. Instead, most would be satisfied to discover a set of concepts that help to understand and support a better path toward organizational improvement or improved national policy. That perspective suggests we should focus on identifying concepts that are relevant to each stakeholder's specific situation. That is, those concepts that the stakeholders find most meaningful (Wright & Wallis, 2019). And, historically, that is what has been done to support successful decisions. For example, an engineer who is calculating the orbit for a communication satellite uses Newton's laws of motion, ignoring Aristotle's other categories (e.g., passion).

We agree with Westerhoff (2005) that there is no absolute, perfect, ultimate, final set of categories or concepts. Instead, the set of concepts within a theory should be chosen by the researcher—at least as a starting point. While every individual and group must have some perspective, a problem arises when different stakeholder groups build their understandings on concepts that have unarticulated overlaps (above section) and/or poorly understood causal connections between them (below section).

Causality

There has been so much written on the topic of causality and its relationships to developing useful theories that it hardly seems worth writing more (see especially Pearl, 2000). However, in conversations with colleagues we routinely hear that they deploy techniques for understanding a situation or supporting clients such as "concept mapping," "bricolage," "storytelling," and "rich description" as substitutes for understanding causal relationships between concepts.

While those techniques are certainly useful to some extent (e.g., supporting generative conversations) research in the science of conceptual systems indicates that representations are more useful for understanding situations and enacting desired change when they (as theories, models, knowledge, etc.) include measurable concepts related with causal connections (Wallis, 2016a).

In short, it seems to be increasingly accepted that theories are more useful when they indicate causal relationships (cf. Ackermann & Eden, 2004; Bauer, Booth, & McGroarty-Torres, 2016; Bryson, Ackermann, Eden, & Finn, 2004; Bureš, 2017; Deegan, 2011; Goodier, Austin, Soetanto, & Dainty, 2010; McGlashan, Johnstone, Creighton, de la Haye, & Allender, 2016; Panetti, Parmentola, Wallis, & Ferretti, 2018; Pearl, 2000; Sloman & Hagmayer, 2006; Waldmann, Holyoak, & Fratianne, 1995).

Part of the reluctance by some to accept causal models may stem from past reliance on purely linear models (for example: resources > activities > results). Encouragingly, in the field of program evaluation (and quite possibly others), there seems to be an increasing dissatisfaction with purely linear models. There seems to be, instead, a growing realization that models with more *com-*

plex causal interconnections between those concepts will provide a more useful representation of the real world and so be more useful for planning and evaluation (Rogers, 2008; Wright & Wallis, 2019). These nonlinear models would show many activities, each of which causes changes in each other, along with complex causal connections between resources and goals.

This perspective is generally accepted within the complexity sciences, where it is reasonably held that *simple* causal connections are suspect; the world is more complex than that. That is, for an abstract example, we would be suspicious of a simple claim that only A causes only B because such a claim would contradict the idea that the world is made of many complex causal interactions. While it may be difficult to unravel those connections, we believe that it is possible. To avoid making decisions (or taking action) based on causal relationships, therefore, is to invite poor results.

Although identifying causal connections may be difficult (Sprites, Glymour, & Scheines, 1993), the benefits are worth the effort because understanding causal relationships improves our *actionable* knowledge (Johnson-Laird, 1980) necessary for effective decision making (Sloman & Hagmayer, 2006) especially in business settings (Axelrod, 1976). Also, understanding causal relationships supports long-term learning in children (Bauer et al., 2016).

For creating diagrams, specifically knowledge maps, causal relationships are presented as arrows between concepts. Such causal connections are also referred to as edges, morphisms, links, and so on. For the present paper, we will generally refer to them as arrows or causal connections.

Orthogonality

To deepen the brief introduction to orthogonality presented in the introduction section, it seems interesting to note that the term is derived from the Latin "orthogonus" meaning "right angled," orthogonal is "pertaining to or involving right angles" (from geometry) and (from statistics) "statistically independent" (OED, 2007: 2029). That is to say, when we are looking at two boxes (or independent variables) and their causal arrows pointing towards a third box (dependent variable), we are more certain that the resulting understanding is valid (and presumably useful) because those two causal variables are independent of one another.

An example from the field of psychology relates to models having been developed to describe three factors of behavior (Finkel, 2014) and the Big Five factors of personality (Saucier, 2002) where researchers understand that the greater degree of orthogonality, by which they mean independence between variables, the more valid the model and the more useful it may be for making predictions. However, the orthogonality of data is often difficult to justify on theoretical grounds (Mishler & Rose, 1997).

Although it may be challenging, in contemplating "how orthogonal" two concepts might be to one another, we push our thinking. And that is the stuff that scientific advances are made of.

Rethinking our assumptions is important for making progress because researchers of the social/behavioral sciences (as well as the humanities) are rarely in a position to conduct repeatable experiments under controlled situations. We, along with our students and clients living in a rich and complex world, require a more pragmatic guide for developing simple understandings into complex theories and eventually creating law-like conceptual structures. We need a deeper understanding—a "why" approach of logics that is different from yet related to a "what" approach of data. We must find a balance between the human desire for parsimony and the need to understand our complex world to reach challenging goals. In the world of policy implementation, there are simply too many variables in our world to completely make sense of it all and gain the policy results we expect (Goggin, 1986). And, variables that are important may not be accessible for measurement (Hayek, 1974).

A perspective from statistics is focused on finding variables that are independent or "uncorrelated" (Ahn, 1997). That is an important idea because it relates to the reliability of data in theory building as well as the construction of useful theories—based on "concatenated" logic structures (Wallis, 2016b). Briefly, a concatenated structure is one with more than one causal arrow pointing toward a concept from other concepts (Figure 1). While orthogonality between variables is a desirable goal variables of "constructs in the real world are rarely uncorrelated" (Ford, MacCallum, & Tait, 1986).

Sterner (2014) suggests that epistemology and ontology are orthogonal to one another. Importantly, he notes that orthogonality is different from opposition. That is, concepts which are orthogonal are 90 degrees apart, they are not opposites. He states, "Much needs to be done, though, before the idea of 'concurrent but orthogonal' can offer a full accounting of the conceptual structures underlying the debates over explanation" (p. 251). Lissack and Graber (2014) suggest that epistemology and ontology are not in opposition, rather they are orthogonal and congruent. Yet, in their explanation, they suggest that the division between the two is not so clear and that the activity around them may exist on some spectrum.

In this interconnected universe, no variables are *truly* independent. So, a better (or, at least an alternative) way to choose them might be to evaluate "how orthogonal" they are; and, for improving theory, continually strive to develop concepts within the context of a theory that are increasingly orthogonal.

Let us go back to a simpler perspective for clarity. In mathematics, an example of orthogonality may be found in any algebraic operation. For example, where A x B = C (or, more concretely, money in investments x rate of return = dividends). If A or B were reduced to zero, C would also be zero. So, A would be considered orthogonal to B. However, because A and C are both measures of money (investment and dividends), they may not be considered completely orthogonal.

A clearly non-orthogonal relationship is seen in additive operations where (for example) A + B = C (or, more concretely, (money received from investments + money received from paychecks = money in bank). A or B could be zero but C could still be non-zero).

For another example, if one were to say, "more human labor and more component parts both combine to create finished widgets" we could more easily see labor and parts as orthogonal to one another. Whatever it may be that we are measuring, the two measures are not the same thing. They are (in some sense) multiplicative instead of additive. There will be no finished widgets if either variable drops to zero. Or, from the perspective of Boolean logic, additive causal streams are "OR" operators while orthogonal ones are "AND" operators (my thanks for that insight to Stankovic, personal communication, email, 7/25/2017).

To summarize the key points of this section, in seeking to create more useful theories, concepts should be related to one another by causal connections, and should be relevant/meaningful to the users of the theory and their situation/context. Importantly, there should be no overlap between concepts (we will discuss this more below). In the following section, we conduct a deeper exploration into orthogonality to find out and clarify how an orthogonal perspective may be used for developing more useful theory.

Exploration

In creating models, it is important to consider how the constituent concepts are selected or understood within some context, and how they are differentiated, combined, and correlated (Gabora, Rosch, & Aerts, 2008) leading to a more useful model.

While orthogonality is (to some extent) used (and to some extent misused) in the literature of the social/behavioral sciences to *describe* relationships between concepts, it is generally ignored as a tool for *evaluating* our theories with any level of objectivity and so for improving them. That is, to create theory that is more useful in practical application. The process of striving towards orthogonality provides a way to re-think concepts and assumptions to bring new and better understanding to topics of study.

Ingo Pies and others use orthogonality to identify new directions of thinking to resolve conflict—to find win-win situations (Beckmann, Hielscher, & Pies, 2014; Hielscher, Pies, & Valentinov, 2012; Pies, 2016). For example, in the conflicting views/goals between environmentalists (seeking to preserve nature) and capitalists (seeking to exploit nature), an orthogonal position might be seen in the new goal of sustainability—something of interest to both sides. Their approach, combined with the systems-based approach presented in the present article, may prove a powerful tool for helping the world to better understand how to understand their world for mutual benefit instead of mutually destructive conflict.

Seabury (1991) notes that the process of abstraction involves identifying similarities and disregarding differences. For example, in developing laws of motion, Galileo learned to ignore the color of his spheres (Oh, 2014) to focus on the more relevant measurement of movement. In contrast, when looking at organizations, we may find ourselves looking at (for example) relationships between two different departments. When, in actuality, they may be abstracted in many alternative categories. To use such a broad description as "department" may disguise the few (yet important) differences while ignoring potentially confusing similarities. This orthogonal perspective provides a tool to help us differentiate those important details.

If we engage an orthogonal perspective "correctly" as we identify and clarify concepts for a theory, we will be ignoring (perhaps removing them from the theory, or re-combining them with other concepts in the theory) those concepts that are less orthogonal to one another and focusing on those concepts that are more orthogonal. For a practical example from gardening, both seeds and soil are made of physical matter. Yet, when creating a theory of gardening, and deciding which should be planted in the other, we focus on the differences between the two—the biological material of one and the nutrients of the other rather than construct a theory focused on how those two are both composed of matter.

While the orthogonality between concepts may seem clear within a simple example from gardening, it becomes devilishly difficult when we are examining a complex social topic such as economics. It may seem obvious that the concepts should relate to economic issues such as interest rates, employment rates, and so on. One thing seems certain; that to understand a more complex phenomena such as an economy, a theory will require more concepts. And, with a growing number of concepts, a greater probability of low orthogonality between some of the concepts.

In a universe where everything seems connected, we must ask if it is even possible for any two concepts is be independent. For example, looking at Ohms Law of electricity there are three concepts;

each of those is dependent from (causally resulting from) the other two; each is what we call transformative or concatenated (Wallis, 2019). Because each concept/variable is transformed by every change in input from other concept/variables, it seems that they "behave" as if they were independent variables even though they are clearly interconnected when one views the complete model. That is, in an experiment, the data might show no correlation between the two "input" variables when viewed from the "perspective" of the output variable, suggesting a better term would be "interdependent variables." Further exploration is beyond the scope of the present paper, but raises interesting possibilities for the philosophy of science.

Because everything is connected, there may be no such thing as perfect, ultimate, or "global" orthogonality. However, it seems useful to differentiate between things in the real world so it seems useful to understand that differentiation as a kind of "local orthogonality." For example, "math" doesn't care if a student balances an algebraic equation, but the professor can differentiate between right and wrong answers on the test. So, there can be learning and improvement.

Perspectival Orthogonality

Hypothetically, some ultimate list of concepts may be possible, when the topic is something at a similar "ultimate" level such as "reality" or perhaps "knowledge." However, "The uncertainty principle states that in an effort to gain precision about individual parts of a system, researchers lose precision in their knowledge of the whole system. For example, researchers can measure the momentum of a particle or its position, but both variables cannot be measured at once." (Monahan & Fisher, 2010: 360). That is to say, the researcher unavoidably interacts with the subject or topic of study. And, for our purpose, that interaction between observer and observed means that concepts will emerge and become part of the theory.

Therefore, while achieving global orthogonality seems impossible (or at least very unlikely), "perspectival orthogonality" still seems possible. And, a potentially useful notion provided that the conditions for finding orthogonality are effectively articulated.

For example, an ocean might be viewed as "raw material" to an uncaring industrialist, as a "subject" to an artist, "beauty" to vacationers, or a "preserve" to an environmentalist. Each perspective will have some orthogonal counterpart. Not only because of the perspective of the viewer, but also according to its use. So, while a shift in perspective may result in some concepts being removed (or included) to a theory, this is more about the relevance or meaningfulness of concepts within a theory according to different perspectives.

For a simple diagrammatic example we may consider a figure depicting two lines, intersecting at right angles (Figure 3). They are orthogonal—and serve as a graphical representation of two causal inputs that are orthogonal. Now, rotate the figure so that the two lines appear to be one. The perceived orthogonality is lost by a change in perspective. Or, for our purpose here, concepts that appear orthogonal in one theory may not appear orthogonal in another theory. The relevant context for concepts is the theory consisting of other concepts.

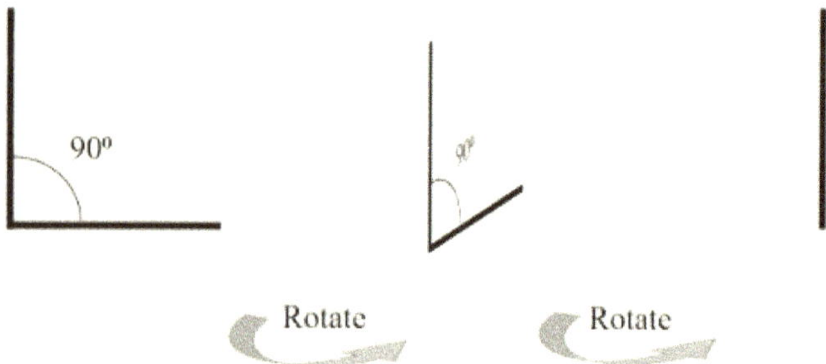

Figure 3 *Shifting/alternative perspectives confounding understanding*

In Figure 4, the center diagram (map) may represent a (presumably, for our purposes here) some optimal theory representing some part of reality. Rotating that view along an axis serves to cause an apparent increase in size of some concepts and a decrease in others as seen in the center map while rotating the figure in the other direction enhances the seeming importance of different concepts (the left and right hand diagrams of Figure 4). Taken further, when a map is rotated sufficiently, some concepts may (seemingly) completely disappear.

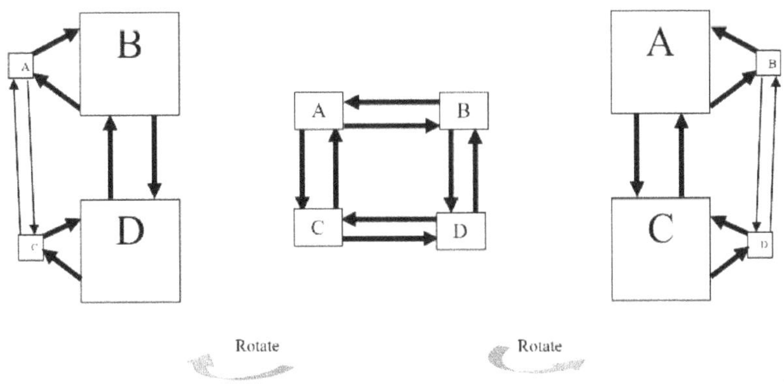

Figure 4 *Shifting/alternative perspectives of an abstract theory*

This graphical representation of perspective helps to show the view of some stakeholders that may be "stuck" in their own perspectives and unable to clearly perceive the larger picture. Such a perspective on perspectives may be extended to include changing views on causal relationships between concepts. Where, for example, one group might identify a causal relationship between two concepts, a second group believes that there is a relationship (perhaps that one influences the other) but not in a causal way, while a third group identifies no relationship at all. This perspectival approach suggests new directions for exploring and understanding how theories may be understood as having different "levels" depending on the perspective from which they are viewed or understood as "nested" within other theories. Such an exploration, however, is beyond the scope of the present paper.

While a perspectival approach opens the door for many alternative views, it should not be interpreted as a laissez-faire process. All parts of a model, from any perspective, should be supported by data (from quantitative and/or qualitative methods). Further, it is important to note that such a perspective is not the end point, once a perspective is found or chosen, it should be followed by orthogonal analysis from that perspective to identify/clarify directions for improving theories.

To summarize, there seem to be two unavoidable aspects that we should consider. The human perspective of choosing concepts that are relevant/meaningful to a topic/situation, and the structural perspective of comparing each concept with the others in the theory to improve orthogonality.

Historical Examples

About two thousand years ago, Plutarch articulated a theory of electrostatic attraction where rubbing a piece of amber was said to cause exhalations that pushed air that then pushed small objects (e.g., hair) toward the piece of amber (Wallis, 2010: 161). An orthogonal perspective would call attention to the overlap between the concepts of pushing air and pushing small objects; that is, both include pushing (which we may refer to as a sub-concept). Then, noting how such an overlap indicates an inappropriate lack of orthogonality between concepts in the theory, this perspective might suggest how the theory could be improved by removing one of those concepts. Such an insight might have inspired philosophers of the time to re-think the concepts of the theory and remove the concept of pushing air (for which they had no evidence anyway, only a strongly held assumption). And, as a result, the philosophers could have accelerated the development of science.

It is worth noting that the relationship noted above is a linear one (exhalations → pushing air → pushing small objects). In contrast, a more complex relationship may be seen in a theory of public health developed during an online conversation between novices and experts.

An orthogonal perspective draws attention to the overlaps between the concepts within the theory. In Figure 5, note that the sub-concept "people" appears in three concepts. Also, note that the concepts of healthy people and sick people may be understood as having an overlap because "healthiness" and "sickness" may be understood as two ends of one scale. That provides an opportunity to re-think or re-consider the concepts; perhaps by combining those two into a single concept such as "level of health in society."

This process may be challenging for some, testing deeply held beliefs and convenient observations. But those convenient measures of social phenomena may easily lead to problems such as economic tragedy (Hayek, 1974). However, we are not looking here for an approach that is comfortable or convenient. We are

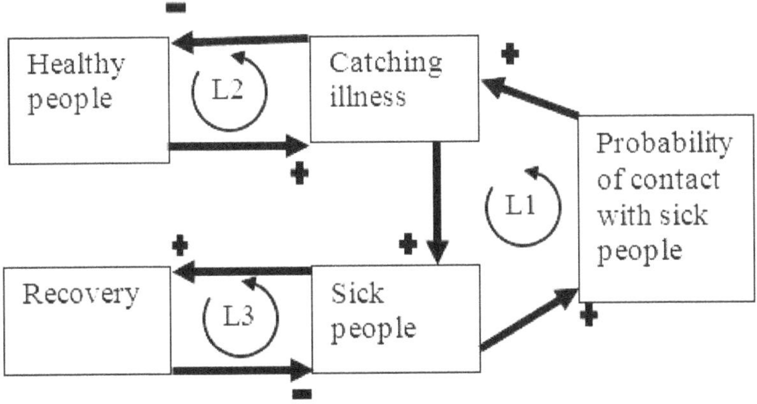

Figure 5 *Theory of public health (Wallis, 2019), (boxes indicate concepts, straight arrows indicate causal connections, curved arrows indicate feedback loops).*

seeking to develop a method that is reasonably accessible to the average researcher for evaluating the level of orthogonality between concepts of a theory and then improve that orthogonality through carefully focused efforts. Challenge is to be expected and embraced.

Historically, this process could have been used to gain insight more rapidly and so accelerate the advance of the natural sciences. Now, it could be used to accelerate the advancement of the social sciences.

A recent interdisciplinary theory of poverty integrated concepts and causal connections from ten source theories (Wallis & Wright, 2019) included 37 concepts. Within those 37, a number were identified with potentially overlapping sub-concepts. For example, one concept "education, skills, resources" seems to overlap with other concepts that may be seen as resources such as "savings," "social capital," and "child care and early education."

This kind of comparison suggests that the concepts should, in some sense, define each other. That is to say, when creating and/or evaluating a theory, we should compare each concept with the other concepts with an eye toward reducing and eventually eliminating the redundancies between sub-concepts. That, in turn, will support the development of theory that is more parsimonious and so more in line with Ockham's razor (cf. Kelly, 2007; Kelly & Mayo-Wilson, 2012).

While re-thinking concepts and revising theories seems likely to support advancements in science, that approach could be more scientific (and so more useful) if we could evaluate theories based on some *level* of orthogonality between all the concepts within a theory. Such a "scale of orthogonality" would provide a useful metric; allowing scholars to compare theories with greater objectivity and so choose the most useful theories for teaching, research, and practical application.

Scale of Orthogonality

Another way to look at it is to ask *how orthogonal* are the concepts of a theory in relationship to one another? What is our measure of differentiation?

Can we develop a "scale of orthogonality?" Where concepts within a theory may be compared to show how orthogonal they are to one another? For example, "water" and "light" both support growth of plants. And, those may be seen as highly orthogonal to one another because removing either one would result in no plant growth. In contrast, if that growth was supported by both "fluorescent light" and "sun light" those two forms of light would not be *quite so* orthogonal to one another because both share a sub-concept (light).

Figure 6 provides an abstract example of a small theory containing three concepts (A, B, C), each with sub-concepts (represented by numbers). Where those sub-concepts are the same, they are represented by the same numbers and so may be said to have some amount of overlap.

This example serves to suggest that we can create a scale of orthogonality; at least in a very simple scale of high/medium/low. Future studies may develop a more sophisticated scale.

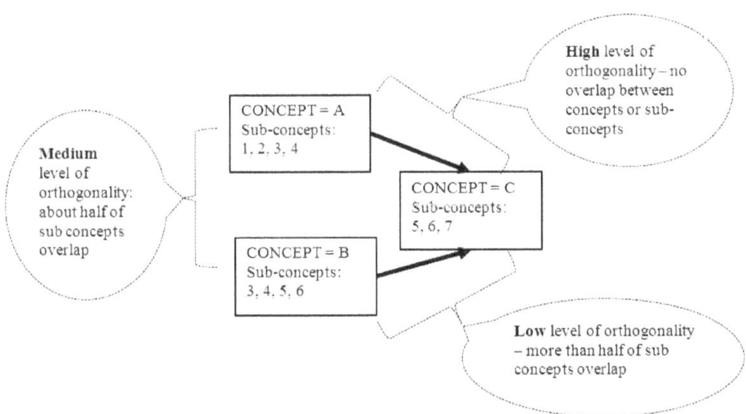

Figure 6 *Concepts and sub-concepts of a simple theory*

Future Directions

From this preliminary exploration, additional paths may be found. In this section, we will identify some of those paths for future research and consider some applications of orthogonality in practice.

In Practice

While some explanations of orthogonality may be complex and nuanced (if not occasionally confusing), some recommendations for using orthogonality in building theories are reasonably direct.

The first and easiest test for orthogonality between causally connected concepts in a theory is to conduct a thought experiment. Consider a conceptual structure where there are two or more input concepts and one output (e.g., Figure 6). If one of those input concepts is reduced to zero, then the resulting "output" concept should also go to zero, if the two inputs are orthogonal to one another. For example, if there is an assembly line where widgets are being made, the assembly requires both parts and labor. If one or the other is removed, the output of widgets will also drop to zero. If the construct passes this "zero-test," it would seem that the inputs are orthogonal.

The second, more nuanced, approach is to identify all the concepts and all their sub-concepts within the theory. Then, re-think and revise them to reduce or eliminate overlaps. For this process, it may be useful to write concepts on large cards, then, on smaller cards or sticky notes attached to or adjacent to each large card, write the various sub-concepts. Then, it becomes a relatively straightforward process to identify the aspects in common across multiple concepts—and so identify what aspects each concept has in common and therefore how orthogonal the concepts are to one another.

Additionally, it is also suggested that collaboration between researchers may lead to more rapid and interesting evaluations of orthogonality. Existing methods for developing theory (e.g., grounded theory, qualitative research) are useful in that they generate many concepts (sometimes defined as categories, themes, etc.). The present paper suggests that those kinds of analyses should be followed by an analysis of the orthogonality between the concepts.

Rather than starting with chosen concepts, and deconstructing them into orthogonal parts (as in the above paragraph, this section), let's return to categories (as discussed in a previous section); or, as they may also be understood, "classifications" (cf. Szostak, 2012). It may be a useful process to consider the concepts of a theory and ask if they may be placed into the same category/classification (e.g., both may be considered "activities"). If so, their orthogonality may be in question and some re-thinking may be indicated.

Again, the re-thinking depends in part on the level of analysis or level of abstraction of the concepts in the theory. So, it might make sense to include two activities if they pass the above zero-test. However, we should not rush to accept a pretense of orthogonality merely for the sake of convenience when re-thinking may lead to significantly useful insights.

Orthogonality may play a useful role in advancing public discourse around large and fuzzy concepts such as national "greatness" (cf. Darwish, Magdy, & Zanouda, 2017) or "transparency in government" (cf. Valentinov, Verschraegen, & Van Assche, 2019). Such notions often grab public attention and seem to become fixed in peoples' minds. As such, some individuals may seem "impervious" to rational debate (Wallis & Valentinov, 2016) which may lead to conflicts between groups; each with their tightly held (yet large and fuzzy) assumptions. Utilizing the orthogonality perspective developed here, it may be possible to deconstruct and reconstruct such complex notions into their interrelated parts.

This orthogonal perspective suggests that some techniques (such as rich descriptions and concept mapping) may be more useful for surfacing insights than generating knowledge. So, we may consider them part of a two-phase process where such techniques are followed by an iterative process of cycling between structural development techniques. Generally, these structural approaches may be understood as IPAx (experimental techniques of the broader Integrative Propositional Analysis methodology) (Wallis, 2019). It is expected that orthogonality will be part of an iterative process, conducted in conjunction with other structural methods such as measures of systemic depth, conceptual breadth, loops (Wallis, 2019) and abstraction (Wallis, 2014) in addition to measures of data and stakeholder relevance (Wright & Wallis, 2019).

Future research

Ontologically, understanding what something "is" (or what dynamics are occurring) may seem to be a straightforward process of making an observation and applying a label. However, we can only gain clarity about what something is by understanding what it is *and* how it causes change. That distinction is what we are striving for in developing more useful theories from an orthogonal perspective.

This exploration has served to surface and develop some preliminary answers to important questions on building more useful theory. More effort is needed before the orthogonal approach is fully developed.

As discussed above, the importance of concatenated structures (e.g., Figure 1) has been established as an improvement over purely linear theories. However, linear theories may serve as a starting point for developing theories of greater systemic structure. When starting with a linear structure, it may be useful in some settings to brainstorm new concepts to add to the theory. In such situations, practitioners and researchers may want to use categories (as discussed above) to support the brainstorming process by purposefully seeking concepts from alternative categories. For example, if some existing theory has three concepts from the business model canvas (Joyce & Paquin, 2016) (e.g., partners, activities, resources), brainstorming might focus on finding concepts in other categories (e.g., value propositions, channels, customer relationship, customer segments, costs, revenues). That would help encourage theory builders to seek what is not there, rather than to repeatedly focus research efforts on what is already in the theory.

With the development of more useful theories, and the synthesis of theories from across and between disciplines, a door may be open for theories which identify recursive relationships between their topics (e.g., global poverty) and their observers (e.g., researchers).

On a deeper level, this orthogonal perspective gives rise to interesting questions around ontology. More specifically, the difference between fact and fiction. Generally, we like to see our theories supported by data. That indicates some relationship between the abstract concepts and the "real world." Yet, it is possible to create a well-structured theory that is entirely fictitious. Indeed, every fiction may be seen as containing something of reality. Even a story about people flying through space faster than the speed of light (clearly a fiction) contains sub-concepts with a clear foundation in reality (people, space). Perhaps some exploration in this direction might yield interesting insights into the nature of reality—or at least our perceptions of "it."

We anticipate that more structured models will lead to greater ability to solve problems. However, we must wonder if greater structure may also indicate tighter focus within a context? For example, grounded theory (cf. Charmaz, 2006) is claimed as a good tool for creating useful theory. However, grounded theory is very context dependent and so the theories/models created should not be considered generalizable. Will increased orthogonality support increased generalizability? A world of research is waiting. A few more questions to be answered are:

- Might we create a "periodic table" of theories based on their orthogonalities?

- What if we were to investigate the potential relationship between the orthogonality of theories and the unanticipated consequences of using those theories in practice?
- Is there a way to understand the *relationships* between concepts as having some orthogonal relationship between them such that they may appear causal to one viewer and non-causal (but somehow influential) to another?
- Can we explore how this kind of orthogonal perspective might be used to identify "missing pieces" of our theoretical puzzles (known unknowns and unknown unknowns)?
- How might the insights and methods presented here be integrated more concretely with those of Ingo Pies to create a more effective process for moving people toward cooperation rather than conflict?

Conclusion

It is worth noting that Ockham's razor is often deployed to suggest that theories should be parsimonious—often interpreted to mean "small" or containing fewer variables or no more than are needed to reach a satisfactory explanation (Kelly, 2007). Unfortunately, we don't know, a priori, how many concepts are needed. Orthogonality supports the development of parsimonious theories by providing a tool for comparing concepts and reducing or eliminating redundancies and clarifying gaps in our theories to more effectively focus research activity.

Lissack and Graber (2014) made interesting progress in their efforts to understand explanations that are sufficient to satisfy. However, we need to be pursuing a larger question. After all, theories of physics in ancient times seemed to satisfy. Yet, they were not highly useful in practical application: Aristotle's theory of motion was not as useful as Newton's. Instead, of satisfaction, we suggest that researchers and practitioners pursue usefulness in theories. As serious scholars, we should push ourselves, and each other, to create theories that are increasingly more useful in practical application; not simply good enough for publication.

In short, orthogonality is one way we understand things as being different, yet connected. As such, it is a critical part of our understanding the world. Importantly, while having a process for rethinking concepts supports the development of theory, having a *rigorous* process suggests a more direct path, so that scholars may make progress more rapidly toward more useful theories.

The focus of the present paper has been mainly on improving theories of the social/behavioral sciences because they have the greatest opportunity for rapid advancement among the increasingly fragmented fields (Donaldson, 1995) before "paradigm wars" cause much greater decline in the perceived legitimacy of our scholarly efforts (McKelvey, 2002). However, it should also be noted that the natural sciences may also benefit from this emerging understanding of orthogonality.

Many forms of information and theory presentation (stories, slogans, etc.) create the pretense of knowledge (Hayek, 1974). Despite, (or because of) the limitations of our human minds, we would benefit from clear, rigorous, direct methods for advancing theory to support advances in practical application.

References

Ackermann, F., & Eden, C. (2004). Using causal mapping - individual and group, traditional and new. In M. Pidd (ed.), *Systems Modeling: Theory and Practice* (pp. 127-143). Chichester, UK: John Wiley & Sons.

Ahn, S. C. (1997). Orthogonality tests in linear models. *Oxford Bulletin of Economics and Statistics, 59*(1), 183-196.

Argyris, C., Putnam, R., & Smith, D. M. (1985). *Action Science: Concepts, Methods, and Skills for Research and Intervention.* San Francisco: Jossey-Bass.

Armstrong, J. S. (1997). Peer review for journals: Evidence on quality control, fairness, and innovation. *Science and Engineering Ethics, 3*, 63-84.

Axelrod, R. (1976). *Structure of decision: The cognitive maps of political elites.* Princeton: Princeton Universtiy Press.

Bateson, G. (1979). *Mind in nature: A necessary unity.* New York: Dutton.

Bauer, J. R., Booth, A. E., & McGroarty-Torres, K. (2016). Causally-rich group play: a powerful context for building preschoolers' vocabulary. *Frontiers in psychology, 7*, 997.

Beckmann, M., Hielscher, S., & Pies, I. (2014). Commitment strategies for sustainability: how business firms can transform trade-offs into win–win outcomes. *Business Strategy and the Environment, 23*(1), 18-37.

Bentz, V. M., & Shapiro, J. J. (1998). *Mindful inquiry in social research.* Thousand Oaks, CA: Sage.

Bryson, J. M., Ackermann, F., Eden, C., & Finn, C. B. (2004). *Visible thinking: Unlocking causal mapping for practical business results*: John Wiley & Sons.

Bunge, M. (2004). How does it work? The search for explanatory mechanisms. *Philosophy of the Social Sciences, 34*(2), 182-210.

Bureš, V. (2017). A Method for Simplification of Complex Group Causal Loop Diagrams Based on Endogenisation, Encapsulation and Order-Oriented Reduction. *Systems, 5*(3), 46.

Charmaz, K. (2006). *Constructing Grounded Theory: A Practical Guide through Qualitative Analysis*. Thousand Oaks, CA: Sage.

Checkland, P. (2000). Soft systems methodology: a thirty year retrospective. *Systems Research and Behavioral Science, 17*(S1), S11-S58.

Dahler-Larsen, P. (2018). Theory-based evaluation meets ambiguity: The role of Janus variables. *American Journal of Evaluation, 39*(1), 6-23.

Darwish, K., Magdy, W., & Zanouda, T. (2017). *Trump vs. Hillary: What went viral during the 2016 US presidential election.* Paper presented at the International conference on social informatics.

Deegan, M. (2011). *Using Causal Maps to Analyze Policy Complexity and Intergovernmental Coordination: An empirical study of floodplain management recommendations.* Paper presented at the System Dynamics Society Conference 2011.

Donaldson, L. (1995). *American Anti-Management Theories of Organization: A Critique of Paradigm Proliferation.* New York: Cambridge University Press.

Dyson, R. G. (2004). Strategic development and SWOT analysis at the University of Warwick. *European Journal of Operational Research, 152*(3), 631-640.

Fink, G. (2014). Type building and mappings in organizational and cultural studies. *European Journal of Cross-Cultural Competence and Management, 3*(1), 93-100.

Finkel, E. J. (2014). The I3 model: Metatheory, theory, and evidence. *Advances in experimental social psychology, 49*, 1-104.

Ford, J. K., MacCallum, R. C., & Tait, M. (1986). The application of exploratory factor analysis in applied psychology: A critical review and analysis. *Personnel psychology, 39*(2), 291-314.

Gabora, L., Rosch, E., & Aerts, D. (2008). Toward an ecological theory of concepts. *Ecological Psychology, 20*(1), 84-116.

Gentner, D., & Toupin, C. (1986). Systematicity and surface similarity in the development of analogy. *Cognitive science, 10*(3), 277-300.

Goggin, M. L. (1986). The" too few cases/too many variables" problem in implementation research. *Western Political Quarterly, 39*(2), 328-347.

Goodier, C., Austin, S., Soetanto, R., & Dainty, A. (2010). Causal mapping and scenario building with multiple organisations. *Futures, 42*(3), 219-229.

Goulding, C. (2002). *Grounded theory: A practical guide for management, business and market researchers.* Thousand Oaks, CA: Sage.

Hammersley, M. (2003, 11-13 September 2003). *Too good to be false? The ethics of belief and its implications for the evidence-based character of educational research, policymaking and practice.* Paper presented at the British educational research association annual conference, Heriot Watt University, Edinburgh.

Hayek, F. (1974). A.(1989),'The pretense of knowledge.' *American Economic Review, 79*(6), 3-7.

Hielscher, S., Pies, I., & Valentinov, V. (2012). How to foster social progress: an ordonomic perspective on progressive institutional change. *Journal of Economic Issues, 46*(3), 779-798.

Hill, T., & Westbrook, R. (1997). SWOT analysis: It's time for a product recall. *Long range planning, 30*(1), 46-52.

Hull, D. L. (1988). *Science as a process: An evolutionary account of the social and conceptual development of science.* Chicago: University of Chicago Press.

Johnson-Laird, P. N. (1980). Mental models in cognitive science. *Cognitive science, 4*(1), 71-115.

Joyce, A., & Paquin, R. L. (2016). The triple layered business model canvas: A tool to design more sustainable business models. *Journal of Cleaner Production, 135*, 1474-1486.

Kaplan, R. S., & Norton, D. P. (1996). Using the balanced scorecard as a strategic management system: Harvard business review Boston.

Kelly, K. T. (2007). Simplicity, truth, and the unending game of science. In S. Bold, B. Löwe, T. Räsch, & J. v. Benthem (eds.), *Foundations of the Formal Sciences V: Infinite Games* (pp. 368). London: College Publications.

Kelly, K. T., & Mayo-Wilson, C. (2012). Causal conclusions that flip repeatedly and their justification. Retrieved from http://arxiv.org/ftp/arxiv/papers/1203/1203.3488.pdf

Kvålseth, T. O. (1985). Cautionary note about R 2. *The American Statistician, 39*(4), 279-285.

Lissack, M. (2019). Understanding Is a Design Problem: Cognizing from a Designerly Thinking Perspective. Part 1. *She Ji: The Journal of Design, Economics, and Innovation, In press.*

Lissack, M., & Graber, A. (2014). *Modes of explanation: Affordances for action and prediction*: Palgrave Macmillan.

McGlashan, J., Johnstone, M., Creighton, D., de la Haye, K., & Allender, S. (2016). Quantifying a systems map: network analysis of a childhood obesity causal loop diagram. *PloS one, 11*(10), e0165459.

McKelvey, B. (2002). Model-centered organization science epistemology. In J. A. C. Baum (Ed.), *Blackwell's Companion to Organizations* (pp. 752-780). Thousand Oaks, California: Sage.

Mishler, W., & Rose, R. (1997). Trust, distrust and skepticism: Popular evaluations of civil and political institutions in post-communist societies. *The journal of politics, 59*(02), 418-451.

Monahan, T., & Fisher, J. A. (2010). Benefits of 'observer effects': lessons from the field. *Qualitative research, 10*(3), 357-376.

Newell, W. H. (2007). Decision making in interdisciplinary studies. *Public Administration and Public Policy, 123*, 245.

OED. (Ed.) (2007) (6 ed., Vols. 2). Oxford: Oxford University Press.

Oh, J.-Y. (2014). Understanding natural science based on abductive inference: Continental drift. *Foundations of Science, 19*(2), 153-174.

Panetti, E., Parmentola, A., Wallis, S. E., & Ferretti, M. (2018). What drives technology transitions? An integration of different approaches within transition studies. *Technology Analysis & Strategic Management, (online)*, 1-22. doi:10.1080/09537325.2018.1433295

Pearl, J. (2000). *Causality: Models, reasoning, and inference.* New York: Cambridge University Press.

Peters, B. G. (2019). *Institutional theory in political science: The new institutionalism*: Edward Elgar Publishing.

Pies, I. (2016). The ordonomic approach to order ethics *Order Ethics: An Ethical Framework for the Social Market Economy* (pp. 19-35): Springer.

Rogers, P. J. (2008). Using programme theory to evaluate complicated and complex aspects of interventions. *Evaluation, 14*(1), 29-48.

Saucier, G. (2002). Orthogonal markers for orthogonal factors: The case of the Big Five. *Journal of Research in Personality, 36*(1), 1-31.

Seabury, M. B. (1991). Critical thinking via the abstraction ladder. *English Journal, 80*(2), 44-49.

Sloman, S. A., & Hagmayer, Y. (2006). The causal psycho-logic of choice. *Trends in Cognitive Sciences, 10*(9), 407-412.

Sprites, P., Glymour, C., & Scheines, R. (1993). *Causation, prediction and search* (First, online ed.). Cambridge, MA: MIT Press.

Sterner, B. (2014). Explanation and pluralism *Modes of Explanation* (pp. 249-256): Springer.

Suedfeld, P., Tetlock, P. E., & Streufert, S. (1992). Conceptual/integrative complexity. In C. P. Smith (Ed.), *Handbook of Thematic Content Analysis* (pp. 393-400). New York: Cambridge University Press.

Szostak, R. (2012). Classifying relationships. *Knowledge Organization, 39*(3), 165-178.

Umpleby, S. (1997). Cybernetics of conceptual systems. *Cybernetics & Systems, 28*(8), 635-651.

Valentinov, V., Verschraegen, G., & Van Assche, K. (2019). The limits of transparency: A systems theory view. *Systems Research and Behavioral Science.*

Waldmann, M. R., Holyoak, K. J., & Fratianne, A. (1995). Causal models and the acquisition of category structure. *Journal of Experimental Psychology, 124*(2), 181-206.

Wallis, S. E. (2010). The structure of theory and the structure of scientific revolutions: What constitutes an advance in theory? In S. E. Wallis (Ed.), *Cybernetics and systems theory in management: Views, tools, and advancements* (pp. 151-174). Hershey, PA: IGI Global.

Wallis, S. E. (2014). Abstraction and insight: Building better conceptual systems to support more effective social change. *Foundations of Science, 19*(4), 353-362. doi:10.1007/s10699-014-9359-x

Wallis, S. E. (2016a). The science of conceptual systems: A progress report. *Foundations of Science, 21*(4), 579-602.

Wallis, S. E. (2016b). Structures of logic in policy and theory: Identifying sub-systemic bricks for investigating, building, and understanding conceptual systems. *Foundations of Science, 20*(3), 213-231.

Wallis, S. E. (2019). *Learning to Map and Act Cybernetically Without Learning Cybernetics: **How Many Loops are Needed to Enable Effective Decisions and Actions?* Paper presented at the American Society for Cybernetics, Vancouver, BC, Canada.

Wallis, S. E., & Valentinov, V. (2016). The imperviance of conceptual systems: Cognitive and moral aspects. *Kybernetes, 45*(9).

Wallis, S. E., & Wright, B. (2019). Integrative propositional analysis for understanding and reducing poverty. *Kybernetes, In Press.*

Westerhoff, J. (2005). Ontological categories: Their nature and significance.

Wright, B., & Wallis, S. E. (2019). *Practical Mapping for Applied Research and Program Evaluation.* Thousand Oaks, CA: SAGE.

Yin, R. K. (1984). *Case Study Research: Design and Methods* (Vol. 5). Beverly Hills, CA: Sage.

Yin, R. K. (2011). *Applications of case study research*: Sage.

Zude, Y., & Yolles, M. (2006, July, 2006). *From Knowledge Cybernetics to Feng Shui.* Paper presented at the International Society for the Systems Sciences, Sonoma State University, Rohnert Park, California.

Chapter 13
Understanding and Improving the Usefulness of Conceptual Systems—An IPA Based Perspective on Levels of Structure and Emergence

Wallis, S. E. (2020). Understanding and improving the usefulness of conceptual systems: An Integrative Propositional Analysis-based perspective on levels of structure and emergence Systems Research and Behavioral Science, in press. https://onlinelibrary.wiley.com/doi/abs/10.1002/sres.2680

My thanks to Michael Lissack and the ISCE conference in Paris 2013 for deep conversations resulting in new insights that led to this paper. Also to Thomas Marzolf, Kent Palmer, and Len Troncale whose conversations and insights proved inspiring and illuminating. Finally, to two anonymous reviewers whose comments, both kind and challenging, led to the development of an improved paper.

Abstract: *Terms like "levels" and "nested" are used to describe relationships between components of conceptual systems (theories, models, etc.). However, they have not been fully explored. This paper investigates levels to better understand how theories are structured; and so, how we may develop more useful theories and models to better support more effective practice. We find a horizontal dimension (represented by causal connections between concepts at one ontological level) and a vertical dimension (represented by connections of emergence between concepts of differing ontological levels). This view of emergence offers a new way to structurally distinguish between conceptual components of a theory, thus supporting a new approach to building theories that better reflects our systemic world. A third, perspectival, approach may be applied to aid in the understanding* *of both dimensions. A typology is proposed as are conventions for diagramming theories, and new criteria for improving the structure of theories.*

Introduction

It is generally accepted that organizations may be seen as having different levels (often presented hierarchically). For example, Lindberg (2004) talks about levels in human/organizational systems as individual, interpersonal, organizational, and community. While those may easily be interpreted as existing on different levels (in some sense) they might also be interpreted as having different numbers of people and interactions—things that are common across all seeming levels. So, despite (and/or because of) interesting arguments and conversations around borders and boundaries between those levels (cf. Midgley & Rajagopalan, 2019) it seems that they are quite permeable because people and interactions may also cross those boundaries. Such complexity enables diverse views without necessarily improving understanding; instead requiring detailed descriptions of varied views with attendant confusion and fragmentation that may accompany such detailed description. Instead, we seek here to understand levels in a way that enables the greatest variety of useful explanatory models requiring the fewest "rules."

Notions such as levels are also commonly found in the systems sciences. First, that systems have different levels (Ashmos, Huonker, & Reuben R. McDaniel, 1998; Pascale, 1999; Tower, 2002). Also, it seems that differences in levels seem to emerge from changes or interactions in (and between) systems (Axelrod & Cohen, 2000; Harder, Robertson, & Woodward, 2004; Moss, 2001; Pascale, 1999). While, in turn, having different levels causes interaction and change (Hunt & Ropo, 2003; Stacey, 1996).

But what constitutes a level? We may casually talk about levels in terms of "how many" people may be involved. For example, a "micro" level might include a couple of dozen people while a "macro" level might include billions (Alexander, Giesen, Munch, Anderson, & Smelser, 1987). Are those "levels" or are they simply numbers on a scale? That fuzziness of understanding (or, perhaps, they may be examples of false clarity) applies also to conceptual systems (theories and models). Thus we may be led to choose poor models and so make poor decisions. How might we gain more clarity about what (or how) one level is understood as emerging from another?

Emergence, as a term, was coined by Lewes in 1875 based on the Latin *emergo*, to arise or to come forth (Vintiadis, 2013). Clayton (2006) provides a nice and concise history of emergence from Aristotle through Hegel, the British Emergentists, and to the near present with twists, turns, and reversals in the way emergence has been understood over the centuries in various disciplines including the mathematical physics of Whitehead, James in psychology, and Morgan in evolutionary biology. Conversations continued in recent decades around what these concepts mean and how they may be applied with authors such as Polanyi (chemistry and biology) and Sperry (neuroscience) and many others with various definitions proposed for a variety of concepts including: strong emergence, weak emergence, causality, downward causation, irreducibility of emergence, properties that may be seen as emerging, epistemology, and ontology.

Among those many perspectives, it is worth noting that "weak emergence is sometimes called "epistemological emergence," in contrast to strong or "ontological emergence" (Clayton, 2006: 8), a perspective supported in the present paper. Also linked with those ideas, Morgan suggests that causality between components occurs on a horizontal level; while, in contrast, there are discontinuities between levels of emergence. Those emergent levels of reality are different from one another in substantial ways (*ibid*: 12). Morgan's lays the foundation for the present piece which builds on his work by providing a way to *measure* those levels as structures of conceptual systems and so support more scientific understanding of levels.

Within emergentism, the notion of causal closure seems particularly relevant for a variety of philosophical debates including the mind-body problem. Some suggest that all physical events are caused by and result from other physical events (causally closed), (cf. Kim, 1993) reductively arguing, for example, that the mental domain is reducible to the physical domain. Others suggest that physical events may be caused by (and/or result from) mental events (Lowe, 2000) and, building on evolutionary theory and complexity perspectives, see such things as meaning, language, and social norms as existing on an apparently different level from the physical (cf. Watanabe & Smuts, 2004). Sparacio (2003) argues that the mental is a higher level, emergent feature, of the physical because it has different properties and so has a different ontological status. Similarly, (Gabbani, 2013) provides an excellent epistemological critique of Kim's causal closure.

Philosophical conjectures, conversations, and concerns continue to this day around issues such as emergence, causal closure, and the mind-body problem. Despite extensive philosophical investigation into the emergentic notion of causal closure, few papers even attempt to provide an effective definition of it; to say nothing of a way to measure it. This is a critical gap that must be addressed if we are to make progress.

Having developed methods for measuring the structure of theories on one level (concepts connected by causal relationships) our research suggests that it may be possible to develop measures of the structure of theory on other levels and so move emergentism a step closer to becoming a science rather than the increasingly fragmented confusion of philosophical expositions.

Wallis (2014a) suggests that we may improve our understanding of theory, and how we may create more useful theory, by understanding where those theories fit on some scale of abstraction and finding how, "theories at one level of scale (e.g., apples) are somehow nested within theories of another scale (e.g., fruit). The connection between levels, then, might be understood as an emergent property" (p. 196). However, that view seems to suggest that there is only one scale from the highly abstract down to the very concrete. While, in contrast, from some hypothetical "satellite-view" we limited humans seem to see or experience and react to levels and such differences (even when we create them ourselves). So, there may be some difficulty in effectively identifying what those levels (and their implied boundaries) are. For example, in much the same way that a CEO may knock on the janitor's door (and visa-versa), levels may seem permeable, insubstantial, or subject to changes in perspective. Nested dolls seem invisible until they are unpacked; at which time they become clear.

Terms such as levels and nested are also applied to conceptual systems and /or blends of physical and conceptual systems. For example, "More recent developments assert that knowledge acquisition is a multilevel nested process" (from Zhao & Anand, 2009) cited in (Tian, Li, & Wei, 2013). And (to the extent that embeddedness might be understood as a kind of level) "Evaluators can thus embed theories within one another in the same way that systems are embedded within each other" (Westhorp, 2012: 411).

While neither of those statements may be positively said to be inaccurate, they do seem to leave out a critical consideration. Specifically, how might we know if we are nesting, embedding, or leveling theories in the best possible way? How do we know that our theories have the same number of levels as our reality?

The implications for the social/behavioral sciences are profound. If we can develop a rigorous, reasonably objective, framework for understanding and creating better theories where the levels of our theories better match the levels of our larger world and policies, that new perspective may serve to accelerate the advance of knowledge, science, and practice in a world that desperately awaits effective answers to wicked problems.

This is important because theories that are merely *said* to have a useful number of levels might be arranged improperly and so provide false or misleading insights. It is one thing to say that one *has* all the parts needed for a system to work effectively, but something very different to say *how* those parts are connected for the system to work effectively. More generally, "There are credible and rather urgent reasons to believe that we must manage and mobilize our knowledge more effectively" (Wexler, 2001: 262) where theories may be understood as a kind of formalized or explicit knowledge. By understanding levels, we may learn how to determine if our conceptual models are complete; and so, stop working on them and start using them to the maximum of their potential benefit (Gonzales, 2001).

Within the present paper, one goal is to better understand "how connected" levels of theories are; and, by providing some measurement of those relationships, provide a clearer path for creating more useful theories for practical application. Another goal is to do that in a way that is more accessible to a wider range of scholars, practitioners, and students for evaluating and improving theories.

This exploration is of particular importance when our field is striving to create theories which are effective representations of systems in the physical world. Because, if physical world systems have levels, those should be reflected in the theory as well. Otherwise, the conceptual system provides only a poor representation of the physical system and becomes relatively useless or misleading. Thus, we may better understand our conceptual systems and so better understand our physical systems (Westhorp, 2012). In contrast, "single level models are likely to overlook relevant interactions" (Bachmann, Engelen, & Schwens, 2016: 310).

This is important because, according to Wexler (2001), knowledge maps, when created and used appropriately, support organizational success and provide a competitive advantage providing economic returns (higher income), structural returns (improved organizational processes, learning, communication, and knowledge), cultural returns (including trust, socialization of new organizational members, shared values, ability to predict and adapt to change), and knowledge returns (improves organizational memory, identifies where research is needed, clarifies distribution of accountability and rewards, shared vocabulary and communication, learning for planning).

The importance of this topic is also reflected in a recent special issue of Chaos (2016), Vol. 26, titled, "Complex Dynamics in Networks, Multilayered Structures and Systems." That issue, however is for the more mathematically inclined. The present paper, in contrast, will be focused on the relational or qualitative aspects of levels, specifically within conceptual systems while drawing on examples from a variety of disciplines.

This investigation has implications for interdisciplinary studies as well because we might see each discipline as on a separate level from the others. Or, perhaps, some disciplines (certainly sub-disciplines) might be viewed as nested within one another. By better understanding those relationships, we may expect to improve the ability of theories to address complex problems requiring multidisciplinary solutions.

If we live in a world where our systems have levels, that world would best be understood and engaged if our conceptual systems also had levels. However, our understanding of those levels seems to be lacking in rigor with the attendant likelihood for confusion. So, in this paper, we seed to discover new insights into the logical systemic structure of conceptual systems with a focus on levels found in theories, plans, and models. With new insights, we anticipate the ability to develop theories that will be more effective in practical application for reaching specified goals.

After providing some definitions, we will explore the notion of levels from a number of different perspectives. Following those, we will provide suggestions for building better theories (more useful in practical application for reaching desired goals).

Definitions

A system may be defined as "*A set of elements in interaction* (von Bertalanffy 1968)" which "...can be either physical or conceptual, or a combination of both." https://sebokwiki.org/wiki/System_(glossary)

This structural similarity between physical and conceptual systems allows us to draw close parallels between them. While one might argue that there are many other ways to categorize systems (e.g., psycho-social, biological) for the present paper, for ease of representation and comprehension, we will use the term "social systems" or "physical systems" (including social systems) to represent all non-conceptual systems (with the understanding that there are still many overlaps and intertwining relationships within systems to be worked out later).

"Generally, a conceptual system is any form of theory, theory of change, theory of the middle range, grand theory, model, mental model, policy, etc." (Wallis, 2016a: 579). A theory may be understood as "An ordered set of assertions" (Weick, 1989: 517) such as a set of sentences, propositions, or hypotheses. Within each proposition may be found conceptual elements (typically nouns) and interactions (typically verbs).

Here, for the sake of clarity, we will present most theories abstractly as diagrams or (more simply) "maps" as is sometimes done for mind maps, concept maps, knowledge maps, maps, diagrams, causal loop diagrams, practical maps, logic models, program models, semantic networks, and so on.

The components of those conceptual systems will be generally referred to as concepts or variables while the connections between concepts will be generally referred to as causal relationships or connections (represented on diagrams as arrows).

In the next section, we will explore a variety of ways that differences in levels have been presented as an orienting background before developing in to emergence in greater depth.

Similarities and Differences

The term "level" has been applied to a variety of systems in a variety of ways. While a comprehensive review of all perspectives is beyond the scope of the present paper (a recent search in Google Scholar™ found over 800 articles mentioning "levels" published in *Systems Research and Behavioral Science*, alone) the present section will provide a range of orienting insights.

Whether we are talking about conceptual systems, physical systems, or a combination of the two, one difficulty is understanding how different parts or levels are related. Sometimes, those relationships seem quite close as in two people standing together, linked by a common language and common interests; they talk and learn and have the possibility of collaborative action. Other times, those relationships seem more distant as in a boiler providing heat within the basement of a high-rise, separated by hundreds of floors from a cell phone antenna on the roof providing telecommunication. Those differences (perhaps "making a difference") is what we are investigating here.

In identifying fundamental systems processes, Friendshuh and Troncale (2012) identified "hierarchy" as an important isomorphy. Within that general term, they include "levels" and "subsystems" and the idea that one system may be "subsumed" by another.

Fractals also seem to exist with self-similar or self-same structure at different levels of scale (Mandelbrot, 1977). Also, Bateson's "dual description" may be thought of as two levels of understanding, combined to create a third. However, drawing on Salthe, McKelvey (2004) suggests that dual descriptions are not sufficient, that three levels of understanding are a minimum.

As an example, Ward (2019) talks about using multiple models—"model pluralism"—to understand complex situations with theories that are at different levels in an effort to counter an apparent trend toward simplistic or "single factor" theories (Ward, 2014). Those levels, however, correspond with levels of analysis including (for understanding psychological depression) physiology, behavior, neurological, psychological, and phenomenological (Ward & Clack, 2019). So, the conceptual system is simply a reflection of the physical system—or more accurately, of the assumptions about the physical system—and so subject to the same potential failings of those assumptions.

Higgins and Shirley (2000) discuss theories as existing on different levels as micro-range, middle-range, grand, and meta level theory. Each with a distinct scope and place on a scale of abstraction. Ward and Hudson (1998) also talk about three levels for theories. However, their understanding of levels seems to correspond with the complexity or simplicity of theories. That is, on the first level, theories describe a situation, on another the theories analyze one factor, and on the third level theories include multiple factors. Research may be used to synthesize theories at lower levels and move them up to higher levels; particularly "theory knitting" (Kalmar & Sternberg, 1988). Leeuw and Donaldson (2015) also suggest theory knitting as a way to bridge differing levels of theory (such as a theory about a policy and a theory about how a policy is made).

While the above perspectives may be useful in some ways, a problem arises when applied to theories at differing levels of abstraction. For a simple example, one theory might address "apples" while another theory addresses "fruit." While there appears to be an overlap between apples and fruit (suggesting the opportunity to knit together two theories containing those two concepts), those concepts differ on a scale of abstraction so should not be located within the same theory lest confusion arise (Wallis, 2014a). For example, it would be awkward to say something like "having more apples causes us to have more fruit" because apples are a kind of fruit; thus suggesting a tautological aspect to the statement. Similarly, because there is an ontic overlap between apples and fruit, they should not be contained in the same theory due to the potential confusion of having a theory with non-orthogonal elements (Wallis, 2020d).

Recognizing that kind of difficulty, in an introduction to a special issue on theory building, Klein, Tosi, and Albert A. Cannella (1999) write about some challenges to building theories including the confusingly large number of available theories, the researcher's skill or comfort at moving between levels of scale, the complexity of larger theories, and the difficulty of finding an appropriate publication. However, their understanding of levels is also related to the systems under analysis (micro, meso, and macro) rather than some levels within the theories themselves. That view is similar to many who are interested in synthesizing theories relating to different levels of physical systems.

The study of systems dynamics (SD) uses a variety of measures to evaluate which *part* of the theory is most influential on the behavior of the system represented by the theory. SD is used to model and study networks of multiple loops. Those "SD models are conceived as closed systems" (Schwaninger, 2011: 8979). However, SD has its share of limitations (including) that the results of those measures are not always intuitively clear and those measure might be difficult to determine in a highly complex model with many feedback loops (Hayward & Roach, 2017). Indeed, the very practice of simplifying the analytical process by focusing on one part of the theory (explained as "nested" in the theory as a whole), may be seen as limiting an SD understanding of the model (Saleh, Oliva, Kampmann, & Davidsen, 2010).

Interestingly, SD looks at the "distance" between nodes as an indicator of their similarity or difference or to show the degree of influence one node has upon another (McGlashan, Johnstone, Creighton, de la Haye, & Allender, 2016). That distance might be thought of as one way to measure the levels of difference between concepts.

It has been noted that most "semiotic networks [as representations of knowledge] have a hierarchical structure, all self-organizing semantic networks actually consist of a number of smaller self-organizing semantic networks across different scales" (Klenk, Binnig, & Schmidt, 2000: 152). However, it is not clear exactly what those levels are, or how they emerge.

Nested

Another, closely related, view of levels is found in the idea that some systems may be understood as "nested" within other systems. For example, a program model (also known as a program theory) may be nested (in some sense) within the larger program or policy. Within a theory, complex concepts may be said to have a number of sub-concepts (Wallis, 2020d) which may be thought of as a nested relationship. Social network analysis (Hawe, Webster, & Shiell, 2004; Neal & Neal, 2013) may be understood as looking at nested and/or networked relationships at multiple levels. Within a story (fictional, or not, or somewhere in-between) sub-plots may be nested within the main story. Emotions may be nested within our bodies and minds. Figure 1 provides one diagrammatic example; there, (Fink, 2019) shows individual scholars nested within research organization, nested in society, nested in the ecological environment.

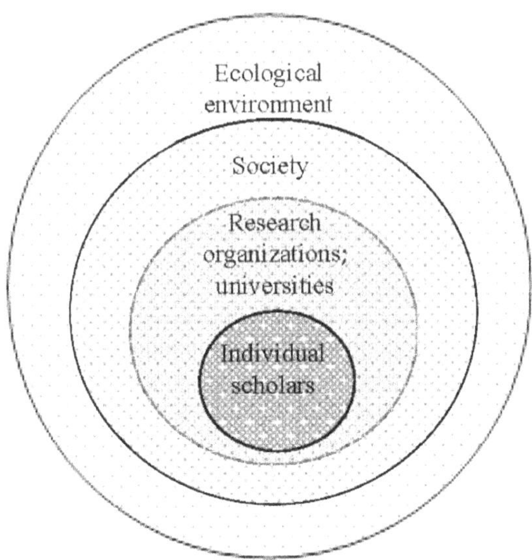

Figure 1 *Scholars nested within universities, nested within society, nested within ecological environment (from Fink 2019)*

While this kind of nesting relationship serves to illustrate some relationships, it also (as with any presentation) de-emphasizes or ignores other relationships because each level is connected, in some way, to other levels. While it is not represented by Figure 1, individual scholars, for an example, are intimately connected with the ecological environment through processes such as living, breathing, and eating.

Like individual people are nested within teams, teams within departments, and departments within organizations, we can also think of knowledge maps (as a kind of conceptual system or theory) as nested within other knowledge maps. For example, a number of stakeholder groups within a city might collaborate to create a knowledge map showing the relationship between major problems they face (e.g., poverty, unemployment, homelessness). While that high-level map might show some linkages between those issues, such a map might not be very useful for a team focused on feeding people on the street. So, each team might create a map that is more relevant to their own level of operation. Such an operational map could be seen as nested within (or at a different level from) the higher-level map as seen in Figure 2.

While these may be seen as nested, they may also be seen as existing at different levels of some organization. So, as noted above, the boundaries are quite permeable or indeterminate, having been created by intuition or biased observation. Therefore, some clarity is needed.

The overlap, or interplay, between physical and conceptual systems threatens to confound us. For example, Ohm's law of electricity (a kind of theory) may be nested within a team of engineers who are hierarchically working under the direction of a person who uses Situational Leadership (a kind of theory) to manage her team, while they are all part of a larger firm within an industry that is partially controlled by an informal group of elites who are using economic theories to advance their collective interests within the broader global bio/socio-economic system.

The general idea of recognizing different systems as existing on different levels or as nested within one another seems to make sense if those systems are disconnected from one another. That

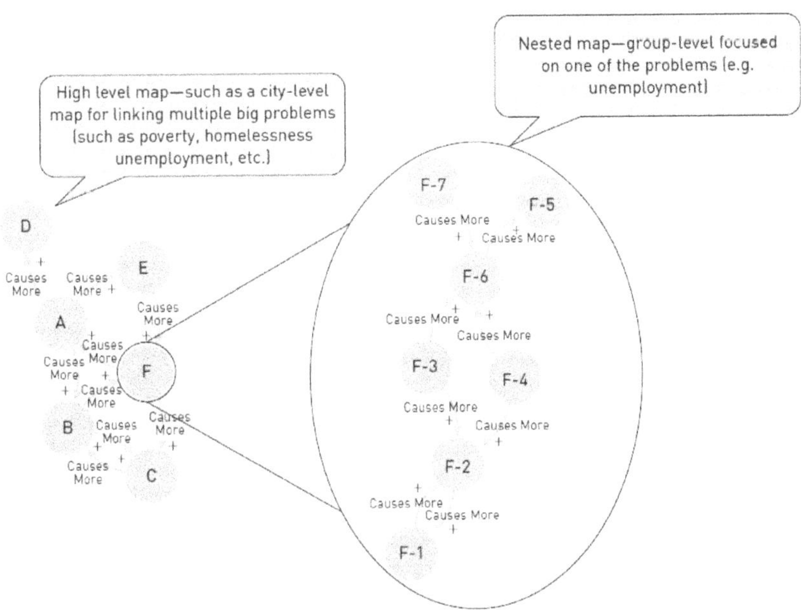

Figure 2 *Nested map within a higher level map—from "Practical Mapping for Applied Research and Program Evaluation" (Wright & Wallis, 2019: 222)*

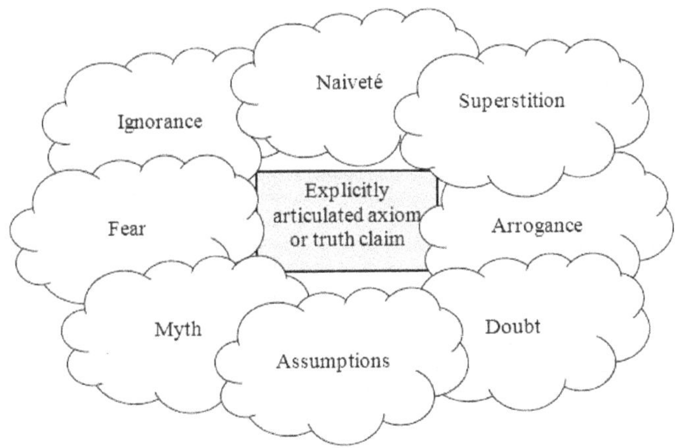

Figure 3 *Box of explicit belief surrounded by cloud of fuzzy tacit beliefs.*

view, however, negates the most important assumption of systems thinking. That a boundary cannot be a point, a line, or a plane in three dimensional space, but a tetrahedron which identifies an area of space that may be seen as separate from another (McNa-

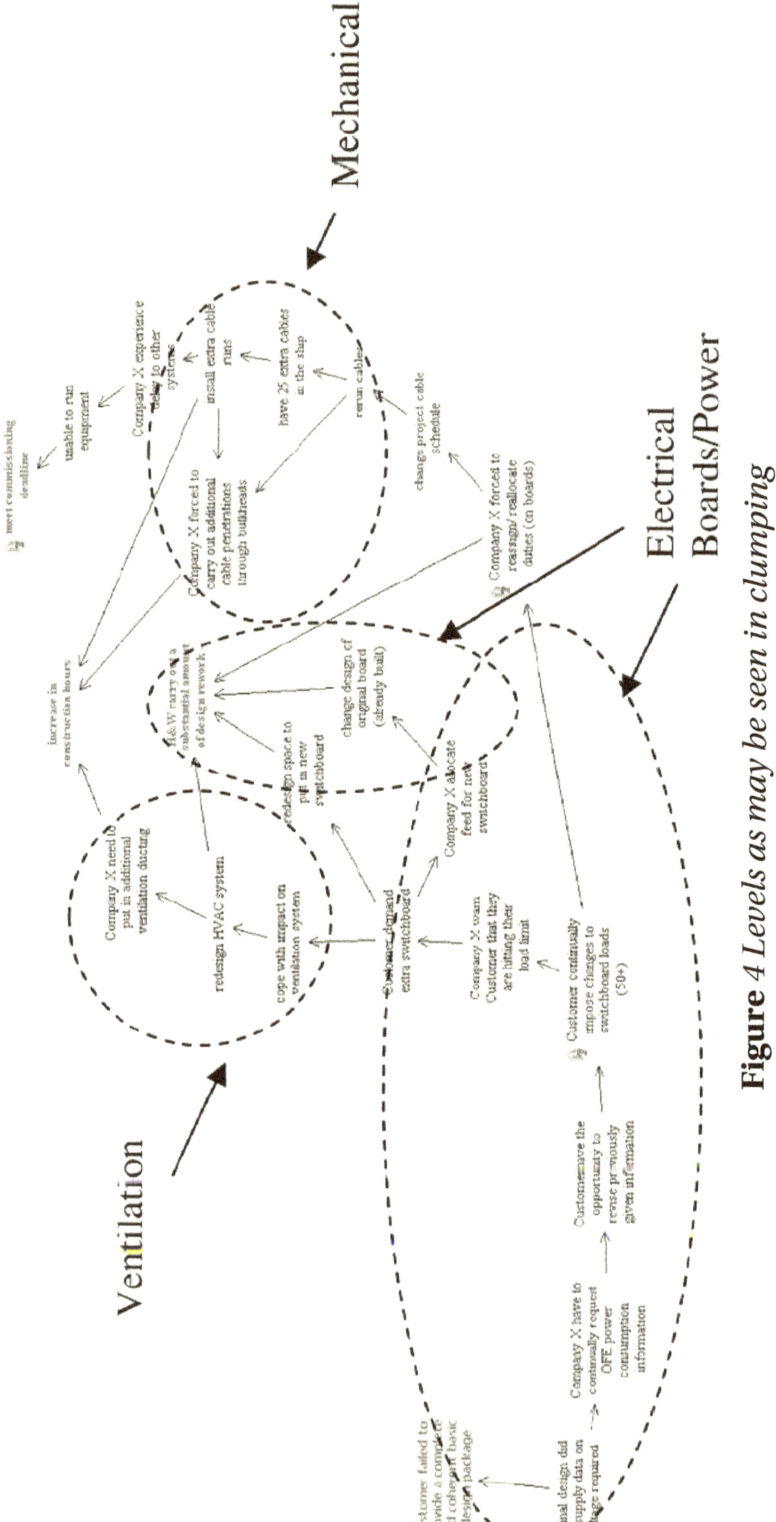

Figure *4 Levels as may be seen in clumping*
(Howick, Eden, Ackermann, & Williams, 2008: 1079)

mara, 2009). Therefore, we must develop a more nuanced understanding of how our conceptual systems are connected with one another, as well as the connections within each system. Not in a simple, reductionist way; rather in a complex systemic way.

Tentative typology

The above collection of perspectives may seem somewhat kaleidoscopic in the many points and counterpoints provided. Some relationships seem *fuzzy* while others seem to exhibit a *gap* which may be represented as a lack of understanding. Other perspectives seem to suggest that levels may be seen as *categories* of things or that some *boundary* may exist between the levels. Another relevant perspective is that there may exist some kind of change or *transformation* that occurs between levels. And, finally, the question may be raised how we can understand complex issues using relatively compact theories; perhaps where complex concepts are *enfolded* with one another.

To clarify the many perspectives, the following sub-sections suggest a tentative typology for levels; each type building on the one before to build our understanding of levels in conceptual systems. While the following is neither comprehensive nor conclusive, we hope it will encourage conversations that will move the field toward greater clarity. More to the flow of the present paper, these types will serve to focus our attention on key perspectives to be developed in the following full section on emergence.

Type 1—"Fuzzy"

On the simplest level, we may see a difference in levels between that which is known and that which is unknown; or, partially known or inferred. This may be understood as the difference between our explicit and tacit beliefs as shown in Figure 3.

Another way to understand a Type 1 conceptual system is as something where one or more concepts have some ontological expression, but are not expressed in a way that identifies epistemological relationship(s) with other concepts. That lack of connection draws perilously close to negating assumptions of systems thinking in that such conceptual systems are presented as non-systemic; that is to say, each concept is not clearly connected to any other. In contrast, a systemic perspective suggests that the existence of one concept implies the search for others.

Type 2—"Gap"

A conceptual system with two (or more) concepts, where no connections have been explicitly identified (or, at least, claimed as one might do in developing research hypotheses), the next type is where a gap between them has been identified.

Based on the assumption that we live in a systemic universe where things are connected, Type 2 conceptual systems suggest a research process where we can use a kind of *gap analysis* (Wright & Wallis, 2019: 162) to indicate which direction to take with our research to better surface those unknowns. Using a practical map of circles (one for each concept) and arrows (for causal connections between the circles), we take our investigatory cue from those gaps (any space between circles that does not have an arrow) as an opportunity to conduct research. In short, if one person claims, for example, "A is true and B is true," another might reasonably ask (or investigate) the potential causal relationship between A and B because a gap exists between A and B.

Type 3—"Categories" or "Implied Connections"

In a categorical view, some concepts within a theory are identified as being categorically similar or somehow related as in Figure 4. Regardless of the connections between concepts that *are or are not* expressed, Type 3 conceptual system carries a clear implication that there are some connections without necessarily explaining what kinds of connections they are.

In Figure 4, much of the text is not visible due to space limitations. The point of the figure is to show that there *are* categories, or clumps (there, expressed as "Ventilation," Mechanical," "Electrical," and "Boards/Power) implying that there are additional connections between the concepts contained within each of those categories.

Again, there is an implied call for research. This time, because the connections (and/or overlaps between the concepts) are more clearly implied so the implications for research are more explicit than in Type 2. Type 3, much as Type 1, provides some improved understanding. However, it is worth noting that categorization may be seen mainly as a kind of indicator for relevance. That is, showing what categories or concepts may seem relevant to a particular person or stakeholder group in a particular situation. For example, managers may be interested in a category of theory called "management theory." Such an approach is not highly effective, however, because it does nothing to tell us which management theory might be best or most useful.

Type 4—"Bounding"

For physical systems, each person's world may be said to be bounded by what they understand and what they interact with so that boundaries are erected for the convenience of the observer / analyzer (Hutchins, 1995). That, however, is "other-bounding" while here, instead, we are looking at "self-bounding."

Metaphorically, we may imagine a village as a grouping of building whose inhabitants interact more frequently with one another than with inhabitants of any other village. Where that village is situated on an island, it might be tempting to say that the boundary of the village is the water which surrounds it. That would certainly provide a convenient demarcation. Instead, however, the perspective presented here suggests that the boundary of the city is better defined by the close relationship of the buildings and systemic interaction of its people.

Any time a set of concepts and their connections are described as a theory or model, a "line" has been drawn around them describing or delineating that as a conceptual system. We can think of a theory as existing within its own boundaries (Wallis, 2010); as being self-bounded or as a self-defined conceptual system. The boundary of such a conceptual system is made of the concepts contained/described within it.

In Figure 4, the entire diagram is presented as a complete theory. The boundary for that theory would be identified as the concepts contained within it. It would be a misuse of that theory, therefore, to apply it (except, perhaps, experimentally) to situations that do not include one or more of those concepts.

That is to say, a theory (or other conceptual system) should not be applied to understand and/or engage a physical system where the concepts of the theory are not reflected in the physical system; a theory of electronics should not be applied to situations where no electronics exist.

Type 5—"Transformative"

No part of a conceptual system is completely cut off from its surrounding systems because a system may be in some ways open and closed at the same time. For example, systems may be open to information but closed to matter (Roth, 2019). A simplistic view might suggest that my liver is closed to the open air. However, my liver still receives oxygen via my red blood cells after they have become oxygenated—after a *transformation* of those cells has occurred.

"The difference in stability of conceptual systems may be understood in terms of systemic structure which may be measured using Integrative Propositional Analysis (IPA), (Wallis, 2016a). Those with higher levels of structure are more open to data but less open to changes in conceptual components. Those with lower levels of structure are more closed to data, but more open to changes in conceptual components" (Wallis, 2020a).

A relatively straightforward way to evaluate levels may be seen in differentiating between linear, non-transformative, concepts. However, such representations would only be found in relatively immature theories. A more useful and sophisticated measure of difference would be between transformative changes. That is, for conceptual systems, we may understand theories as having different levels where the path of causality is interrupted by transformation.

For example, returning again to Figure 2 showing one map nested within another, the causal path from A to E is direct, with no transformation. While, in contrast, the causal path from F-7 to F-5 must pass through F-6 which transforms the nature of F-7. Naturally, from a systems perspective, we would look with suspicion on claims that more A causes more E in a linear fashion. However, it serves as an adequate example of causal distance.

Alternatively, we might look at that same figure and see the "simple" concepts or non-transformative concepts (A, D, E, F-1, F-3, F-4, F-5, F-7) as existing on one level of understanding and the transformative concepts (B, C, F, F-2, F-6) as existing on another.

So, another type of boundary is seen where each concept within a theory is structurally defined by the other concepts. For example, in Ohm's Law, the measure of volts is defined by the measures of amps and ohms; each is transformative (Figure 6) from the other two.

Type 6—"Unfolding/Enfolding"

Another approach suggests a kind of enfolding in the structure of theories. Enfolding may be seen as similar to developing "subroutines." Rozin (1976), for example, suggests that as children learn they are also organizing their knowledge into subroutines which makes mental tasks more efficient. Unfolding may be understood as a more rigorous/formal version of "unpacking." For example, using a simple phrase such as "I wrote the document on the cloud" refers to the use of personal memory, cognition, and typing skills along with a vast array of technological systems including computers, telecommunication equipment, and servers (note, each of those things/processes mentioned may also be unpacked).

Here, however, we look at folding and unfolding in a more formal/rigorous way. For example, in the structure: $A \rightarrow B \leftarrow C$, we might see B as the enfolded concept while A and C are the unfolded concepts. Importantly, that is a "transformative" structure (Wallis, 2016b). Indicating that the combination of A and C have resulted in something new.

For a more concrete example from physics, velocity is equal to the distance travelled over a length of time ($V=\Delta d/\Delta t$). To lessen the cognitive load, one might simply ask, "How fast are you going?" instead of asking "what is your change in distance divided by your change in time?" Answers such as "I'm doing sixty" are likely to be clearly understood. The concepts of time and space are enfolded into the concept of velocity.

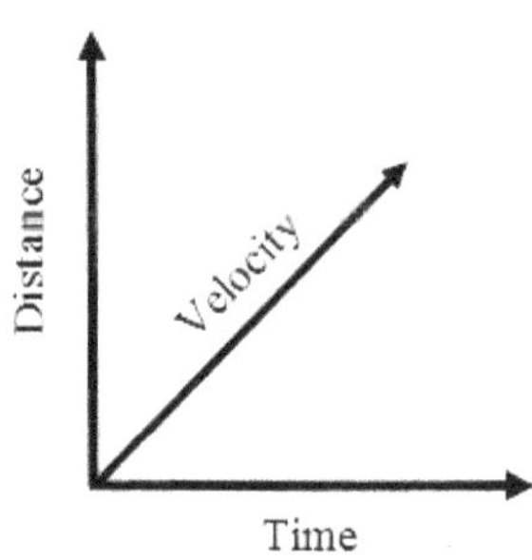

Figure 5 *Velocity, Time, and Distance as folds of one another*

For a more complex example, consider one of Einstein's ideas. According to (Okun, 1989) for a body at rest: $E^2 - p^2c^2 = mc^2$

Where E = energy, c = the speed of light, m = mass, and p = momentum. That momentum has a formula of its own, expressed as: $p = v \cdot E/c^2$

We might see the second formula (for momentum) as nested in the first. We may also see "p" as the enfolded version of $v \cdot E/c^2$. It gives pause to imagine how many things might be enfolded in our assumptions; just waiting to be unfolded so that we may develop a more nuanced and useful understanding of them.

By understanding how concepts may be enfolded and unfolded, we may develop theories that are parsimoniously easier to use while still retaining the requisite variety needed for dealing with complex situations.

Levels of Emergence

While the above section and sub-sections provide an orienting backdrop on a number of ways we might interpret levels, this section takes a much deeper dive into how we may understand different levels based on the idea of emergence.

Briefly, according to Page, there are three kinds of emergence. Simple emergence is obtained by combining parts in an orderly way to reach a predictable outcome. Weak emergence is less predictable than the simple, while strong emergence describes unexpected results that are completely unpredictable (https://sebokwiki.org/wiki/Emergence).

This view dovetails neatly with Sillitto *et al.* (2017) who suggest that a system may be defined as "A collection of possibly interacting, related components that exhibits emergence" (p. 13).

It is reasonable to infer that when we see more emergence occurring, we can be increasingly certain that we have a better understanding of some system. Although, on a side note, we should specifically disregard general claims such as "everything is emerging all the time" because that does not provide a good/useful explanation of what is emerging, or from what the emergence is occurring, or where we might explore to generate greater understanding.

When viewing a conceptual system, simple emergence may be seen in a concatenated logic structure as in Figure 2. Where concept F-6 (as one example) is transformative—concatenated from F-7, F-3, and F-4. In that theory, F-6 is a predictable outcome. For a simple example, in a factory setting, having more workers, more raw material, and more tools will cause the production of end-product widgets.

While such a process seems reasonably clear, understanding unanticipated outcomes is not. Because, it seems, there are two reasons for such outcomes. First, because a theory is structurally weak; the second because it is structurally strong. A structurally weak theory may have too few concepts (and and/or too few causal connections between them) to adequately represent the relevant situation in the physical world. Using a simple theory to understand a complex situation will lead to frequent surprises.

On the other end of the structural scale are highly systemic theories; the best of which have all of their component concepts included in loops (Wallis, 2019) and all of their concepts are transformative—concatenated from other concepts (Wallis, 2016a) such as Ohm's Law (I=E/R). Using Ohm's Law as a theory (represented diagrammatically in Figure 6), it may be seen that each concept, by itself, is perfectly predictable—but only if the other two are known. For example, one can calculate the resistance of a circuit if one knows the voltage and the current. The high level of systemic structure makes the theory highly useful in practical application.

Importantly, therefore, we may say that one emergent property of theory is the usefulness of that theory for understanding and engaging physical world phenomena.

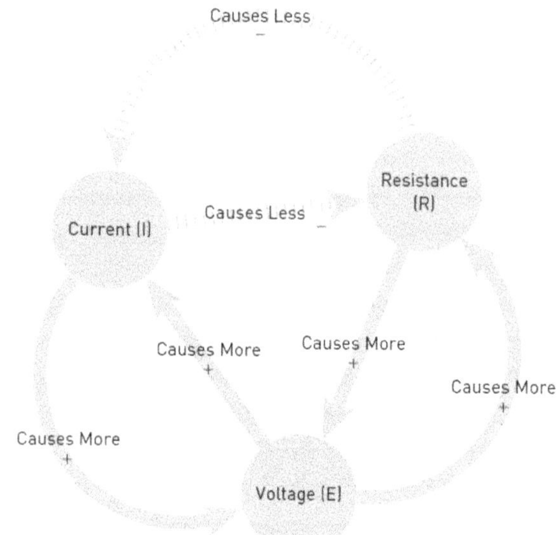

Figure 6 *Ohm's Law as a practical map (Wright & Wallis, 2019: 152)*

With such a highly systemic theory (supported by sufficient data and applied in relevant situations) surprising or unanticipated outcomes will be minimal or nonexistent for the concepts included within the theory. Over the long run, however, the unanticipated outcomes are much more substantial. Because the theories are so useful in practical application, engineers are able to design computers, cell phones, cars, planes, and the plethora of technology that surrounds us. Those vast revolutions in technology and lifestyle emerged as a result of applying the many highly structured conceptual systems to change physical systems.

It is important to note that "Emergent outcomes, by definition, occur at a different *level* of a system than the level in which the original interaction occurred" (Westhorp, 2012: 411, emphasis added). Because current, resistance, and voltage all appear to be at the same level, we can understand emergence as something happening on a different level; represented diagrammatically in Figure 7.

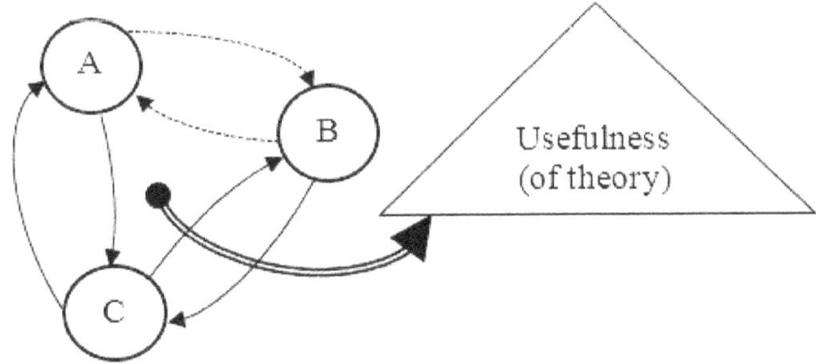

Figure 7 *Abstract example of highly systemic theory with one emergent property (usefulness) shown*

Where objective research may lead to highly systemic theories such as Ohm's Law, there is also a subjective component about the context in which the theory may be applied. "We can combine the benefits of objective study from the modern era with the benefits of subjective study from the postmodern era by adopting a systems approach " (Laszlo & Laszlo, 2009: 7).

To summarize, there are two different mechanisms at work here. The short-term predictable and the long-term emergence. Where predictability may be thought of as a function *within* the interacting concepts at a single ontological level, emergence occurs between concepts at one level and results in something at a *different* ontological level.

We may see concepts with causal connections between them as being ontologically similar; perhaps on some scale of abstraction (Wallis, 2014a) and/or some scale of orthogonality (Wallis, 2020d). While, in contrast, the properties/things/concepts/components that are emergent seem to be on ontologically different level than the concepts from which they emerged.

This perspective suggests a scale of emergence roughly analogous to the one proposed by Page (and generally accepted by SEBOK). At the lowest end, a collection of concepts at varying levels on some scale of abstraction and/or orthogonality. Using such conceptual systems as knowledge is unlikely to result in predictable outcomes and similarly shows little opportunity for emergence. At the highest end, we would see the most emergence from a set of concepts, all closer to the same level of abstraction, orthogonal from one another, transformative (concatenated from other concepts in the theory) and all connected through causal loops.

While the present example has identified one concept (usefulness) as an emergent property, there may be other concepts emerging. The key point of this preliminary investigation is to suggest that theories should be constructed of concepts with clear delineation between levels on two dimensions. The horizontal dimension having concepts of the same ontological level linked with causal connections; and the vertical dimension having concepts of differing ontological levels linked with connections representing relationships related to emergence; it is also worth noting that the higher level concept emerges only from a highly systemic theory as measured using IPA (cf. Wallis & Wright, 2019), not from a single or sets of disconnected conceptual or components (as may be seen in lower level "types" in the above section).

This is an important perspective for developing better theories because, "...in an absolute sense, the emergent products do not exist as such. They do, however, exist in the alternative (although deeply—complexity—related) macro-level description..." (Richardson, 2007: 7). That is to say, emergence seems to occur at every level. "Indeed, in SoSPT the entire range of observed natural systems from galaxies to societies are linked sets of hierarchical levels portrayed in an unbroken series of emergence events" (Friendshuh & Troncale, 2012: 14).

In short, anything and everything may be viewed as emergent. Parts of theories along with concepts and logical inferences, all exist in our minds and the literature as a jumbled interacting mass, from which more systemic theories may occasionally emerge.

Perhaps; however, that does not mean we are doomed to be lost in a maelstrom of relative realities. It is still possible, and even desirable, to assign levels based on our perspectives. Importantly, however, we must recall that it is our *understanding* of emergence and ontological perspectives that we are concerned with here.

If we accept (tentatively, awaiting additional research and empirical validation) that we can avoid the conundrum of infinite regress in attempting to find some foundational perspective, that it is not "turtles all the way down" (cf. Meulenberg-Buskens, 1997) but instead, it is emergence all the way down, we can stop being distracted by "the turtles" and find a better focus for improving our theories.

Whether something appears as an emergent phenomena, or not, depends on one's perspective. For conceptual systems, that may become more confusing as one theory may contain many perspectives—raising the need for very careful theory building to ensure that each theory stays on its appropriate level of emergence by focusing only on those perspectives that are of that same *level* of emergence (same ontological level).

The ability to clarify emergent properties from a systems theoretic perspective helps to resolve the tricky business of understanding our world because we don't always know what we are looking at. For example, one might look at a group of people who work together and say it is a "team." However, in some views of systems thinking, one might "draw a line" around anything and call it a system. So, calling a certain group of people a team is a kind of "shorthand" for all that is understood to be (and, quite possibly, not understood to be) a team. For example, according to https://en.wikipedia.org/wiki/Team "As defined by Professor Leigh Thompson of the Kellogg School of Management, "[a] team is a group of people who are interdependent with respect to information, resources, and skills and who seek to combine their efforts to achieve a common goal".1 Yet, when we look at a group of people we do not see that interdependence, their resources, or skills. So, instead, it might be more useful to think of "team" as an emergent property of those interdependent people.

For a hypothetical example, in Figure 8, we might imagine that concepts A, B, and C represent Number of People, Interdependency, and Resources—all causally interacting in a way that results in the emergence of Team (or, perhaps, "team-ness") as Emergent Property 1.

The insights in this section suggest new standards for representing theories. Such standards are needed to provide greater clarity, replacing the many ad-hoc ways diagrams have been presented. Here, we draw a distinction instead between connection #1 and #2 presented in Figure 8 and tentatively suggest that theories should include only causal connections *within* ontological levels and only emergent connections *between* ontological levels.

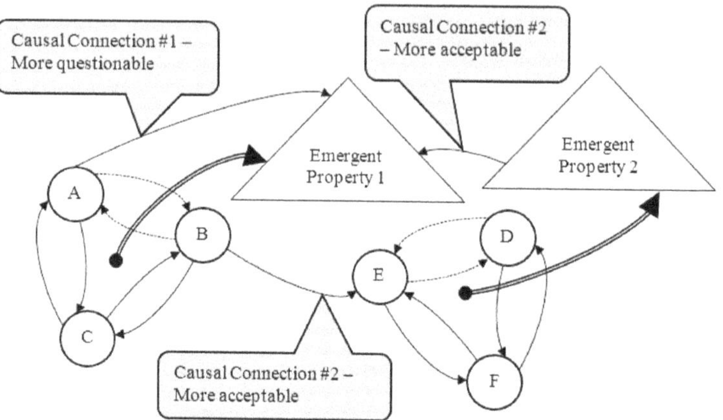

Figure 8 *Correct and misleading representations of causal connections within and between levels/scales of emergence*

In keeping with our understanding of a highly complex universe it may be noted that an individual concept does not emerge from another individual concept; rather emergence occurs from the interaction of multiple concepts. Although, it is certainly possible (indeed, seems quite likely) that multiple concepts may emerge from multiple concepts. The difficulty lies in sorting them out in a way that their causal and emergence relationships provide us with useful theories for practical application.

Diagramming Conventions

To more clearly communicate the understandings of scholars in an increasingly fragmented field, a set of standards is suggested here. Naturally, this is only a starting point; but one that scholars, practitioners, and editors might discuss at conferences, eventually leading to general agreement for the advancement of the field.

As Figure 9 shows, the shape of each conceptual component of a diagram might be changed to indicate its level on some scale of emergence. Here, triangles are seen as emergent properties of circles, rectangles as seen as emergent properties of triangles, and 3-D rectangles are seen as emerging from 2-D rectangles. Lines (with arrows indicating direction) may be shown as:

- Solid = "causes more" (may also have a "+" sign at head of arrow);
- Dashed = "causes less" (may also have a "-" sign at head of arrow);
- Double = "emergence";
- Shaded (with descriptive label) = non-causal relationship.

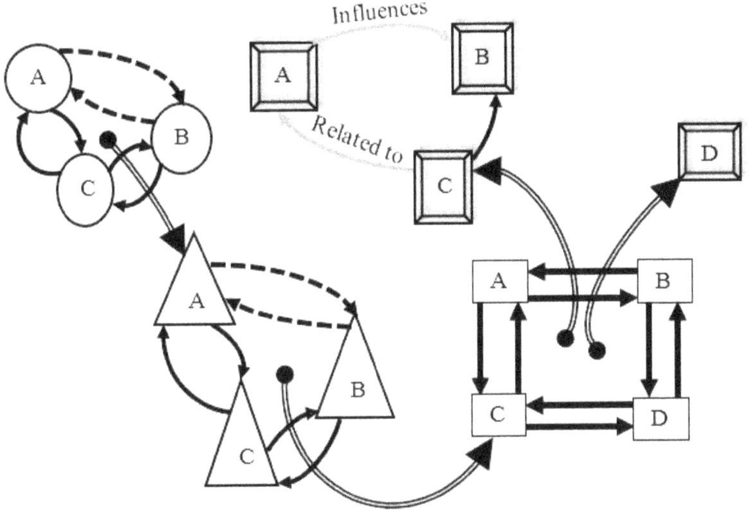

Figure 9 *Proposed norm for representing four levels of emergence within a single diagram*

The insights presented in this section offer fundamentally new thinking about the phenomenon of emergence with profound implications for improving our conceptual systems and accelerating our science to better understand and address the many problems facing our planet and our civilization.

As suggested by the typology developed in the previous section, emergence is not the only way we can recognize levels of difference. In the next section, we will delve into a perspectival approach to understanding levels in conceptual systems.

Levels of Perspective

The notion of perspective has been deeply explored in the literature across multiple disciplines. Common wisdom across those perspectives on perspectives seems to suggest that we should accept that there are many perspectives, each one important to some stakeholders, and that communities of stakeholders should strive to find common ground through techniques such as dialog and "dot voting." As those topics are adequately explored elsewhere, the present paper instead strives to find new ways of understanding perspectives by looking at the *structure* of our conceptual systems.

Diagrammatically, let us begin with the idea that a theory may be represented by boxes (for concepts) and arrows (for causal connections). That theory represents some situation that is shared by different stakeholder groups. Each stakeholder group might look at that theory from a different perspective; and, from each unique perspective, see their priorities as the parts of the whole that are perspectivally closer (Wallis, 2020d). Indeed, some concepts may appear so close and large that other concepts may appear to be "hidden" behind them. Thus, much as some stories or narratives, some parts of the theory are emphasized while others are de-emphasized. The same may also be true of causal connections between the concepts; where some connections are concealed and others emphasized depending on the viewers' perspectives.

For a diagrammatic example, please consider Figure 10 where the central diagram represents the complete view of some situation, and the diagrams on either side represent the views of different stakeholder groups operating at different levels of the overall society.

The diagram on the left of Figure 10, we might say, represents the view of the BD group while the one on the right represents the AC group. Each prioritizes their own while minimizing the views of the other. Now, let us re-orient the diagrams further.

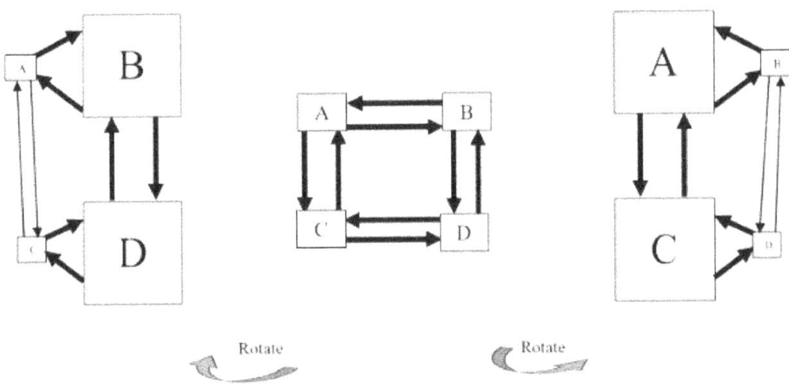

Figure 10 *One theory, rotated to emphasize different perspectives and so concepts - from (Wallis, 2020d)*

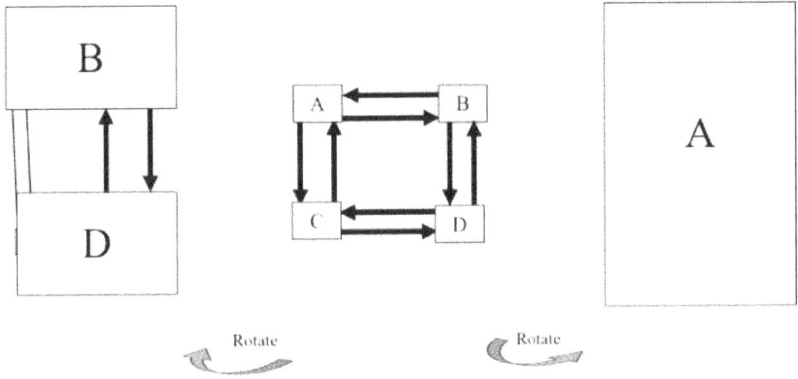

Figure 11 *Conceptual rotation of a single model to represent sub-groups' increasingly narrow perspectives and so foci.*

Figure 11 suggests that the BD group is so focused two concepts that they have lost sight of the others. Except, in the diagram on the left of Figure 11, for a slight hint that there *may* be another concept. There are also connections (thin lines) but they are not seen as causal arrows. Again, that knowledge is hidden. The A group (on the right hand side of Figure 11) is so focused on their one idea/goal/concept that they see nothing else. This kind of simple model suggests a kind of imperviance (Wallis & Valentinov, 2016a) to thinking that may be overcome only with difficulty.

Operationally, this suggests a difficulty with the common approach to prioritizing where organizational participants brainstorm ideas, then use dot voting to determine where to focus their collective efforts. Such an approach may help focus the group on a few concepts (perhaps the group's goals), but that seems reductionist because it is done at the cost of concealing other concepts and causal connections. This, in turn, suggests that a better approach to developing and clarifying knowledge would be to use causal knowledge mapping which would include a kind of gap analysis (Wright & Wallis, 2019: 162) to develop *interdependent* goals.

This view of views takes on a particular importance when attempting to discern causal relationships between concepts. For example, looking at Figure 12, from the perspective of concept A, concepts B and C may be seen as one level (or causal link) removed from concept A; while concept D is two levels removed.

Consider a stakeholder group for whom A is perspectivally near and D is far. So far, that they do not see or recognize D (that lack of recognition represented by shading). Nor do they recognize the causal links directly related to D (B↔D and C↔D). So, instead of an accurate understanding, the stakeholder group may incorrectly infer that *other* causal connections exist because their interactive effects on visible levels (B, C) as indicated by the shaded dashed arrows. Naturally, such inferences are likely to be inaccurate because A does not have data about D, nor does A understand the causal logical connections between D and other components. Thus, A "fills in" the blank spot on the map with assumptions of causal connections.

Such a "knowledge fog" may apply also (perhaps more so) to differences between ontological levels of emergence. So much so, that with sufficient distance, events may appear random or unpredictable.

Thus, we have reductionist thinkers (e.g., Kim, 1993) striving to explain the mind as reducible to the body because they are using only causal connections to explain connections that would more be more appropriately explained with connections representing emergence.

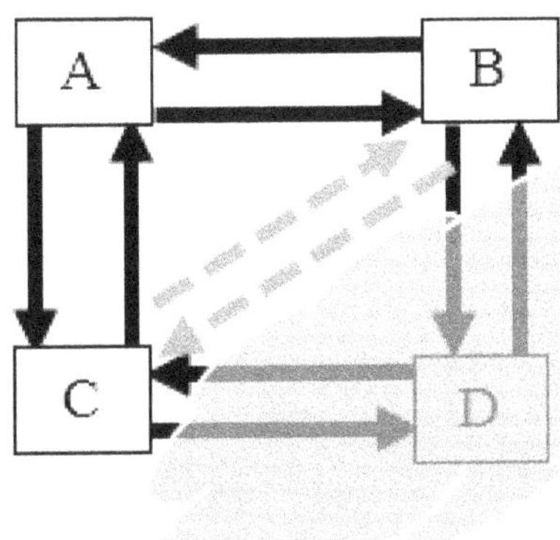

Figure 12 *Inferred causal connections (dashed arrows) based on lack of understanding of actual causal connections*

For a more concrete example, one stakeholder group may fill a gap in understanding with the idea that deities are causing lightning (instead of naturally occurring atmospheric conditions). Their mistaken assumption may, in turn, lead to behaviors that are counter-productive (e.g., performing ritual sacrifices to appease some deity instead of installing lightning rods). This perspective on perspectives has profound implications for policy, planning, knowledge use, and decision making.

For a more practical example, Figure 13 shows ten theories of poverty synthesized into one (Wallis & Wright, 2019). This represents a shared understanding of poverty according to the ten stakeholder groups (from five academic disciplines and five organizations across the political spectrum).

However, in contrast, the view from only one of those groups may be seen in Figure 14.

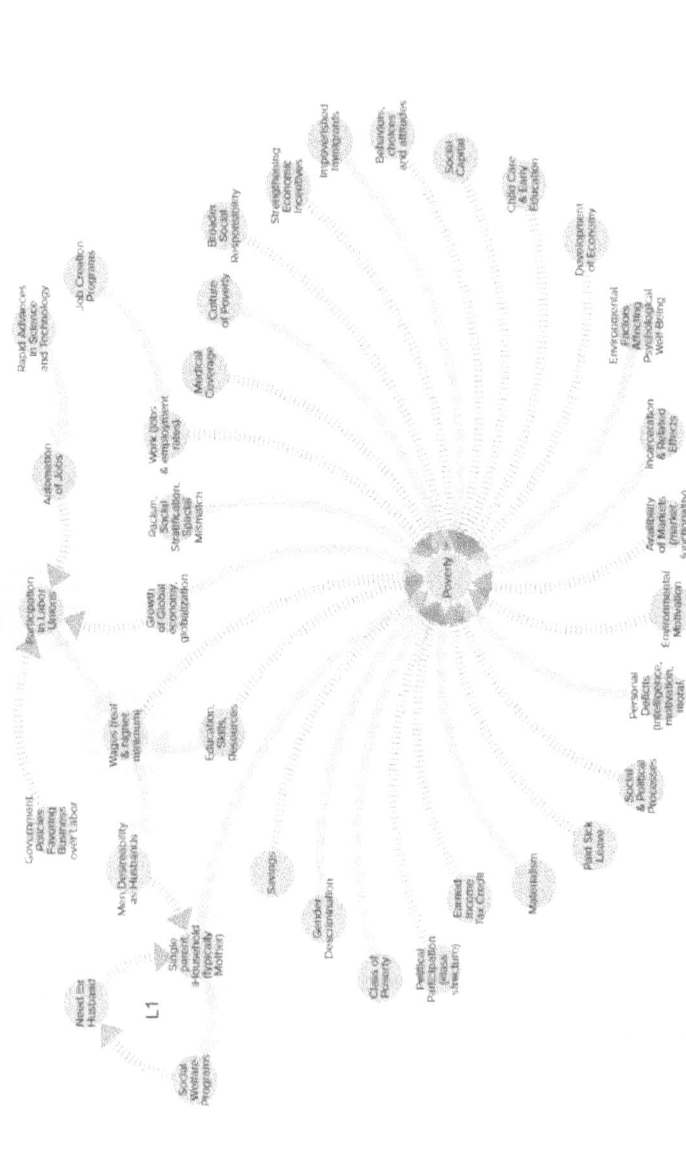

Figure 13 *Interdisciplinary synthesis of 10 theories of poverty (Wallis & Wright, 2019). To read details more clearly, please visit: https://kumu.io/Steve/caps2018*

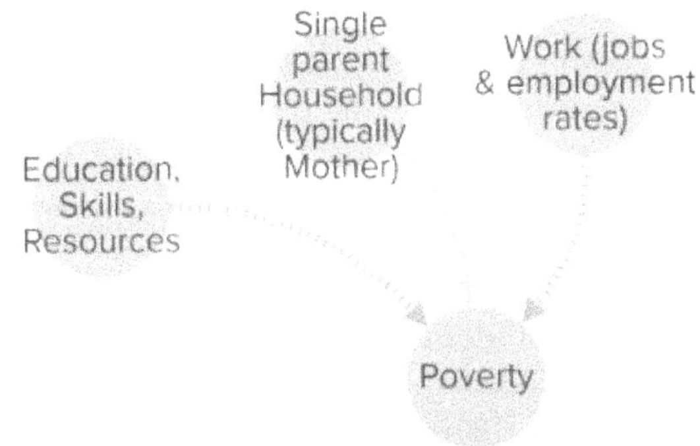

Figure 14 *Understanding of poverty from one stakeholder group's perspective (concepts and connections are also in Figure 13)*

Believing that the problem is fully understood, this group (with their single stakeholder perspective) may begin to take action on the problem of poverty. They may also argue *against* the actions of other stakeholder groups believing that the other groups have misrepresented the situation. In contrast to either view or all views, the present paper along with studies on structure cited through this paper suggests a range of actions to develop more effective theories/policy models as conceptual systems to enable more successfully addressing the issue and collaborating with other stakeholder groups.

Hutchins (1995) uses the term "complexity shifting" to describe how a group shifts the difficulty or complexity of a problem to an expert (for whom the problem appears less complex) or to a tool (that may be used to resolve the problem more easily). By using the techniques as discussed in previous sections, we are shifting the complexity of our theories to better address seemingly impenetrable problems of society and policy. These techniques may also suggest how we *improperly* reduce our cognitive load (in the short run) by ignoring theory and making decisions based on limited "moral" perspectives, simplistic theories, false assumptions, or even random choice.

Such improper reductions, however, are unlikely to result in desirable situations; they are not sustainable (Wallis & Valentinov, 2016a, 2016b, 2017). The view of Figure 14 is only about 10% of the view of Figure 13 reflecting a much more limited understanding of the poverty problem. In short, while perspective-based approaches to identifying levels (e.g., listing priorities or describing hierarchies) may have some benefit for improved understanding, they are as likely to conceal as reveal.

This perspective on perspectives emphasizes the need to include multiple stakeholder views when developing theories, policies, strategies, and other conceptual systems. This relates to dual descriptions (Bateson, 1979) and conversations to build a socially constructed reality (Dent & Umpleby, 1998). This is important because it is the observer who determines the complexity of the system (Zelmer, Allen, & Kesseboehmer, 2006). Or, to put it another way, "A 'system' is a set of variables selected by an observer" (Umpleby, 2009: 4). So, it may be more accurate to state that observers, consciously and/or unconsciously through their perspectives, determine the *appearance* of complexity.

Those differing views may lead to conflict. Especially if the views represent simple conceptual systems. For example, two groups in the same nation may each insist that their shared nation move toward a goal that excludes the other group's goal; thus leading to conflict.

Adding perspectives to expand the number of concepts in the theory stands in contrast to "normal science," which tends to suggest investigating the fewest alternatives with an eye to choosing "the best" one. A systems/structural perspective in contrast, suggests expanding the shared model. So, instead of looking at a simplistic model of two concepts (A, B) and pushing to a choice of "A or B," we should seek to improve the model by adding C which may be a direction that will benefit more levels of society (Beckmann, Hielscher, & Pies, 2014; Hielscher, Pies, & Valentinov, 2012).

This diagrammatic approach to understanding perspectives, and how those perspectives affect our understanding across and between levels of conceptual and social systems, may be useful for consultants and other change agents in explaining to their clients how we often do not understand the limits of our own understanding.

Before moving on, it is worth stressing some kcy concepts from this section. While all perspectives may be useful to some extent, they (as conceptual systems like theories, models, strategic plans, etc.) should be thought of as starting points for further exploration. Adding more concepts to those perspectives will improve them when the concepts are of the *same* ontological level and when causal connections between the concepts are identified. More importantly for this new perspective on emergence, concepts on *differing* ontological levels should be linked with connections showing emergence; however, such connections are more valid when they are understood as emerging from a highly systemic conceptual system as measured using IPA. This process, applied with a reasonable amount of rigor and careful consideration seems likely to impel our field to new ways of thinking and set the stage for accelerating the advance of science.

Future Directions

As this paper is somewhat exploratory in nature, it is as much about the future as the past. While the limitations of the human mind are insurmountable (Wallis & Valentinov, 2016b), we may expand those limits through the use of theories, computers, and collaboration. However, those also have limits. But by understanding our limits we may learn to exceed them by gaining additional, perhaps orthogonal, insights (Pies, 2016; Wallis, 2020d). In this section, we will identify a few directions for future research and practice.

Generally, scholars interested in the various methods of mapping (e.g., concept maps, mind maps, causal maps) should strive to develop quality standards for their respective methods (cf. Wexler, 2001) so users may differentiate between high and low quality maps within each method. This, in turn, will support comparisons and improvements between methods. It may be, for example, that a high quality concept map (with descriptive, rather than causal, connections between the concepts) may prove to be more useful to planners than a low quality casual map (which includes only causal connections). These kinds of efforts will support continual improvement of all methods and the advancement of the field.

Importantly, for synthesizing theories, this paper suggests that we should first strive to identify concepts at each level of emergence, then, link concepts into theories according to their levels of emergence, rather than try to force-fit concepts between levels.

The existence of levels suggests that it may be worthwhile to measure the *distance* between levels according to one or more ways suggested here (e.g., emergence, transformation, time). Such a measure may be useful in developing a periodic table of theories (Feuer, 1995; Wallis, 2014c) to further support improvements in theory and practice. The most basic approach would be to measure or quantify the levels between concepts within theories; this may be considered an IPAx (experimental technique within the larger family of Integrative Propositional Analysis methods) approach based on system dynamics.

For example, in Figure 2, C may be seen as being four causal steps removed from D suggesting some measure of structure for the theory overall. In contrast, F4 is only two causal steps removed from several other concepts—suggesting higher structural cohesion. However, F3 and F4 have no causal connections between them—suggesting a lack of structural closeness. In greater contrast, in Ohm's Law (Figure 6) each concept is only one causal step away from each of the other two concepts.

We may anticipate that as theories evolve from lesser to greater usefulness that there will be fewer steps between the concepts (which implies also that a greater percentage of the concepts will be connected via causal pathways). Such a measure may serve as an indicator of the *completeness* of the theory and so guide scholars toward the development of better theory and guide practitioners and professors in the choice of more useful theories. Similar approaches may be developed to evaluate the range of vertical levels of emergence contained in the theory. And, both may be used someday to develop a periodic table of theories or perhaps some algebra of emergence.

Another important advance to be considered is the automation of these processes for evaluating and advancing theory. While current artificial intelligence technology can support coding of texts, it has not yet reached the level where in can replace the need for human sense-making. If peer reviewed papers could be analyzed by an artificial intelligence, the propositions clearly delineated, and the results then run through the process of IPA and IPAx, we may advance the social/behavioral sciences at an astonishingly rapid pace with attendant benefits for humanity.

Also, this structural view hints at a new approach for finding and/or managing "unknown unknowns" (cf. Stefano, Camillo, & Riccardo, 2014). Because, practically speaking, something that is unknowable to one person may be known to another. So, by looking at the systemic structure of each level, and determining the emergence of each level, we may see where something is "missing" from the diagram (and, at what level of emergence it is missing from, and so where to look to find it). Thus the practice of gap analysis (noted in above sections) may be applied to evaluate and understand what is missing between ontological levels of emergence in much the same way that it is now applied to understand what is missing (and so to suggest the opportunity for research) between concepts within the same ontological level un-connected by causal linkages.

Future investigations may suggest changes to our understanding of conceptual sub-structures. That is, emergence may be thought of as a sixth "logic structure" in addition to the existing five (atomistic, linear, branching, concatenated, and circular) (Wallis, 2016b). A seventh structure may be seen as existing (set, in some sense, between atomistic and linear), where multiple concepts exist, but are connected with non-causal arrows.

A "static" theory (where a "snapshot" of a theory is captured on a page) may be combined with "dynamic" theories of action (e.g., action research) even though the dynamic theory may also be a snapshot on a page. This may result in a world-bridging metatheory where iterations of intertwined theories and their related activities are developed. Those metatheories may pave the way for a much deeper level of understanding—where iterations of activities change the theoretical "rules" of iterations.

Unaddressed here, but worthy of additional research, is the idea of downward causation (cf. Emmeche, Køppe, & Stjernfelt, 2000). What affects do emergent properties have on submergent properties? Might the emergent concepts affect the submergent concepts as a whole system? Individual components? The causal connections between them? Much remains to be explored.

Returning our focus to causal closure, (Kineman, 2012), in discussing Rosen's perspectives, notes that a key question to consider relates to what might be considered as inside compared to what is outside the system. With a physical structure, it is relatively easy to decide who is inside or outside a building. With the fragmented nature of our conceptual ecology, however, the difference is not always so clear (for example, in the above typology, "fuzzy" conceptual systems. Much like a pile of boards does not have an inside or outside in the same way that a house has, we can see what is in or out of a conceptual system only when it has a high level of structural closure. As such, we may use IPA (cf. Wallis & Wright, 2019) to measure "how structured" and therefore "how closed" a conceptual system is.

For example, in Figure 2, the nested map has an IPA Systemicity (measure of causal logic structure) of 0.27 (two concatenated concepts divided by seven total concepts) so may be said to be 27% closed (or, conversely, 73% open). The high level map is 50% closed, while Figure 6 is 100% closed. The goal here is not to close the conversation; rather, to move it forward by providing new tools and insights.

Measuring closure, at least in the realm of conceptual systems, provides a new technique for pushing our thinking to new and more productive directions for addressing such deep philosophical issues as conceptual closure and the mind-body problem. Especially as moving toward closure on one dimension pushes us toward emergence on another. It is not sufficient to claim that the mind emerges from the body. Such a presentation is too sparse to be of much use. Instead, enough concepts and causal connections for the physical must be provided to justify a claim of emergence. In short, this approach will help us to understand the role of emergence in conceptual systems as a path to developing better representations of our physical systems.

This is also a step forward toward and (perhaps) a resolution of the micro-macro problem (Alexander *et al.*, 1987). That problem is presented in the Sorites paradox, or the paradox of the heap. From one perspective, we may ask, "If you have a heap of sand, and remove one grain of sane at a time, when does the heap stop being a heap?" Or, we may ask, "How many individual grains of rice are required in order to have a heap of rice?" While there are many ways to parse the answer from a component perspective, another way to think of it is as a problem of understanding emergence—or emergent dimension properties. Or, to say it in another way, grains and heaps are on different ontological levels. They are of different universes and so cannot be measured with the same yardstick. One might measure the number of grains by counting them, or the number of heaps by counting them; but heaps may also be measured according to their height and the slope of their sides (which is determined in part by the material of which the heap is made). Otherwise, one might just as well ask, "How many two dimensional squares can you fit into a three dimensional box?" Such a question is meaningless (and, worse, useless) unless you understand the three dimensional construct as an emergent property of two dimensional constructs interacting or connected at right angles.

Some differences may be separated by distance while other may be separated by time in a complex multilayered way (Tateo & Valsiner, 2015). Awareness of a "time horizon" for connectedness of systems seems to be important as we attempt to address simple problems (e.g., the operation of a thermostat) and more complex problems (e.g., global social evolution, advances in technology, and such). Some theories, particularly statistical maps, seem to suggest that causal change occurs instantaneously. That may seem to be the case for some theories/laws of physics; however, at the human/social scale very little or nothing happens that quickly! Other theories, particularly some computer models, have explicitly built-in time scales to show how much time must pass between an action and a result. Given that there are some unexpected outcomes that occur within moments while others emerge over centuries, the study of time differences does not merely "call for more study," it practically screams.

It seems difficult to see the trees (as a submergent property) and the forest (emergent property), at the same time. Human individuals have difficulty comprehending the world more than one or two levels removed from their immediate perspectives. From a perspectival view, the CEO has little understanding of the daily knowledge required and deployed by the line workers. The voter has little understanding of candidates for public office. This has profound implications for the developing a just society (Wallis

& Valentinov, 2016b). The present paper suggests that there are significant implications for working with maps of different stakeholder groups. For example, using a map made in the board room may create confusion if it is used as a reference for communicating with a supervisor working in the boiler room. When each level map is more (someday, completely) systemic, so that each level is rigorously understood as emerging from the other (not only in some fuzzy metaphorical sense, but actually, objectively verified as such) we will be able to more effectively communicate between levels of human organizations. The implications raised here might be explored in relationships to levels of emergence within and between social/organizational systems. Imagine, for example, an organizational chart based on levels of emergence, rather than a hierarchy of power and (supposed) control? A challenging notion!

Scholars often have addressed, developed, and presented theories in the form of text. Sometimes confusing, that process may support the process of "cherry picking" (Wallis, 2016b) and lead to moral conflicts as scholars argue over small pieces of the puzzle pulled from multiple larger texts. We must be careful not to confuse the piece with the puzzle. Instead, our field should purposefully move beyond the postmodern perspective of generating ever-increasing number of concepts and theories, and move instead toward synthesizing those theories. If we are to bring unity to the many disciplines and fields of science, we may want to begin with our own.

The above summaries, contributions, reflections, insights, and conversations suggest a scale of relative usefulness for conceptual systems and the diagrammatic maps that represent them. The lowest level is a set of concepts which are poorly understood ontologically. There would be little or no empirical connection or "correspondence" between each of the concepts and something that is measurable or measured in the physical world. For example, one might say that there is an "essence" of a cat. Without a measure, "essence" is at the lowest level of understanding or practical usefulness. The second level, and slightly more useful, is one where there are measurable concepts but having few or no connections between them. For example, a list of terms (e.g., cat, mouse). The third level would be a set of concepts with conceptual connections between them. For example "cats chase mice." Here, "chase" is the conceptual connection between the two actors. Fourth, are theories with measurable concepts having causal connections between them. For example, "Having more cats will cause there to be fewer mice). At the fifth level, we have concepts at differing ontological levels (with some causal connections between concepts at the same level). The sixth, and perhaps highest level, are those conceptual systems which include measurable concepts, causal connections, and emergent connections to concepts of differing ontological levels. For example where, "The interaction between cats and mice will lead to their co-evolution." Here, coevolution is the emergent property.

From a structural perspective, that is for developing theories of improved structure for improved usefulness in practical application, there now exists a range of techniques within the broader structural methodology. In that methodology, Integrative Propositional Analysis (IPA) may be used to evaluate the Complexity (that is "simple complexity," complicatedness, or the number of concepts within the theory) and Systemicity (measure of systemic structure) of theories (Wallis, 2016a) and to integrate/synthesize theories within and between disciplines (Wallis, 2014b, 2020c) with objectivity and rigor.

Theoretical models may also be evaluated to improve the level of relative abstraction between concepts (Wallis, 2014a) and to identify the core and belt of theories (Wallis, 2014c) along with the sub-structures of concepts which may be used to show what parts of the theory may be more useful in application and which require more research (Wallis, 2016b). Another technique is to identify and increase the number of loops within each theory and increase the percentage of concepts included in those loops (Wallis, 2019). Additionally, concepts within conceptual systems should be evaluated and revised to improve the orthogonality between them (Wallis, 2020d). The present paper adds to this suite of tools the idea that we may evaluate, separate, and measure the levels between causally and emergently connected concepts to better understand and advance theory.

It is worth mentioning that as each technique is applied, it may change that theory's rating/score according to one or more other techniques. Therefore, scholars should cycle between these techniques; along with improving theory through the other two (non-structural) dimensions of knowledge by improving the quantity and quality of the supporting data and the relevance or meaning to the users of the theory (Wright & Wallis, 2019).

Conclusion

In the world of physical (including social) systems it is easy to apply labels—to claim that we have identified important levels of difference. However, it is sometimes difficult to see the value of those levels and to adjudicate between conflicting claims (e.g., the benefits of flat vs. hierarchical organizational structures) because of our inescapable biases and because we have not identified immutable "laws" of the social world. The conceptual world offers an abstract snapshot, frozen in time, and governed by the fewest possible rules which themselves are based in well-founded assumptions of systems thinking, as a way to re-think and so better understand our physical world.

Generally, by using the above structural techniques to understand, evaluate, synthesize, and improve conceptual systems, scholars may make more rapid progress in developing theories that are more useful in practical application. Such progress is necessary for overcoming our individual and collective ignorance (Wallis & Valentinov, 2016a) and developing more sustainable theories (Wallis & Valentinov, 2017) that are more easily modeled for experimentation and insight (Wallis & Johnson, 2018) for a more just (Wallis & Valentinov, 2016b) and sustainable world within and beyond our organizations (Wallis, 2020b).

Understanding causality, emergence, and perspectives through diagrams of theory provides benefits to those working at the highest level of human aspiration. For example, the idea that, "Liberty consists of being able to do anything that does not harm others: thus, the exercise of the natural rights of every man or woman has no bounds other than those that guarantee other members of society the enjoyment of these same rights." (The Declaration of the Rights of Man and of the Citizen of 1789, cited in Fink, 2019). However, even well-intentioned individuals may harm others out of ignorance. That ignorance is a lack of knowledge and a function of the levels of that knowledge. That is to say, if one's theory has gaps, has levels where something is missing, it may be suggested that the person should strive to develop a better theory to inform their action lest they cause unintentional harm instead of optimizing their ability to generate sustainable and purposeful good.

This paper has presented (diagrammatically in Figure 12) a structural understanding of the ease with which causality may be *improperly* inferred. To resolve such problems, we suggest three general approaches for improving conceptual system such as models and theories which may be understood as representations of useful knowledge (Wright & Wallis, 2019). First, adding perspectives from additional stakeholder groups to expand the scope of the theoretical map (rather than deploying reductionist science). Second, triangulation through a variety of research methods to gather more data relating to the theory. Third, the use of IPA / IPAx methods developed in this and other papers to evaluate and improve the structure of the theory.

Fractal organization theory suggests that levels of hierarchy in organizational systems may be counterproductive and that organizations should, instead, adopt flatter organizations based on hierarchies of emergence (cf. Raye, 2014). Here, we find similar insights suggesting that theories should be created on (at least) two dimensions. First, a horizontal dimension where components are concepts of the same ontological level with causal connections between them. Second, a vertical dimension where components are concepts of different ontological levels with emergent connections between them. More specifically, recalling that concepts of a lower ontological level must be a highly systemic group from which a higher level concept emerges—one concept does not emerge from one concept. Because this structure of conceptual systems seems to mirror the structure of natural systems, this perspective enables us to create the largest variety of *useful* theories following the fewest number of rules.

With the present approach, we may measure causal closure and emergence; so, we have new tools for advancing conversations around emergence that will enable us to more effectively resolve philosophical questions such as the mind-body problem and other conundrums because we can now better understand theories on two dimensions of structure (horizontal causality and vertical emergence). This bold new approach to advancing knowledge and science is particularly useful because it may be used more pragmatically to support the creation of theories, policies, and other conceptual systems, that better reflect the systemic world in which we live.

References

Alexander, J. C., Giesen, B., Munch, R., Anderson, W. T., & Smelser, N. J. (eds.). (1987). *The Micro-Macro Link*: University of Chicago Press.

Ashmos, D. P., Huonker, J. W., & Reuben R. McDaniel, J. (1998). Participation as a complicating mechanism: The effect of clinical Professional and middle manager participation on hospital performance. *Health Care Management Review, 23*(4), 7-20.

Axelrod, R., & Cohen, M. D. (2000). *Harnessing Complexity: Organizational Implications of a Scientific Frontier*. New York: Basic Books.

Bachmann, J.-T., Engelen, A., & Schwens, C. (2016). Toward a better understanding of the association between strategic planning and entrepreneurial orientation—The moderating role of national culture. *Journal of International Management, 22*(4), 297-315.

Bateson, G. (1979). *Mind in nature: A necessary unity*. New York: Dutton.

Beckmann, M., Hielscher, S., & Pies, I. (2014). Commitment strategies for sustainability: how business firms can transform trade-offs into win-win outcomes. *Business Strategy and the Environment, 23*(1), 18-37.

Clayton, P. (2006). Conceptual foundations of emergence theory. *The re-emergence of emergence: The emergentist hypothesis from science to religion*, 1-31.

Dent, E. B., & Umpleby, S. (1998). Underlying assumptions of several traditions in systems theory and cybernetics. In R. Trappl (Ed.), *Cybernetic and Systems '98* (pp. 513-518). Vienna, Austria: Austrian Society for Cybernetic Studies.

Emmeche, C., Køppe, S., & Stjernfelt, F. (2000). Levels, emergence, and three versions of downward causation. *Downward causation. Minds, bodies and matter*, 13-34.

Feuer, L. S. (1995). Varieties of Scientific Experience Emotive Aims in Scientific Hypotheses.

Fink, G. (2019). Freedom of science. *Kybernetes, in press.*

Friendshuh, L., & Troncale, L. (2012). *SoSPT 1.: Identifying fundamental systems processes for a general theory of systems*. Paper presented at the International Society for Systems Sciences (ISSS), San Jose, California, USA.

Gabbani, C. (2013). The Causal Closure of What? An Epistemological Critique of the Principle of Causal Closure. *Philosophical Inquiries, 1*(1), 145-174.

Gonzales, R. M. (2001). *Completeness of Conceptual Models Developed using the Integrated System Conceptual Model (ISCM) Development Process*. Paper presented at the INCOSE International Symposium.

Harder, J., Robertson, P. J., & Woodward, H. (2004). The spirit of the new workplace: Breathing life into organizations. *Organization Development Journal, 22*(2), 79-103.

Hawe, P., Webster, C., & Shiell, A. (2004). A glossary of terms for navigating the field of social network analysis. *Journal of Epidemiology & Community Health, 58*(12), 971-975.

Hayward, J., & Roach, P. A. (2017). Newton's laws as an interpretive framework in system dynamics. *System Dynamics Review, 33*(3-4), 183-218.

Hielscher, S., Pies, I., & Valentinov, V. (2012). How to foster social progress: an ordonomic perspective on progressive institutional change. *Journal of Economic Issues, 46*(3), 779-798.

Higgins, P. A., & Shirley, M. M. (2000). Levels of theoretical thinking in nursing. *Nursing Outlook, 48*(4), 179-183.

Howick, S., Eden, C., Ackermann, F., & Williams, T. (2008). Building confidence in models for multiple audiences: The modelling cascade. *European Journal of Operational Research, 186*(3), 1068-1083.

Hunt, J. G. G., & Ropo, A. (2003). Longitudinal organizational research and the third scientific discipline. *Group & Organizational Management, 28*(3), 315-340.

Hutchins, E. (1995). *Cognition in the Wild*: MIT Press.

Kalmar, D. A., & Sternberg, R. J. (1988). Theory knitting: An integrative approach to theory development. *Philosophical Psychology, 1*(2), 153-170.

Kim, J. (1993). *Supervenience and mind: Selected philosophical essays*: Cambridge University Press.

Kineman, J. J. (2012). R-Theory: A Synthesis of Robert Rosen's Relational Complexity. *Systems Research and Behavioral Science, 29*(5), 527-538.

Klein, K. J., Tosi, H., & Albert A. Cannella, J. (1999). Multilevel theory building: Benefits, barriers, and new developments. *Academy of Management review, 24*(2), 243-248.

Klenk, J., Binnig, G., & Schmidt, G. (2000). Handling Complexity with Self-Organizing Fractial Semantic Networks. *Emergence, A Journal of Complexity Issues in Organizations and Management, 2*(4), 151-162.

Laszlo, A., & Laszlo, E. (2009). The systems sciences in service of humanity. *Systems Science and Cybernetics-Volume II*, 32.

Leeuw, F. L., & Donaldson, S. I. (2015). Theory in evaluation: Reducing confusion and encouraging debate. *Evaluation, 21*(4), 467-480.

Lindberg, C. (2004, 12/17/2004). [Conversation on the use of complexity theory in personal and professional practice].

Lowe, E. J. (2000). Causal closure principles and emergentism. *Philosophy, 75*(4), 571-585.

Mandelbrot, B. B. (1977). *Fractals: form, chance, and dimension* (Vol. 706): WH Freeman San Francisco.

McGlashan, J., Johnstone, M., Creighton, D., de la Haye, K., & Allender, S. (2016). Quantifying a systems map: network analysis of a childhood obesity causal loop diagram. *PloS one, 11*(10), e0165459.

McKelvey, B. (2004). Toward a 0th law of thermodynamics: Order-creation complexity dynamics from physics and biology to bioeconomics. *Journal of Bioeconomics, 6*(1), 65-96.

McNamara, C. (2009). The geometry of thinking *Systems Science and Cybernetics* (Vol. 1): EOLSS Publishers Co Ltd

Meulenberg-Buskens, I. (1997). Turtles all the way down?—On a quest for quality in qualitative research. *South African Journal of Psychology, 27*(2), 111-115.

Midgley, G., & Rajagopalan, R. (2019). Critical Systems Thinking, Systemic Intervention and Beyond. *The Handbook of Systems Science. New York: Springer.*

Moss, M. (2001). Sensemaking, Complexity and Organizational Knowledge. *Knowledge and Process Management*, 217.

Neal, J. W., & Neal, Z. P. (2013). Nested or networked? Future directions for ecological systems theory. *Social Development, 22*(4), 722-737.

Okun, L. B. (1989). The concept of mass. *Physics today, 42*(6), 31-36.

Pascale, R. T. (1999). Surfing the edge of chaos. *Sloan Management Review, 40*(3), 83.

Pies, I. (2016). The ordonomic approach to order ethics *Order Ethics: An Ethical Framework for the Social Market Economy* (pp. 19-35): Springer.

Raye, J. (2014). Fractal organization theory. *Journal of Organizational Transformation & Social Change, 11*(1), 50-68.

Richardson, K. A. (2007). On the Relativity of Recognizing the Products of Emergence and the Nature of the Hierarchy of Physical Matter. *Worldviews, Science and Us: Philosophy and Complexity: University of Liverpool, UK, 11-14 September 2005*, 117.

Roth, S. (2019). The open theory and its enemy: Implicit moralization as epistemological obstacle for general systems theory. *Systems Research and Behavioral Science, 36*(3), 281-288.

Rozin, P. (1976). The psychobiological approach to human memory. *Neural mechanisms of learning and memory*, 3-46.

Saleh, M., Oliva, R., Kampmann, C. E., & Davidsen, P. I. (2010). A comprehensive analytical approach for policy analysis of system dynamics models. *European Journal of Operational Research, 203*(3), 673-683.

Schwaninger, M. (2011). System dynamics in the evolution of the systems approach. *Complex Systems in Finance and Econometrics*, 753-766.

Sillitto, H., Dori, D., Griego, R. M., Jackson, S., Krob, D., Godfrey, P., . . . McKinney, D. (2017). *Defining "system": A comprehensive approach.* Paper presented at the INCOSE International Symposium.

Sparacio, M. (2003). Mental realism: rejecting the causal closure thesis and expanding our physical ontology. *PCID, 2*, 8.

Stacey, R. D. (1996). *Complexity and Creativity in Organizations.* San Francisco: Berrett-Koehler Publishers, Inc.

Stefano, A., Camillo, C., & Riccardo, O. (2014). *Policy modeling as a new area for research: perspectives for a systems thinking and system dynamics approach.* Paper presented at the Proceedings of the Business Systems Laboratory 2nd International Symposium.

Tateo, L., & Valsiner, J. (2015). Time breath of psychological theories: A meta-theoretical focus. *Review of General Psychology, 19*(3), 357-364.

Tian, Y., Li, Y., & Wei, Z. (2013). Managerial incentive and external knowledge acquisition under technological uncertainty: a nested system perspective. *Systems Research and Behavioral Science, 30*(3), 214-228.

Tower, D. (2002). Creating the Complex Adaptive Organization: a Primer on Complex Adaptive Systems. *OD Practitioner, 34*(3).

Umpleby, S. A. (2009). Ross Ashby's general theory of adaptive systems. *International Journal of General Systems, 38*(2), 231-238.

Vintiadis, E. (2013). Emergence.

Wallis, S. E. (2010). Toward a science of metatheory. *Integral Review, 6*(Special Issue: "Emerging Perspectives of Metatheory and Theory").

Wallis, S. E. (2014a). Abstraction and insight: Building better conceptual systems to support more effective social change. *Foundations of Science, 19*(4), 353-362. doi:10.1007/s10699-014-9359-x

Wallis, S. E. (2014b). Existing and emerging methods for integrating theories within and between disciplines. Organizational Transformation and Social Change, 11*(1), 3-24.

Wallis, S. E. (2014c). A systems approach to understanding theory: Finding the core, identifying opportunities for improvement, and integrating fragmented fields. *Systems Research and Behavioral Science, 31*(1), 23-31.

Wallis, S. E. (2016a). The science of conceptual systems: A progress report. *Foundations of Science, 21*(4), 579–602.

Wallis, S. E. (2016b). Structures of logic in policy and theory: Identifying sub-systemic bricks for investigating, building, and understanding conceptual systems. *Foundations of Science, 20*(3), 213-231.

Wallis, S. E. (2019). *Learning to Map and Act Cybernetically Without Learning Cybernetics: How Many Loops are Needed to Enable Effective Decisions and Actions?* Paper presented at the American Society for Cybernetics, Vancouver, BC, Canada.

Wallis, S. E. (2020a). Commentary on Roth: Adding a conceptual systems perspective. *Systems Research and Behavioral Science, 37*(1), 178-181.

Wallis, S. E. (2020b). Integrative Propositional Analysis for Developing Capacity in an Academic Research Institution by Improving Strategic Planning. *Systems Research and Behavioral Science, 37*(1), 56-67.

Wallis, S. E. (2020c). The missing piece of the integrative studies puzzle. *Interdisciplinary Science Reviews, 44*(3-4).

Wallis, S. E. (2020d). Orthogonality: Developing a structural/perspectival approach for improving theoretical models. *Systems Research and Behavioral Science, in press*. doi:10.1002/sres.2634

Wallis, S. E., & Johnson, L. (2018). Using Integrative Propositional Analysis to Understand and Integrate Four Theories of Social Power Systems. *Journal on Policy and Complex Systems, 4*(1), 169-194.

Wallis, S. E., & Valentinov, V. (2016a). The imperviance of conceptual systems: Cognitive and moral aspects. *Kybernetes, 45*(9).

Wallis, S. E., & Valentinov, V. (2016b). A limit to our thinking and some unanticipated moral consequences: A science of conceptual systems perspective with some potential solutions. *Systemic Practice and Action Research, 30*(2), 103-116.

Wallis, S. E., & Valentinov, V. (2017). What Is Sustainable Theory? A Luhmannian Perspective on the Science of Conceptual Systems. *Foundations of Science, 22*(4), 733-747. doi:10.1007/s10699-016-9496-5

Wallis, S. E., & Wright, B. (2019). Integrative propositional analysis for understanding and reducing poverty. *Kybernetes, 48*(6), 1264-1277.

Ward, T. (2014). The explanation of sexual offending: From single factor theories to integrative pluralism. *Journal of Sexual Aggression, 20*(2), 130-141.

Ward, T. (2019). Why theory matters in correctional psychology. *Aggression and Violent Behavior, 48*, 36-45.

Ward, T., & Clack, S. (2019). From symptoms of psychopathology to the explanation of clinical phenomena. *New Ideas in Psychology, 54*, 40-49.

Ward, T., & Hudson, S. M. (1998). The construction and development of theory in the sexual offending area: A metatheoretical framework. *Sexual Abuse, 10*(1), 47-63.

Watanabe, J. M., & Smuts, B. B. (2004). Cooperation, commitment, and communication in the evolution of human sociality. *The origins and nature of sociality*, 288-309.

Weick, K. E. (1989). Theory construction as disciplined imagination. *Academy of Management review, 14*(4), 516-531.

Westhorp, G. (2012). Using complexity-consistent theory for evaluating complex systems. *Evaluation, 18*(4), 405-420.

Wexler, M. N. (2001). The who, what and why of knowledge mapping. *Journal of Knowledge Management, 5*(3), 249.

Wright, B., & Wallis, S. E. (2019). *Practical Mapping for Applied Research and Program Evaluation*. Thousand Oaks, CA: SAGE.

Zelmer, A., Allen, T., & Kesseboehmer, K. (2006). The nature of ecological complexity. *Ecological Complexity, 3*, 171-182.

Footnote

1. https://en.wikipedia.org/wiki/Team.

Chapter 14
Evaluating and Improving Theory Using Conceptual Loops—A Science of Conceptual Systems (SOCS) Approach

Wallis, S. E. (2020). Evaluating and improving theory using conceptual loops: A science of conceptual systems (SOCS) approach. Cybernetics and Human Knowing, 27(3).

My thanks go out to two anonymous reviewers; their carefully considered comments and detailed feedback served to generate a more complete and coherent paper.

Abstract: *In the field of cybernetics, it is generally accepted that a better theory is one that presents more conceptual loops (often represented diagrammatically with boxes and arrows) which may represent processes in the world of physical, biological, and social systems. However, that assumption has not been adequately tested. This preliminary study evaluates loops and other conceptual structures found in theories drawn from two disciplines, and stages of evolution within those disciplines, that show theories evolving over time from lesser to greater usefulness. Comparing the loop structures with the usefulness of the theories, this paper provides recommendations for measuring the structure of theories as tools for developing theories that are more useful in practical application. Because theories are ubiquitous across academic disciplines and fields of practice, this research is expected to support multi/inter/trans-disciplinary scholars and practitioners.*

Introduction

The purpose of this paper is to conduct and present a preliminary exploration of causal loop structures (conceptual loops, or more simply, loops) found within some theories with an eye to developing a method for using loops as one indicator for the quality (effectiveness or usefulness) of theories. That measure, in turn, is expected to suggest one or more directions for accelerating the advance of theories in all sciences, with a particular focus here on improving theories of the social/behavioral sciences. Such advance is expected to provide improvements in practice for the benefit of humans and humanity. Such an approach is called for because present theories have proved insufficient for understanding and resolving the vast problems facing our peoples and our planet.

Theory may be understood as a representation of knowledge, potentially understood as "knowledge_of" or "knowledge_for" (Glanville, 2009). As such, this paper responds to Glanville's call to explore similarities and differences between the two by exploring how less useful knowledge (knowledge_of) may have fewer loops than knowledge which is of greater use for understanding situations and enabling successful change to reach desired goals (knowledge_for).

Here, theory may be understood as a set of assertions (Weick, 1989) or propositions. Each proposition includes one or more concepts relating to things in the physical world (including physical, biological, and social systems) and may include descriptions or explanations of how those things are related; "a declarative sentence expressing a relationship among some terms" (Van de Ven, 2007: 117). Much like what Popper would call an axiomatic system, model, or set of hypotheses (Popper, 2002: 53).

Building on systems thinking (as understanding relationships between things), each theory may be understood as a kind of "conceptual system" (Wallis, 2016a) existing in a conceptual world or ecology of knowledge (Disessa, 2002) where (for our purposes) the things in the system are concepts and the relationships between them are causal connections. As will be explained in greater depth below, a conceptual system is useful, in some way, for understanding the world of physical systems; and, for taking action with some expectation of successfully reaching goals. Examples of conceptual systems include theories, models, mental models, program theories, policy models, large scale theories, middle range theories, laws, strategic plans, and so on. Those terms will be used interchangeably in the present paper, although we will mostly refer to them as theories.

For clarity, those theories may be presented as diagrams—with boxes containing concepts (each representing something in the physical world) and arrows between the boxes providing a representation of presumed or expected causal relationships. There are many forms or types of diagrams from making simple "Venn Diagrams" to complex "GIGA-Mapping" (cf. Sevaldson, 2011). While each form may have its benefits, our focus will be on diagrams where measurable concepts (as nodes or variables) are connected with arrows (representing causal relationships) as a long history of research has shown those to have measurable structures which may be correlated with their usefulness in practical application (Wallis, 2018, 2020b).

Those diagrams may be referred to as knowledge maps; or, more simply, maps. Figure 1 shows a simple map representing the number of predators and prey in an ecosystem. With more predators, there will be a decline in the number of prey (indicated by the minus sign on the causal arrow). With an increase in the number of prey, the increased food source will lead to an increase in the number of predators.

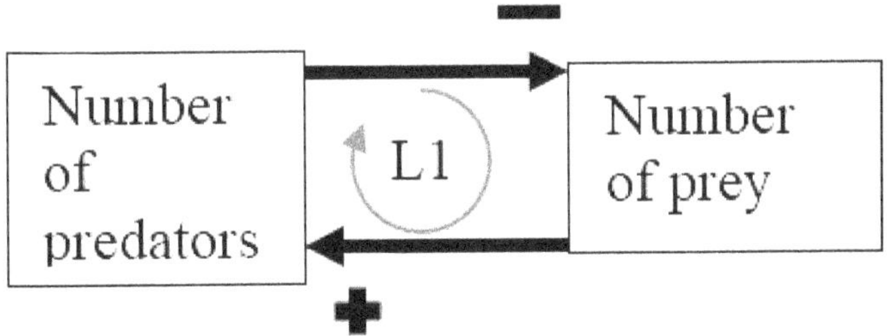

Figure 1 *A simple loop*

Here, it is important to recognize that the focus of this paper is on the conceptual systems (theories, models) rather than the underlying situations of the physical systems. We are focused on theories themselves as representations of "how the world works," regardless of the source of the theory (personal experience, collaborative deliberation, formal academic analysis, etc.). As such, any theory may be situated in another meta-level of understanding where the theory emerges from, and is used in, practice (which, although important, is not the focus of the present paper).

While calls for advancing theories have been heard since the dawn of science, progress in the social/behavioral sciences has been slow. Previous calls for "more data" have not proved highly effective in advancing the social/behavioral sciences. Available data is increasing exponentially, but the quality of our theories remains staid (Wallis, 2015a, 2015b). Even "big data" (Russom, 2011; Wamba, Akter, Edwards, Chopin, & Gnanzou, 2015) while interesting and useful on some level has not served to significantly advance and so fulfill the promise of the social/behavioral sciences.

With that lack of efficacy, it has been said that organizational theory has failed (Burrell, 1997; Lewis & Kelemen, 2002) as has social theory (Appelbaum, 1970). Policies (informed by policy theories/models) seem to rarely reach their stated goals (Light, 2016; Wallis, 2011, 2016a) often leading to "worse outcomes than the previous status quo" (Beaulieu-B & Dufort, 2017: 1). Many examples of unanticipated consequences from purposeful policy actions and short-sighted practices may be found with Sterman (2012: 24).

Instead of finding improvements in theory, with attendant improvements in practice suggesting a scientific revolution (Kuhn, 1970), recent decades have seen something more akin to a celebration of postmodernism; an explosion of insights. For better or worse, those many insights and associated theories have been viewed as part of the problem of disciplinary fragmentation. That process may be seen where scholars use a variety of terms; perhaps to increase the novelty and attractiveness of their work (Hovorka, Birt, Larsen, & Finnie, 2012). Such fragmentation (and confusion) seems common across the social/behavioral sciences (cf. Bakker, 2011; Ledoux, 2012; Lewis & Kelemen, 2002; Newell, 2007) and perhaps beyond. Fragmentation may be seen where each scholar has great expertise in a small area so that each expert seem like a novice to other scholars (Lipton & Thompson, 1988).

It seems, therefore, that we are in need of a new way to develop our theories, and so our sciences. To that end, this paper will conduct and present a preliminary exploration of the causal loop structure of conceptual systems (conceptual loops) to answer this research question: How might we more rigorously evaluate loops presented in theories to support our ability to develop theories that are more useful in practical application?

A Structural Path Forward

Because theories are ubiquitous across disciplines, and those theories have a kind of measurable internal structure, it is anticipated that the same general ideas about structure may be applied to theories of all sciences. By measuring changes in the structure of theories as they evolve over time from lesser to greater usefulness, we may be able to identify what kinds of changes support improvements. With such insights, we may be able to accelerate the development of our theories and our sciences to more rapidly and effectively understand and resolve the wicked problems of the world.

Past studies of theories have used Integrative Propositional Analysis (IPA) to evaluate their structures based on the number of concepts ("breadth" or "simple complexity") and a measure of causal connectivity ("depth" or "systemicity"). And, so far, the more structured theories seem to be more useful than theories that are less structured understanding situations and reaching goals (Wallis, 2016a). We will delve into IPA in greater detail below.

While loops have been identified as a useful sub-structure within theories (Wallis, 2016b), IPA has not had a specific method for evaluating loops. The present paper is a step forward to remedying that lack because, aside from a general understanding that "more loops are better," no clear guide seems to exist as to how many loops are required in a theory to make it better (more useful for understanding situations and enabling effective change).

In the present paper, that IPA stream of research is continued with a focus on the structure of loops within theories with the idea that by building theories with more loops it is expected that theories will become more useful in practical application for understanding situations and taking effective action to reach goals.

To clarify, this process (and this paper) is not about finding examples from the natural sciences that may be applied through metaphor to gain insights elsewhere (although those may be interesting). Nor is this paper about seeking some "consensus among stakeholders" where scholars (somehow) agree to use a common theory regardless of its efficacy. Instead, in the present paper, we will look at the internal structure of theories for clues as to what makes them more efficacious or useful for understanding and resolving problems in the social/physical world.

It is worth noting that, for the present paper, theories are the subject of analysis. That is, by counting the structural components (e.g., concepts, causal connections, loops, etc.), we gain in data that will help us to better understand how our theories are structured.

Systems, Dynamics, Loops, and Maps

In this section, we briefly cover some background perspectives to orient readers of various disciplines to the science of conceptual systems (SOCS).

Systems thinking suggests a way of understanding the world by understanding interactions (Sterman, 2012). "Systems thinking is about understanding these interactions and interconnections and identifying the leverage points to solve the problems in the organization." (Vemuri & Bellinger, 2017: 2). Generally, systems thinking encompasses a variety of more specific perspectives including cybernetics and systems dynamics (Schwaninger, 2015).

One key idea of systems thinking is that we expect to identify similar patterns, or isomorphic similarities, (Friendshuh & Troncale, 2012) across differing fields of study. And, by finding such similarities, improve our understanding within fields, along with our interdisciplinary understanding (among many fields). Such understanding improves our ability to take effective action and reach our goals. These isomorphies are visible across a vast range of theories (conceptual systems) most clearly as concepts and causal connections between them that exist in so many theories. Another pattern, although less frequently found in theories of the social/behavioral sciences, is the causal loop. Loops (of course) are perhaps most studied in the field of cybernetics and systems dynamics.

Emerging in the 1940s, the field of cybernetics (overlapping with systems sciences) has been gaining interest, particularly in Asia and Europe (Umpleby, Wu, & Hughes, 2017). Applied across many (if not all) disciplines, cybernetics is "concerned with circular causal mechanisms in biological and social systems" (Umpleby, Wu, & Hughes, 2017: 34). Those kinds of loops may be easily seen in a simple form where your home thermostat switches the furnace on when the house is too cold and off when the house is warm enough. Here, we extend that definition to include "conceptual systems."

The study of cybernetics is based primarily on ideas of feedback, control, and guidance (Schwaninger, 2015). "Feedback is used in virtually all scientific fields, from physics, chemistry and ecology to the social and economic sciences, as a principle for the explanation of system behavior and the design of systems" (Schwaninger, 2015: 56). Because loops are seemingly ubiquitous (as are theories) across the sciences, we may be able to use insights gained from identifying causal conceptual loops in theories (which may be different from feedback loops) of one science to better understand and develop theories in another science. Particularly, if we are able to identify the role played by loops within the causal structure of theories.

Within the broader study of systems, Systems Dynamics (SD) has evolved to be, "A methodology and discipline for the modeling, simulation and control of dynamic systems. The main emphasis falls on the role of structure and its relationship with the dynamic behavior of systems, which are modeled as networks of informationally closed feedback loops between stock and flow variables [boxes/concepts]" (Schwaninger, 2011: 8974). It is worth noting that, "Linking feedback loops and system behavior is part of the foundation of system dynamics, yet the lack of formal tools has so far prevented a systematic application of the concept, except for very simple systems" (Kampmann, 2012: 370).

In SD, experts use complex processes and formulae to simplify complex theories and find leverage points where policy actions might have the greatest effects (Oliva, 2015; Saleh, Oliva, Kampmann, & Davidsen, 2010). Even for rigorous processes, some element of judgement is required, along with some computational expertise (Oliva, 2015: 208). Another SD approach is to identify the "dominant" loops within a theory (Duggan & Oliva, 2013; Hayward & Roach, 2017). Which, while interesting, requires higher mathematical skills than are easily accessible to many researchers.

Those methods, to the best of this author's knowledge, are not used to compare theories to show which ones might be the more accurate or useful interpretation of the physical world. And, for many scholars and practitioners, that level of expertise may not be readily accessible. One goal of the present paper, in contrast, is to understand how we might evaluate the whole model in a way that balances rigor with accessibility.

"The SD method relies mostly on the enumeration of feedback loops." (Wakeland & Medina, 2010: 17) but, "How many feedback loops does a given system have? Unfortunately, there are no general formulas for finding the loops for a given system, short of direct enumeration." (Kampmann, 2012: 374); a problem that exists in parallel for identifying loops in the physical world, as well as identifying loops on a map. Both may be highly complex and subject to misunderstanding.

For clarity, it is important to repeat that this article will be investigating loops found in theories (conceptual loops) rather than the many different kinds of loops that may be found in physical or social systems. While we may interpret different kinds of systems as having different kinds of feedback (e.g., verbal or electronic), we look at conceptual systems differently—focusing on conceptual loops.

To put it another way, we are investigating the conceptual systems which *represent* those other systems. Such *representation* may be mechanical and simplistic, regardless of the complexity of the physical/social system represented. For example, a social system that is highly complex and non-linear may be represented by a theory that is simple and linear. Of course, such a linear representation is thought to be less useful than a conceptual system that contains multiple loops. Theories with more loops are generally expected to provide a more accurate, and therefore more useful, representation of complex, nonlinear social systems.

Importantly, in this study, we are learning how to analyze the structure of our conceptual systems to see how useful a representation they provide for understanding and enabling change in those physical/social systems. So, before taking action (for example to change a social system), we will have an indicator for whether that action will be successful, based on the structural qualities of the theory used to inform or support that change.

To help understand the structure of our theories, we use diagramming and other related forms of visualization. These are central to design thinking and design practice supporting creativity and synthesis (Sevaldson, 2011).

In the present paper, we are seeking to find more accessible techniques that may be easily acquired and applied by scholars and practitioners across a variety of scientific disciplines to improve their theories and accelerate the development of their fields. Optimally, such a process will allow the field to identify better theories that may be the focus of more advanced forms of research.

Three streams of research supporting a structural perspective and the usefulness of maps

As noted above, a conceptual system is a set of interrelated propositions—a theory—reflecting actionable knowledge and/or an understanding of how the world works. Briefly, to provide an orienting background, the science of conceptual systems may be found in at least three streams of research. From the field of political psychology, the measure of "Integrative Complexity" (IC) is applied to texts (typically found in personal communications or public speeches). IC may be defined briefly as a measure of conceptual complexity including the differentiation and integration of concepts held by individuals and/or groups that enable more effective information processing and decision making (Suedfeld & Tetlock, 1977). Research in that stream shows that individuals and teams whose texts reflect a higher level of structure are generally more successful at reaching their goals (Suedfeld & Rank, 1976; Suedfeld, Tetlock, & Streufert, 1992; Wong, Ormiston, & Tetlock, 2011). Suedfeld (2010) provides a 30-year history of that stream.

From the field of education, the "systematicity" stream is generally focused in the classroom, where students with a better understanding of the course material are said to have knowledge with greater structure. Drawing on Fodor and Pylyshyn, Phillips and Wilson (2014) suggest that, "Systematicity is a property of cognition whereby the capacity for some cognitive abilities implies the capacity for certain others." That capacity is demonstrated when students create concept maps showing terms they've learned, along with arrows linking those terms, and lines showing how the terms might be related in some kind of hierarchy. An expert then scores the concept map to determine the students' level of mastery (reflecting capacity for cognition on the topic of study). Novices tend to create simplistic maps with low levels of structure, while experts create complex maps with high levels of structure (Novak, 2002, 2010). Such advancement is also seen from the perspective of psychology where, "According to Piaget's developmental theory of learning, individuals' assimilate external events and accommodate them to develop a mental structure the facilitates reasoning and understanding" (Gray, Zanre, & Gray, 2014: 30).

While the IC and systematicity perspectives support the general notion that having more structured knowledge is more useful in practical application for understanding situations and making decisions to reach specified goals, neither was developed to evaluate formal theories. So, we will not be using those methods to study theories in the present paper (future research may explore the overlap between them).

Instead, we will focus on a third stream of research, emerging from the systems sciences. "Integrative Propositional Analysis" (IPA) is used primarily for analyzing formal theories such as those as presented in peer reviewed literature, IPA is a methodology used to evaluate and determine the structure of theories based on the number of concepts and the number/direction of causal connections between them (Wallis, 2016a). IPA is a relatively new methodology, which suggests a limit to its potential usefulness. Also, while IPA has been used to show a general correlation between the structure and usefulness of theory that relationship has not been quantified with great accuracy.

In general, IPA has shown that with more structure, theories are more useful for decision making in practical social/physical world situations. Because theories, with greater and lesser levels of structure, are ubiquitous across all sciences, IPA has been applied to successfully evaluate theories from many disciplines including physics (Wallis, 2010), psychology (Wallis, 2015b), sociology (Wallis, 2015a), policy (Wallis, 2011), strategic planning for academic institutions (Wallis, 2020b), applied social research (Wright & Wallis, 2019), and others. IPA has also proven useful for integrating theories within and between disciplines for improving the structure, and so the potential usefulness, of those theories (Wallis, 2014b, 2018, 2020c) enabling greater sustainability and morality while avoiding the limits of our own human cognition (Wallis & Valentinov, 2016a, 2016b, 2017).

Despite the clear importance of causal loops for an understanding of how the world works (evidenced by their importance in multiple fields of study), research on mental models shows that few incorporate any loops. Axelrod (1976) found virtually no loops processes in the cognitive maps of political leaders. Dörner (1980, 1996) found that people tend to think in single-strand causal series and had difficulty in systems with side effects and multiple causal pathways, much less loops. Sweeney and Sterman (2007) found limited recognition of feedback processes among both middle school students and their teachers.

That lack of loops may be related to cultural limitations. Where "Western" thought tends to be more linear than Asian thought (Wallis, 2016b). African logic (reflected in the concept of Ubuntu) also tends to be more circular, representing a higher level of interconnectedness or interrelatedness for understanding organizations and contexts (Chilisa, Major, Gaotlhobogwe, & Mokgolodi, 2015; Wallis, 2019). Whatever the underlying reason(s), we can improve our ability to understand real-world systems through our conceptual systems by creating diagrams such as maps, mind maps, concept maps, and so on.

Benefits of Mapping

Mapping supports clarity of understanding (Hovorka *et al.*, 2012) and collaboration (Bryson, Ackermann, & Eden, 2016) at least in part because maps help alleviate the limits of human cognition (Bureš, 2017). Mapping is also beneficial because it helps us to integrate or synthesize diverse perspectives more easily and effectively. This is important because, "Collaboration and cross-sector collaboration have therefore emerged as hallmarks of the new approach in which public managers frequently must work jointly with nonprofit organizations, business, the media, and citizens to accomplish public purposes" (Bryson *et al.*, 2016: 912) which is important because "collaboration is usually difficult and success is hardly assured" (p. 912). The turn towards visual mapping also supports clarity, accountability, trust, and coordination through interdependent goals because assumptions can be "on the table" visible to all (Wallis & Wright, 2015).

As abstract representations of physical world systems, it is general understood that, "The truth and quality of the models [theories] used hinges primarily on the correct match of the model [theory] with the concrete ('real') [physical] system in focus" (Schwaninger, 2015: 572). A key assumption of the present paper is that our best (most useful) representations of the physical world, our laws of physics, include causal loops as may be seen in Figure 4—a causal loop representation of Newton's second law. We may reasonably expect that developing loops in theories of other sciences will similarly lead to improvements in the usefulness of those theories thus improving practices and practical success in many fields.

Before moving on to compare maps of theories, some clarity in terminology may be useful because many of our fragmented fields use different terms when referring to the same (or very similar) things. For this paper, we will use the term "map" to refer to a theory presented as a diagram (which might also be referred to as a loop diagram, graph, or directed graph). Links and edges will be referred to as causal arrows or more simply "arrows." And, we will use the term "concept" (or "box" on a map) instead of terms such as nodes, points, variables, or vertices.

It is also worth noting that our focus will be exclusively on causal theories. That is, where the map shows how changes in one or more things cause changes in one or more other things. That causal perspective is foundational to scientific understanding (Pearl, 2000). Indeed, "[S]tudies of learning, attribution, explanation, reasoning, judgement and decision making all suggest that people are highly sensitive to causal structure" (Sloman & Hagmayer, 2006: 408).

Generally, as noted above, the perspective from the systems sciences seems to suggest that having more loops in a theory will provide a better map; a more useful representation of the physical world. While this seems intuitively correct, it does not seem to be supported by empirical studies; relying, instead, on illustrative examples of how systems work in the physical world (e.g., thermostats and furnaces). And, how they fail when a loop becomes disconnected (such as a thermostat failing to switch on its furnace because of a lack of communication between the two). The present paper will focus on addressing that lack of empirical evidence to find out the extent to which our assumptions may be warranted.

Methods for Measuring Loops and Structure

While it seems important to look for and identify loops (Deegan, 2011: 21) the focus seems to have been on simply identifying or finding what loops exist (or seem to exist), in the physical world, rather than relating the quality and quantity of loops with the usefulness of the theory. In this section, we review a variety of perspectives to see how we might better understand and evaluate loop structures. These measures will inform the present methodology, which will be used to evaluate the structure of theories to help understand what strategies we might use to accelerate the improvement of our theories.

In the field of business studies, Senge, Kleiner, Roberts, Ross, and Smith (1994) show how loops may be used to understand challenging situations. They provide various arrangements of loops relating to organizational issues (e.g., "limits to growth," "fixes that backfire," and "tragedy of the commons"). And, those patterns suggest where intervention might have some successful results. Those patterns typically include only two to three loops (*ibid.*: 150); while, in contrast, the overall process encourages creating maps with more loops to provide a more useful understanding of complex situations. What is missing from that publication is some reasonably objective measure of "how many" or "what kind" of loops might be optimal. Instead, the authors recommend that a group stop adding loops "after a few interdependencies become apparent" (Senge *et al.*, 1994: 163) and instead of continuing to improve their understanding, instigate a conversation about the emerging themes of the map and implications for action.

What 'becomes apparent' for one, does not necessarily mean apparent for all. What might happen when one faction desires to take action while another faction desires continued exploration? What if one person or faction holds a hidden agenda that is not reflected on the map? Such situations provide the opportunity for conflict; so, having more objective methods for evaluating the usefulness of maps may provide some beneficial guidance to practitioners and researchers.

In systems dynamics (SD) "Network analysis provides a suite of quantitative techniques that can summarize the structure of a network and quantify the importance of its elements. Understanding the structural features of a network as a whole can provide key insights into the ease or difficulty by which information, influence, or physical matter flow through the network" (McGlashan, Johnstone, Creighton, de la Haye, & Allender, 2016: 2). Using an SD perspective, we might ask how influential each concept might be on the map as a whole; how central it is or how closely connected each concept might be compared with the other concepts on the map (McGlashan *et al.*, 2016). Those might be thought of as "leverage points" from a structural perspective (Wallis, 2020b; Wright & Wallis, 2019), where a small amount of effort may lead to a large amount of change. Again, while estimating the importance of certain concepts, the SD approach does not seem to provide strong guidance for enumerating loops.

In cybernetics, Umpleby (2018) suggests building maps with a combination of positive and negative loops (basically, loops whose effects balance one another, one causing increase and another a decrease in some concept/box). If most or all the loops are mutually supporting, it would suggest that there will be "runaway" effects. For example, consider a charity whose map shows then that conducting fundraisers will generate more money and that the money raised can be used to conduct more fundraisers. If the theory were valid, they could set such a plan in motion and soon control all the money in the world. As that is a highly unlikely scenario, there must be something missing from the map; one or more balancing loops that limit the charity's ability to raise funds (e.g., "donor fatigue" resulting from a large number of fundraising events). While identifying balancing effects seems potentially interesting, there is no quantitative evaluation of the number of loops or how they may be balanced.

Seeking to include all the concepts of the theory within those loops, another approach is to identify and enumerate independent loops sets (ILS), with a focus on the "shortest loops possible" (Oliva, 2004: 322). Such an approach seems reasonable; first, because by comparing the number of concepts included in all loops, we can identify what percentage of the total number of concepts in the theory are accounted for. That is to say, it seems reasonable to assume that the better theories will have a higher percentage of their concepts as part of one or more loops. That assumption, as yet, seems to be unfounded. Second, by focusing on the shortest loops, we may reduce the confusion found in looking at large and complex theories that may have longer loops (including more concepts) and/or multiple overlapping loops. That is not to say that the longer loops are meaningless or irrelevant, only that this approach may be used as a starting point to making complex maps more comprehensible. And, of greater importance for evaluating maps, focusing on the smallest loops may provide a more objective way to view theories so that differing scholars may reach the same or more similar conclusions about the similarities and differences in structure between theories.

Bureš (2017) suggests it may be useful to consider the, "multi-input and multi-output (MIMO) variables [concepts/boxes]" (p. 5). That is, concepts with more than one arrow pointing inward and more than one pointing out or away from each concept/box. Similarly, Integrative Propositional Analysis (IPA) suggests that theories should be created with at least two inputs for each concept on the map (cf. Wallis & Wright, 2019). A concept with two arrows pointing directly at it is said to be concatenated. And, like a dependent variable influenced by two or more independent variables, is held to be better understood. So, a map with more concatenated concepts will provide greater understanding and so be more useful. While these methods evaluate the structure of the map, neither approach is directly used for the evaluation of loops.

Delving more deeply into IPA, from Wallis (2015b: 371), Integrative Propositional Analysis is a six-step process: 1. Identify the propositions within a theory. 2. Diagram the causal relationships

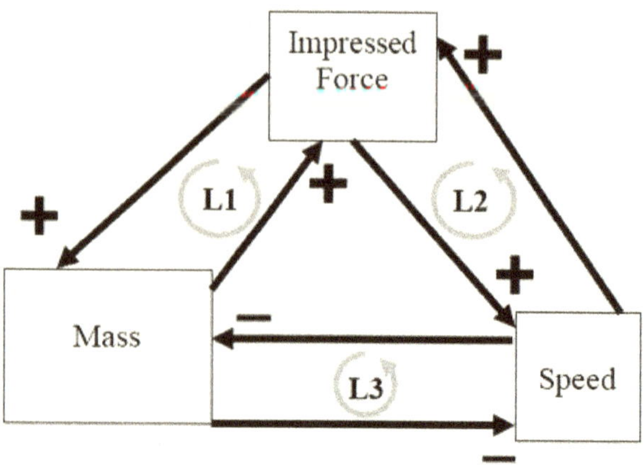

Figure 3 *Force according to Buridan (Impressed force = mass x speed)*

between the concepts within the propositions. 3. Combine those smaller diagrams where they overlap to create a larger, integrated, diagram. 4. Count the total number of concepts to determine the (simple) Complexity of the theory. 5. Identify and count the concatenated concepts (those that have more than one causal influence). 6. Calculate the Systemicity of the theory by dividing the number of concatenated concepts by the total number of concepts.

From that structural perspective, the measure of Systemicity may be used as an indicator of the map's potential usefulness for choosing actions and so reaching goals. IPA is also limited in that it does not consider underlying data. Thus, a theory may have a high level of structure even though it has no counterpart in the physical world. Other measures must be used to ascertain the validity of the data supporting the theory.

Please note that the measure of "systemicity" from IPA is different from the above-mentioned approach of "systematicity." The difference is not important to the present paper as we will use only systemicity here.

Although not all of the above measures expressed or implied specifically suggest or require enumerating loops, they do apply to structure and so we will apply a variety of them to the present study of evaluating theories. This will provide methodological plurality enabling a more effective comparison of the theories under analysis and the methods used. Including:

- The number of smaller loops (two concepts);
- The number of larger loops (three or more concepts);
- Percent of all concepts included in loops;
- The number of fully positive loops;
- The number of fully negative loops;
- The number of loops combining both positive and negative connections;
- The ratio of positive/negative signs on causal connections;
- Number of DIDO or MIMO concepts;
- IPA measure of concatenated concepts;
- Number of leverage points, and;
- IPA measures of complexity and Systemicity.

In the following analysis, we use these measures to analyze the structure of theoretical models from multiple perspectives as their structures evolved over time from lesser to greater usefulness. This is expected to help us see how their structures changed; and so, what inferences we might make for evaluating and improving our theories of the present to develop more useful theories in the future.

Subject of Analysis: Evolving Theories

In this section, we present in map (diagrammatic) form a number of theories for analysis using the methods and measures described in the previous section. First, we will present a series of theories from physics, followed by a series of theories from the health sciences. In each series, the theories may be understood as evolving from lesser to greater usefulness in practical application. The examples will be followed by the results of the analysis, discussion, and recommendations for research and practice.

For each figure, concepts are represented as boxes with causal relationships as arrows (with positive or negative influences indicated), between them. Loops are noted as lighter shaded, circular, arrows and enumerated L1, L2, etc.

Physics: From Aristotle to Newton

Focusing on the nature of "force" in physics, Stinner (1994) also examined the evolution of a theory through history. Drawing on Stinner's text, Aristotle's version of force is presented in Figure 2.

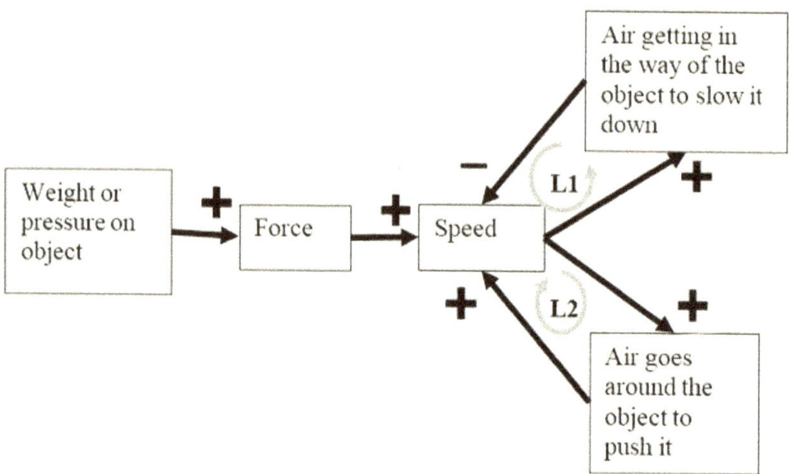

Figure 2 *Aristotelian conceptualization of force*

While Figure 2 has two loops, it is worth noting that the loop involving air going around the object to push it was not supported by data; that was merely a conjecture to account for why the object did not stop immediately when pressure or force was removed.

Moving on through the evolution of the theory, according to Stinner (1994), in the middle ages, John Burdian had come up with another version (Figure 3). For Burdian, when the impressed force is removed, some force (somehow) would continue to push the object, but would eventually die out. Again, this part of the theory was based on conjecture.

While Burdian's version may seem to make more sense than Aristotle's, it still did not provide sufficient explanation of the concepts to launch rockets into orbit (or, more appropriately for the time, predict the trajectory of catapult projectiles).

Figure 4 shows a causal diagram of Newton's famous Second Law (F=ma) with what might be called a "triadic" structure of loops. That law is entirely supported by data. And, it has proved highly useful for (among other things) launching satellites which facilitated global communication, weather observation, and more. In short, more useful than the theories presented in Figures 2 and 3.

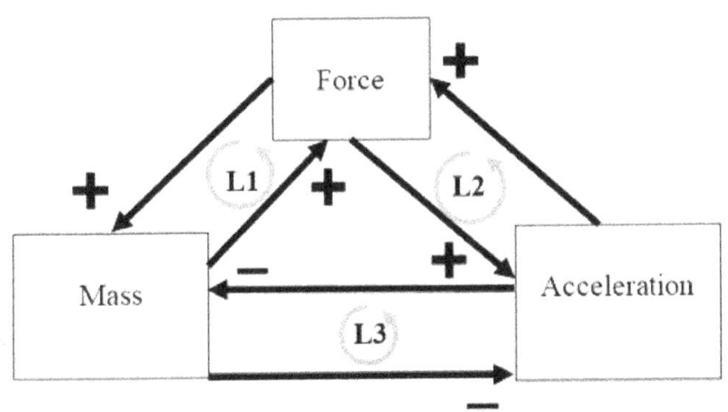

Figure 4 *Newton's Second law (F=ma)*

In comparing Figures 3 and 4, it is worth noting that they are structurally identical (same number of boxes, same kinds of connections). Figure 4 (Newton) includes "acceleration" instead of Burdian's "speed" (Figure 3). Also, Figure 3 includes "Impressed force," instead of Newton's "Force." That notion of impressed force (and related assumptions) were not supported by data. So, scholars seeking to improve their theories should keep in mind that they must be supported by data—structure alone will not suffice.

Returning to our focus on structure, it is worth noting that an alternative explanation of loops might be perceived for Figures 3 and 4. That is, using Figure 4 as an example, one loop may be seen passing through all three concepts while a second loop may be seen passing through all three concepts in the opposite direction. Those loops are ignored in favor of the smaller loops because the two large loops appear to be in parallel with each other and could conceivably be seen as redundant. In contrast, the three loops containing only two concepts each are clearly not redundant. We would not consider Figure 4 to have five loops because that would involve counting each arrow twice—which would introduce manifold uncertainties in evaluating structure.

The sequence of above maps (Figures 2, 3, and 4) generally suggests a progression from lesser to greater understanding and usefulness. The results of the previously described analyses are summarized in Table 1 according to the various measures of structure from the previous section.

Public Health: Progression of a Conversation

As may be seen in the previous sub-section, historical examples of theories may be found in the natural sciences (especially physics), from which we may see the evolution of theories from ancient times to the present. And, with that advancement, we may identify structural components and so identify and infer relationships between the structure of the theories and the usefulness of those theories. Such a history, however, does not yet seem to be available for the social/behavioral sciences. So, for the purpose of the present analysis, we will look at a set of theories drawn from an online conversation where (through conversational turns) the map was improved. Although incomplete, the four theories in Figure 5 may be seen as representing an evolution in knowledge or understanding; with presumed evolutionary improvement in usefulness of that theory for decision making for taking action and reaching goals.

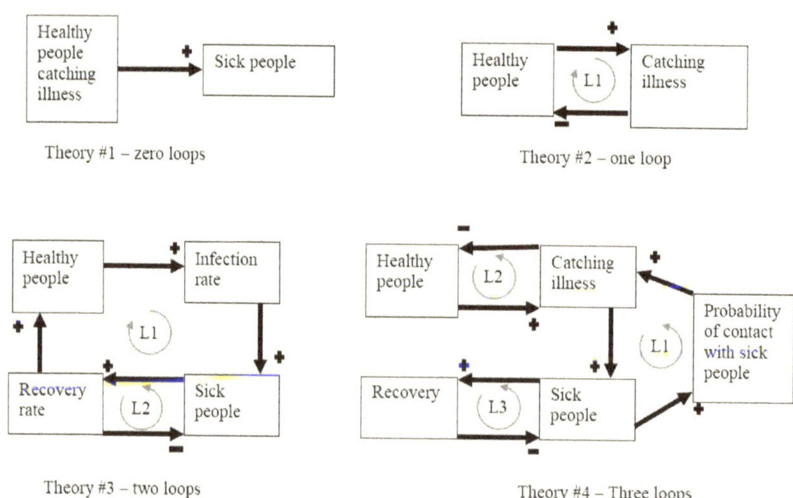

Figure 5 *Progressively valid theories of public health*

Each theory in the series from Theory #1 to Theory #4 may be understood as an improvement of the previous theory. That is, each map represents a deeper knowledge or understanding of the situation as each version built upon the previous through a conversation among experts. The results of the analytical methods (from the previous section) applied to the theories in Figure 5 are summarized in Table 2.

Results of Analysis

In this section, we summarize the results of analysis from the methods previously developed and applied to the theories of the previous section. Those results are shown in Tables 1 and 2, and explained below.

Figure	Small loops (two concepts)	Larger loops 3+ concepts	% Concepts included in loops	Completely positive loops	Completely negative loops	Mixed sign (+ / -) Loops	Sign ratio + : -	DIDO or MIMO Concepts	IPA – Concatenated concepts	Leverage Points	IPA: Complexity / Systemicity
#2 - Aristotle	2	0	60%	1	0	1	5:1	1	1	1	5 / 20%
#3 - Burdian	3	0	100%	2	1	0	4:2	3	3	3	3 / 100%
#4 - Newton	3	0	100%	2	1	0	4:2	3	3	3	3 / 100%

Table 1 *Summary of results: Evolving theories of physics*

Looking at Tables 1 and 2, we begin with the far left hand column which indicates the theory under analysis in order from top to bottom of increasing usefulness for in explanation and application.

The first column of data (after the column listing the theories) shows the number of loops where those loops contain only two concepts (and two causal connections between them). Key to the present study, we do indeed see that the number of loops increases as the theories improve. This serves as an important validation for a key assumption of cybernetics and systems thinking.

The second column looks at larger loops; those with three or

Figure 5	Small loops (two concepts)	Larger loops 3+ concepts	% Concepts included in loops	Completely positive loops	Completely negative loops	Mixed sign (+ / -) Loops	Sign ratio + : -	DIDO or MIMO Concepts	IPA – Concatenated concepts	Leverage Points	IPA: Complexity / Systemicity
Theory 1	0	0	0%	0	0	0	1:0	0	0	0	2 / 0%
Theory 2	1	0	100%	0	0	1	1:1	0	0	0	2 / 0%
Theory 3	1	1	100%	1	0	1	4:1	0	1	2?	4 / 25%
Theory 4	2	1	100%	1	0	2	5:2	2	2	2	5 / 40%

Table 2 *Summary of results: Evolving theories of public health*

more concepts. Interestingly, while there are none of these for theories of physics, we see one for two of the theories from public health. This may indicate an interesting question for future research. Should the better theories of the natural sciences have no larger loops while the theories of the social/behavioral sciences have more?

In the third column, of data we see the percentage of concepts included within loops. Again, there is a general increase in this percentage as the theories improve suggesting that this may be a useful measure for evaluating the potential usefulness of theories.

The fourth column identifies the number of completely positive loops in each theory. That is, those loops where both or all of the arrows reflect positive causality (e.g., more of A causes more of B). Here again, we see a general increase suggesting that this may be a useful measure of improvement in theories.

Next, we see the number of completely negative loops in the fifth column. That is, those loops where both arrows show a negative causal relationship (e.g., more of A causes less of B). Interestingly, while we see an increase in the physical theories of Table 1, the theories of public health in Table 2 show none at all. This will be discussed below in the section titled, "Pluses and minuses."

In the sixth column of data, we have the number of loops for each theory where the causal arrows include both positive and negative causation. Interestingly, the theories of physics in Table 1 show a decrease while the theories of public health in Table 2 show an increase in these kinds of loops. This will be discussed below in the section titled, "Pluses and minuses."

The seventh column considers the ratio of signs on the causal arrows for each theory. That is, the number of arrows representing positive and negative causation. Here, we see an interesting progression; as the theories evolve toward greater usefulness, the ratios also seem to evolve. This may be another area for future study.

The eighth column indicates the number of concepts for each theory that is DIDO (dual-in, dual-out) or MIMO (multiple-in, multiple-out) referring to the number and direction of causal arrows to and from the concepts. This approach (as noted above) has been hypothesized as a potential measure for the usefulness of theories; and the present study provides some preliminary support for that hypothesis.

In the ninth column, the number of IPA concatenated concepts are shown. Briefly and simply, those are the concepts with two causal arrows pointing at them. Much as a concept is better understood when it has two independent variables pointing at it, these concatenated concepts may serve as a proxy—a very simple measure—for the explanatory power of each theory. As expected, there is a general increase in the number of concatenated concepts as theories improve.

In the tenth column, we look at the leverage points for each theory. A leverage point is a concept that is included in (at the intersection of) two or more loops. While we see an increase in leverage points in line with the improvement of the theories, Theory 3 in Table 2 presents an issue. The two loops of that theory overlap so that both loops include two concepts ("Recovery rate" and "Sick people"). This confusion and potential resolution will be addressed in the section below titled, "Reducing redundancies."

In the last (11th) column, measures of IPA are shown. These are included because IPA has served as an effective measure of systemic structure, and so predictor of usefulness, in theories. This serves as a check between existing methods and the methods being developed in the present paper. There, complexity (simple complexity) shows the number of concepts in each theory. Systemicity, is the measure of systemic structure. As expected, there is a general trend toward increasing values in both these measures. The exception being in Table 1 where the Aristotle's theory contains more concepts than later versions. This may be seen as an indicator of increasing parsimony which occurs as theories evolve. Table 2 shows no increase in parsimony suggesting that this theory is as yet not fully mature.

To summarize this section, and note an important contribution of this paper, these structural measures all seem to be effective for evaluating the potential usefulness of conceptual systems (theory, models, etc.) because they seem to correlate (at least roughly) with the increased usefulness of the theories. And, as such, any of these measures may be used to point the way to constructing theories that are more effective for understanding and resolving the wicked problems of the world. Because this is a preliminary study, there are many opportunities for future research.

Summary and Recommendations

Here, we have been seeking to develop techniques that may be accessible to a wide variety of scholars and practitioners across the scientific disciplines to rigorously and objectively evaluate and improve their theoretical models from a structural perspective and accelerate the development of their fields. While this is a preliminary study, a few additional insights may be found.

Importantly, this study provides empirical support to the general assumption that our theories will be more useful when they include more loops; and, to the idea suggested by Oliva (2004) that we have better theories when more of the theory's concepts are included in loops. Also, this study supports the more specific suggestion by Umpleby (2018) that a more realistic and useful theory will include positive and negative loops that "balance" one another.

Also, generally, these results support the validity of multiple methods and techniques that might be used to evaluate the structure of theories; and, therefore, may be used as a guide to improve the usefulness of our theories. More specifically, this study supports IPA measures of complexity and systemicity as indicators for the structural quality of theory (Wallis, 2016a; Wallis & Valentinov, 2016b; Wright & Wallis, 2019). This study also supports DIDO (dual input, dual output) and MIMO (multiple input, multiple output) as indicators for theory quality (Bureš, 2017). IPA and DIDO/MIMO can and should be applied to accelerate the advance of scientific fields by comparing and synthesizing theories to develop theories that are more useful in practical application.

Third, in comparing theories of physics and public health, it is worth noting that in both fields theories are more useful when they have more leverage points. Indeed, in the best of the present examples (Newton's Second Law), all the concepts are also leverage points. That seems to suggest that we can build better theories by creating them with more leverage points. Recalling, of course, that these are "structural" leverage points (concepts held in common by two or more loops) rather than the more "intuitive" leverage points of Meadows (1999). For a discussion on this distinction please see Wallis (2020b).

Fourth, the idea of abstraction seems to be relevant here. Not, however, in the sense that a theory is an abstraction of the physical world; rather, that a theory may be more effective or useful when its component concepts are all at a similar level on some scale of abstraction (Wallis, 2014a). While Newton's concepts of force, mass, and acceleration seem to be similar on some scale of abstraction, Figure 6/Model 4, contains concepts that are clearly at different levels of abstraction. We have some concepts that are more concrete (number of sick people) and others that are more abstract (*probability* of contact with sick people). It may be worth exploring how each of those might be "shifted" to a more common level on some scale of abstraction. Alternatively, it may be useful to separate them into two different theories—each working at its own level of abstraction. More research is needed in this area to clarify the potential benefits of this approach to abstraction.

Naming the Method: IPA "x"

To summarize, the present research suggests a range of measures that may be used for evaluating the structural quality of theories thus supporting accelerated advancement in many fields of science. For those using Integrative Propositional Analysis (IPA) to evaluate and improve the structure of and usefulness of theories and strategic plans according to their Breadth/Complexity and Depth/Systemicity, the following new measures are suggested on a provisional or "experimental" basis (IPAx).

- Loops should be counted according to the shortest possible path. Parallel (redundant) loops should not be counted.
- Theories with more loops are expected to be more useful in practical application than theories with fewer loops.
- Theories with more leverage points are expected to be more useful than theories with fewer.
- Theories with a higher percentage of concepts included in loops are expected to be more useful than theories with a lower percentage.
- Loop/sign balance—Theories are expected to be more useful when the ratio of positive to negative loops (or signs) is closer to 2:1.

While preliminary in nature, these results suggest that scholars and practitioners can and should use the above measures to evaluate the structure of theories for choosing which one(s) to use in practice and as a guide to improving the usefulness of theories through research (by conducting research focused on improving one or more of the measures).

These insights should prove useful for the System Dynamics community as an approach for quantitatively differentiating between models according to their structure. Further research may be interesting in this area focusing on (for example) the use of stock-and-flow diagrams to represent theories.

Insights, Conversations, and Future Research

In this section, we will go into greater depth on a few insights that may be gained from a careful parsing of the data. Those, in turn, will lead to suggestions for additional research.

Pluses and minuses

Umpleby (2018) suggests that loops should balance. That is, positive and negative loops should intersect to avoid unrealistic understanding of a situation. And, in both the health sciences and in physics, we see that kind of structure. However, it is interesting to note that we see two different kinds of balancing structures. In Newton's law, there are three dyad loops. Two that have both their arrows representing positive causation while the third has both its arrows representing negative causation. In contrast, the theories of public health have self-balancing loops with one positive and one negative arrow each.

The end result for both seems to be a balancing effect on the concepts of the theory. So, Umpleby's suggestion seems to stand. However, the difference does raise questions. Is there an optimal structure for theories in which the balance is "within" dyads (as in public health theories), or is the optimal balance "between" dyads (as in Newton's Law)? Or, is one more appropriate for physics while the other more appropriate for the social/behavioral sciences (and, if so, why)? Additional research may lead to new insights into the construction and development of more effective and useful theories.

An alternative approach also presents itself if we step back from a loops perspective to more simply evaluate the signs on the arrows (representing positive and negative causation). Enumerating the number and type of signs for each theory map of our analysis provides a sign ratio (note in Tables 1 and 2). Recognizing, again, that this is a preliminary study, it seems that theories with sign ratios closer to 2:1 are more likely to be useful in practical application.

Rather than simply providing "a combination" as Umpleby suggests, the present results advance the science by suggesting that for a theory to provide the most realistic and useful representation of the world, it should include a ratio of two positive loops to one negative loop according to Newton's law (Figure 4) and a similar ratio for more advanced theories in the social sciences such as Theory 4 in Figure 5. More study is needed to better understand how we might evaluate loop ratios.

This structural perspective of balancing loops allows researchers who are developing theories or strategic plans, the ability to "see the invisible" and identify directions for additional research and practice that will accelerate the development of the theory toward greater usefulness—more effective application in practice. For an abstract example, if a theory is presented where all of the theory's loops are all pointing in one direction (e.g., all are positive), the researcher should start looking for additional concepts and causal connections pointing in the other direction.

Reducing redundancies

Another issue arises with counting the number of loops where it may be suggested that the shortest loop is best (Oliva, 2004, 2015; Saleh *et al.*, 2010). And, indeed, that is what we see in Newton's Law (along with other laws of physics that are highly useful in practical application).

In contrast, in evaluating the number of loops and leverage points for Theory 3 (Table 2, Figure 4), some confusion arose. The redundant loops made it unclear just how the theory should be evaluated.

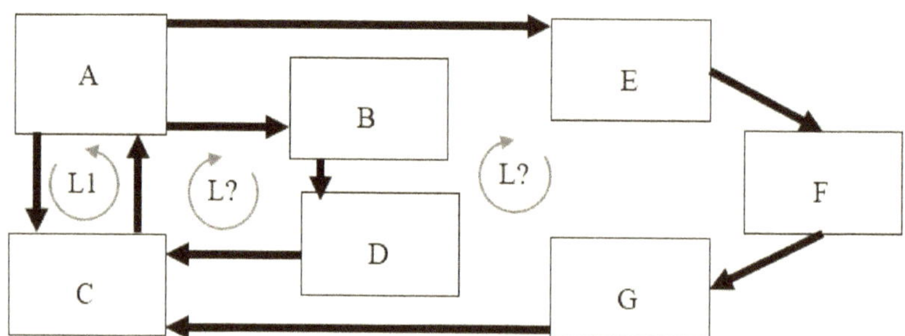

Figure 6 *Redundant loops and redundant leverage points*

As an abstract example, Figure 6 shows the L-1 (A-C) loop as the shortest path (fewest connecting arrows). In contrast, loops formed by B-D and by E-F-G are longer (more connecting arrows included in each loop) and in parallel with the A-C loop. For evaluating the number of loops, this suggests that Figure 6 should be seen as having only one loop.

Another "clue" here is that both A and C seem to be leverage points for all three parallel paths. While more research is needed, it seems intuitively strange, or at least confusing, that there are two leverage points connecting three loops. It is unclear which "one" leverage point is best; which one might be "the" leverage point. To improve the quality and clarity (and so, according to a structural measure, the usefulness) of our maps, it seems that only one leverage point should exist at the intersection of two loops. Although, with more complex theories, it may be reasonable that one leverage point may exist at the intersection of three or more loops. More research is needed to support or refute that assumption.

In comparison to the confusion of loops in Figure 6, Figure 7 shows three distinct loops (L1, L2, L3) because there are no arrows that might be considered parallel, redundant or overlapping—only overlapping concepts. Those overlapping concepts, existing at the intersection of two or more loops, seem to be valid leverage points. A is a leverage point at the intersection of L1 and L2, C is a leverage point at the intersection of L1 and L3.

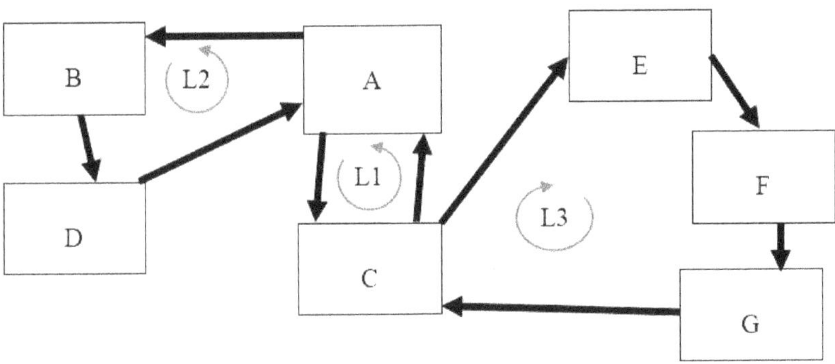

Figure 7 *Three loops with two leverage points*

That process may be understood as a way to reduce the complexity of a theory without unduly sacrificing its explanatory capacity. Following Stinchcombe (1987), Wright and Wallis (2019, Appendix A), suggest reducing linear redundancy by removing intermediary concepts from a causal chain. For example, the linear causal chain: A → B → C → D, could be represented as A → D.

With this structural perspective, loops are more easily differentiable, yet still integrated. And, while more research is needed, those measures seem reasonable for evaluating theories.

Loop of loops

Where a simple causal loop (for example "A causes A" or perhaps "A causes B and B causes A") may appear to be a tautology and so not an appropriate and falsifiable part of a theory (Popper, 2002), it is worth noting that having multiple interconnected simple loops is a different situation. With multiple measurable and falsifiable propositions, we have a valid theoretical model (Priem & Butler, 2001). This kind of structure is seen in Figure 4 (Newton's Second Law) where three "dyad" loops (each containing two concepts) are connected in an overarching "loop of loops" where each concept is a leverage point at the intersection of two loops.

Because we may reasonably expect theories of the social/behavioral sciences to contain more concepts (as social situations are more complex), it may be interesting to explore and create theories of that structure with more concepts, abstractly represented in Figure 8.

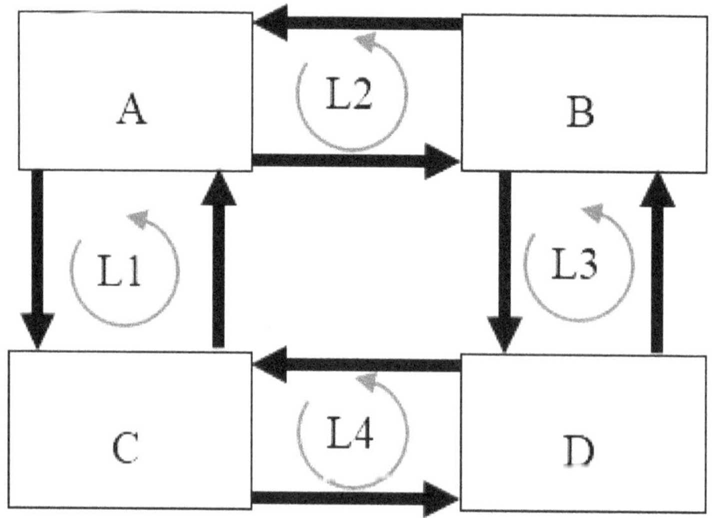

Figure 8 *A loop of loops containing four dyad loops connected in a larger loop*

Larger theories (with more concepts/boxes) are possible when adding more loop dyads creating a larger loop of loops. Again, this is a tentative hypothesis and suggests the opportunity for future research.

Perhaps the most important area for future research is to shine a light on the general assumption presented here of the usefulness of the theories under analysis. The present measures will have greater validity when they are more closely correlated with some objective and rigorous evaluation of that usefulness; perhaps quantified by the number of accurate predictions.

Future studies might also expand the analysis to include more theories overall; and, more specifically, more histories of theories from a broader range of fields. Additionally, future studies should investigate theories of greater complexity. Future studies might consider a range of theories with a variety of loops structures as they are applied in practice over an extended span of time. For example, one set of public service agencies might apply theories of change that are relevant to their missions, but have no loops, while another set of similar agencies apply theories of change which include many loops and leverage points. Alternatively, some organizations providing counseling services might use practices based on theories with few or no loops while another set of organizations provide services based on theories containing multiple loops. The differences in cases may prove rather illuminating provided there is substantive differences in the actual practices followed by those organizations; and, that those practices are clearly based on the theories.

As this is a preliminary exploration, limits of space, time, and other resources means that there are many areas and ideas that are not covered here, that could (and should) be studied in future research.

The notion of "coupled" systems could be explored because, in contrast to the present analysis of relatively isolated systems, "In many social systems the micro, meso and macro level dynamics are coupled, meaning that they cannot be studied or modified in isolation" (Johnson, Fortune, & Bromley, 2017). Future research might also consider the stability of systems (cf. Umpleby, 2018) based on the number and/or ratio of positive and negative loops, in addition to the openness of the system (or systems—linking conceptual, biological, and others). Indeed, having in mind a kind of "template" of loops may help to integrate/synthesize theories from disparate fields to generate theories that are more useful still.

Another direction for future research would be to consider whether the conceptual system is decomposable or not—including the notion of top down (or holographic) feedback (cf. Diniz & Handy, 2004) as well. Indeed, a very wide range of concepts from cybernetics and systems thinking may be researched for their relevance to conceptual systems (e.g., feedback, feed-forward, self-organization, recursions, reflection). It may also be considered how a complex social system may include multiple interwoven levels (Wallis, 2020d) suggesting a highly complex web between physical, biological, and conceptual systems.

The above results also suggest a balancing perspective to Ashby's so-called 'law' of requisite complexity. Instead of looking merely at "how many" components are in a system, as Ashby might, the present paper suggests looking at the number and direction of loops within a conceptual system as a way to evaluate the quality of that conceptual system for education and practice in pursuit of understanding and resolving real physical world problems.

More speculatively, a better measure for the quality of a theory may not lie in the number of concepts or the number of loops. Rather, it may be in "how closed" the conceptual system is. That is to say, our "optimal" theory may be the one (with however *many* loops) where all concepts are included in loops, all loops are of the shortest path, and all loops have positive and negative balance, and all loops are connected with leverage points. This may be understood as a SOCS version of thermodynamic systems in the physical world and how they might be better understood by exploring flows of information and energy. Additional research would certainly lead to interesting insights in this area (cf. McKelvey, 2004). It may be, as suggested in a response to Roth (2019), that "Those [conceptual systems] with higher levels of structure are more open to data but less open to changes in conceptual components. Those with lower levels of structure are more closed to data but more open to changes in conceptual components" (Wallis, 2020a: 180).

Future philosophical exploration might consider how this kind of approach might change the way we think about "truth" when "usefulness" may be a more appropriate term. Because, "Systems sciences defy classification as either epistemology or ontology. Rather, it is reminiscent of the Greek notion of gnosiology concerned with the holistic and integrative exploration of phenomena and events." meanings (Laszlo & Laszlo, 2009: 7).

Conclusions

The present paper has investigated a small set of evolving theories found in physics and public health to find what measures of structure might indicate which theories are more useful in practical application. As a preliminary study, this paper has presented a set of measures that may be used for evaluating the potential usefulness of theories; thus pointing a structural path for improving them. The limited sample size and the exploratory nature of this study limits the certainty of the conclusions that may be drawn. We are closer to the start of this area of study than the end.

This paper has achieved its purpose of conducting a preliminary exploration of causal structures of conceptual loops. And, in doing so, answer the research question of how we might more rigorously and objectively evaluate the conceptual loops presented in theories and models. By better understanding loop structures in theories, we may now reasonably expect to be able to build better theories; and, importantly, understand how much knowledge they hold and so make better decisions around the need for additional research or advancing toward action.

Reflecting on the comparison from the Glanville's (2009) perspective comparing knowledge_of (representing the relatively limited knowledge of those with superficial understanding based on outside observation) and knowledge_for (representing deeper, more nuanced, and more useful knowledge for enabling change based on insights from the inside), the present paper suggests that the distinction is not a binary one. Rather, it seems we can evaluate our theories, regardless of their origin, according to their structure to determine the potential usefulness in practical application. Theories on the knowledge_of end of the scale will have lower scores of IPA's Complexity and Systemicity, DIDO/MIMO, along with lower IPAx scores for loops and inclusions.

The research question for this paper asked: How might we more rigorously evaluate loops presented in theories to support our ability to develop theories that are more useful in practical application? This is important because, without a rigorous, objective approach to improving our theories, we risk developing overly simplistic theories that merely reinforce our biases.

This paper has answered the research question by using multiple methods to evaluate the structure of theories as they evolved from lesser to greater usefulness. Those methods include existing/proven methods such as IPA, along with suggested measure including the enumeration of conceptual loops which had been suggested, but not previously proven, to be useful for evaluating theories as a way to measure the potential usefulness of theories from a structural perspective. The results seem to correlate, showing that measuring the number and structure of conceptual loops within theories does indeed provide a useful tool for evaluating the potential usefulness of conceptual systems such as theories and models.

This is important for researchers, professors, and practitioners because they can use these measurements of structure (including conceptual loops) to show the validity of theories with greater confidence, objectivity, and rigor. That means professors can (and should) choose to teach theories that score higher on one or more of these measures. Practitioners can (and should) have their practices informed by theories with higher scores. Similarly, practitioners involved in strategic planning activities should encourage their clients to develop knowledge maps, theories of change, and similar representations of knowledge that have higher scores. For researchers (including students), the present results are also important because they suggest a path for more rapidly improving theories. For an abstract example, a researcher may begin by conducting a literature review on some topic of interest. And, finding a theory which contains few or no loops, can focus his or her research efforts on finding those causal connections that will "close the loop" and so develop a theory with more loops. And, more generally, researchers can (and should) develop theories with higher scores if they wish those theories to be more useful in practical application.

It should be noted that the present study is limited because it is focused only on causal relationships. It may be that there are a range of qualitative measures of those connections which may provide useful insights into structure. However, it should be understood, that each of those qualitative perspectives comes with a set of assumptions that must also be addressed—potentially leading to an infinite regress without a helpful conclusion.

While the focus of this paper has been on structure, it should also be noted that We gain greater confidence in the usefulness of our theories when they are supported by more qualitative and quantitative empirical data and when the theory is supported by a consensus of rational opinion across the field and perhaps beyond (Senge & Forrester, 1980). In short, knowledge is most useful in practical application when theories are supported on three dimensions of *data* (quantity and quality), *relevance* (to a broader range of stakeholders), and *structure* (Wright & Wallis, 2019). Improving theory on those three dimensions provides a guide to researchers who may want to "build" a "science accelerator" in their institutions to more rapidly advance their field of study for the benefit of all.

References

Appelbaum, R. P. (1970). *Theories of Social Change.* Chicago: Markham.

Bakker, J. I. H. (2011). North Central Sociological Association Presidential Address: Pragmatic Sociology: Healing the Discipline. *Sociological Focus, 44*(3), 167-183.

Beaulieu-B, P., & Dufort, P. (2017). Introduction: Revolution in Military Epistemology. *Journal of Military and Strategic Studies, 17*(4).

Bryson, J. M., Ackermann, F., & Eden, C. (2016). Discovering collaborative advantage: the contributions of goal categories and visual strategy mapping. *Public Administration Review, 76*(6), 912-925.

Bureš, V. (2017). A Method for Simplification of Complex Group Causal Loop Diagrams Based on Endogenisation, Encapsulation and Order-Oriented Reduction. *Systems, 5*(3), 46.

Burrell, G. (1997). *Pandemonium: Towards a Retro-Organizational Theory.* Thousand Oaks, California: Sage.

Chilisa, B., Major, T. E., Gaotlhobogwe, M., & Mokgolodi, H. (2015). Decolonizing and indigenizing evaluation practice in Africa: toward African relational evaluation approaches. *Canadian Journal of Program Evaluation, 30*(3).

Deegan, M. (2011). *Using Causal Maps to Analyze Policy Complexity and Intergovernmental Coordination: An empirical study of floodplain management recommendations.* Paper presented at the System Dynamics Society Conference 2011.

Diniz, L., & Handy, F. (2004). *Holographic Structures Creating Dynamic Governance for NGOs.*

Disessa, A. A. (2002). Why "conceptual ecology" is a good idea *Reconsidering conceptual change: Issues in theory and practice* (pp. 28-60): Springer.

Duggan, J., & Oliva, R. (2013). Methods for identifying structural dominance.

Friendshuh, L., & Troncale, L. (2012). *SoSPT 1.: Identifying fundamental systems processes for a general theory of systems.* Paper presented at the International Society for Systems Sciences (ISSS), San Jose, California, USA.

Glanville, R. (2009). A (cybernetic) musing: Design and cybernetics. *Cybernetics and Human Knowing, 16*(3-4), 175-186.

Gray, S. A., Zanre, E., & Gray, S. (2014). Fuzzy cognitive maps as representations of mental models and group beliefs *Fuzzy cognitive maps for applied sciences and engineering* (pp. 29-48): Springer.

Hayward, J., & Roach, P. A. (2017). Newton's laws as an interpretive framework in system dynamics. *System Dynamics Review, 33*(3-4), 183-218.

Hovorka, D. S., Birt, J., Larsen, K. R., & Finnie, G. (2012). *Visualizing the core-periphery distinction in theory domains.* Paper presented at the ACIS 2012: Location, location, location: Proceedings of the 23rd Australasian Conference on Information Systems 2012.

Johnson, J., Fortune, J., & Bromley, J. (2017). Systems, networks, and policy *Non-Equilibrium Social Science and Policy* (pp. 111-134): Springer, Cham.

Kampmann, C. E. (2012). Feedback loop gains and system behavior (1996). *System Dynamics Review, 28*(4), 370-395.

Kuhn, T. (1970). *The Structure of Scientific Revolutions* (2 ed.). Chicago: The University of Chicago Press.

Laszlo, A., & Laszlo, E. (2009). The systems sciences in service of humanity. *Systems Science and Cybernetics-Volume II*, 32.

Ledoux, L. (2012). Philosophy: Today's manager's best friend? *Philosophy of Management, 11*(3), 11–26.

Lewis, M. W., & Kelemen, M. L. (2002). Multiparadigm inquiry: Exploring organizational pluralism and paradox. *Human Relations, 55*(2), 251-275.

Light, P. C. (2016). Vision + Action = Faithful execution: Why government daydreams and how to stop the cascade of breakdowns that now haunts it. . *Political Science and Politics, 49*(1), 5-26.

Lipton, P., & Thompson, N. (1988). Comparative psychology and the recursive structure of filter explanations. *International Journal of Comparative Psychology.*

McGlashan, J., Johnstone, M., Creighton, D., de la Haye, K., & Allender, S. (2016). Quantifying a systems map: network analysis of a childhood obesity causal loop diagram. *PloS one, 11*(10), e0165459.

McKelvey, B. (2004). Toward a 0th law of thermodynamics: Order-creation complexity dynamics from physics and biology to bioeconomics. *Journal of Bioeconomics, 6*(1), 65-96.

Meadows, D. (1999). Leverage points: Places to intervene in a system, link.

Newell, W. H. (2007). Decision making in interdisciplinary studies. *Public Administration and Public Policy, 123*, 245.

Novak, J. D. (2002). Meaningful learning: The essential factor for conceptual change in limited or inappropriate propositional hierarchies leading to empowerment of learners. *Science education, 86*(4), 548-571.

Novak, J. D. (2010). Learning, creating, and using knowledge: Concept maps as facilitative tools in schools and corporations. *Journal of e-Learning and Knowledge Society, 6*(3), 21-30.

Oliva, R. (2004). Model structure analysis through graph theory: partition heuristics and feedback structure decomposition. *System Dynamics Review: The Journal of the System Dynamics Society, 20*(4), 313-336.

Oliva, R. (2015). Linking structure to behavior using eigenvalue elasticity analysis. *Analytical methods for dynamics modelers*, 207-239.

Pearl, J. (2000). *Causality: Models, reasoning, and inference*. New York: Cambridge University Press.

Phillips, S., & Wilson, W. H. (2014). A category theory explanation for systematicity: Universal constructions. *The architecture of cognition: Rethinking Fodor and Pylyshyn's systematicity challenge*, 227-249.

Popper, K. (2002). *The logic of scientific discovery* (K. Popper, J. Freed, & L. Freed, Trans.). New York: Routledge Classics.

Priem, R. L., & Butler, J. E. (2001). Tautology in the resource-based view and the implications of externally determined resource value: Further comments. *Academy of Management review, 26*(1), 57-66.

Roth, S. (2019). The open theory and its enemy: Implicit moralization as epistemological obstacle for general systems theory. *Systems Research and Behavioral Science, 36*(3), 281-288.

Russom, P. (2011). Big data analytics. *TDWI best practices report, fourth quarter, 19*(4), 1-34.

Saleh, M., Oliva, R., Kampmann, C. E., & Davidsen, P. I. (2010). A comprehensive analytical approach for policy analysis of system dynamics models. *European Journal of Operational Research, 203*(3), 673-683.

Schwaninger, M. (2011). System dynamics in the evolution of the systems approach. *Complex Systems in Finance and Econometrics*, 753-766.

Schwaninger, M. (2015). Model-based Management: A cybernetic concept. *Systems Research and Behavioral Science, 32*(6), 564-578.

Senge, P., & Forrester, J. W. (1980). Tests for building confidence in system dynamics models. *System dynamics, TIMS studies in management sciences, 14*, 209-228.

Senge, P., Kleiner, K., Roberts, S., Ross, R. B., & Smith, B. J. (1994). *The Fifth Discipline Fieldbook: Strategies and Tools for Building a Learning Organization*. New York: Currency Doubleday.

Sevaldson, B. (2011). GIGA-Mapping: Visualization for complexity and systems thinking in design. *Nordes*(4).

Sloman, S. A., & Hagmayer, Y. (2006). The causal psycho-logic of choice. *Trends in Cognitive Sciences, 10*(9), 407-412.

Sterman, J. D. (2012). Sustaining sustainability: creating a systems science in a fragmented academy and polarized world *Sustainability science* (pp. 21-58): Springer.

Stinchcombe, A. L. (1987). *Constructing social theories.* Chicago: University of Chicago Press.

Stinner, A. (1994). The story of force: from Aristotle to Einstein. *Physics education, 29*(2), 77.

Suedfeld, P. (2010). The cognitive processing of politics and politicians: Archival studies of conceptual and integrative complexity. *Journal of personality, 78*(6), 1669-1702.

Suedfeld, P., & Rank, A. D. (1976). Revolutionary leaders: Long-term success as a function of changes in conceptual complexity. *Journal of Personality and Social Psychology, 34*(2), 169-178.

Suedfeld, P., & Tetlock, P. (1977). Integrative complexity of communications in international crises. *Journal of conflict resolution, 21*(1), 169-184.

Suedfeld, P., Tetlock, P. E., & Streufert, S. (1992). Conceptual/integrative complexity. In C. P. Smith (Ed.), *Handbook of Thematic Content Analysis* (pp. 393-400). New York: Cambridge University Press.

Sweeney, L. B., & Sterman, J. D. (2007). Thinking about systems: Student and teacher conceptions of natural and social systems. *System Dynamics Review: The Journal of the System Dynamics Society, 23*(2-3), 285-311.

Umpleby, S. (2018). *Identifying stability or instability in economic systems: Circular and linear analyses of financial crises.* Financial University under the Government of the Russian Federation.

Umpleby, S., Wu, X.-h., & Hughes, E. (2017). Advances in Cybernetics Provide a Foundation for the Future. *International Journal of Systems and Society (IJSS), 4*(1), 29-36.

Van de Ven, A. H. (2007). *Engaged scholarship: A guide for organizational and social research.* New York: Oxford University Press.

Vemuri, P., & Bellinger, G. (2017). Examining the use of systemic approach for adoption of systems thinking in organizations. *Systems, 5*(3), 43.

Wakeland, W., & Medina, U. E. (2010). Comparing discrete simulation and system dynamics: Modeling an anti-insurgency influence operation.

Wallis, S. E. (2010). The structure of theory and the structure of scientific revolutions: What constitutes an advance in theory? In S. E. Wallis (Ed.), *Cybernetics and systems theory in management: Views, tools, and advancements* (pp. 151-174). Hershey, PA: IGI Global.

Wallis, S. E. (2011). *Avoiding Policy Failure: A Workable Approach.* Litchfield Park, AZ: Emergent Publications.

Wallis, S. E. (2014a). Abstraction and insight: Building better conceptual systems to support more effective social change. *Foundations of Science, 19*(4), 353-362. doi:10.1007/s10699-014-9359-x

Wallis, S. E. (2014b). Existing and emerging methods for integrating theories within and between disciplines. *Organizational Transformation and Social Change, 11*(1), 3-24.

Wallis, S. E. (2015a). Are theories of conflict improving? Using propositional analysis to determine the structure of conflict theories over the course of a century. *Emergence: Complexity and Organization, 17*(4), 1-17.

Wallis, S. E. (2015b). Integrative Propositional Analysis: A new quantitative method for evaluating theories in psychology. *Review of General Psychology, 19*(3), 365-380.

Wallis, S. E. (2016a). The science of conceptual systems: A progress report. *Foundations of Science, 21*(4), 579–602.

Wallis, S. E. (2016b). Structures of logic in policy and theory: Identifying sub-systemic bricks for investigating, building, and understanding conceptual systems. *Foundations of Science, 20*(3), 213-231.

Wallis, S. E. (2018, July 22-27). *Actionable knowledge mapping to accelerate interdisciplinary collaborations for research and practice.* Paper presented at the International Society for the Systems Sciences (ISSS), Corvallis, OR.

Wallis, S. E. (2019). Improving our theory of evaluation through an African-made process: An ubuntu answer to the decolonizing question. *Administration Publica, 27*(4).

Wallis, S. E. (2020a). Commentary on Roth: Adding a conceptual systems perspective. *Systems Research and Behavioral Science, 37*(1), 178-181.

Wallis, S. E. (2020b). Integrative Propositional Analysis for Developing Capacity in an Academic Research Institution by Improving Strategic Planning. *Systems Research and Behavioral Science, 37*(1), 56-67.

Wallis, S. E. (2020c). The missing piece of the integrative studies puzzle. *Interdisciplinary Science Reviews, 44*(3-4).

Wallis, S. E. (2020d). Understanding and improving the usefulness of conceptual systems: An IPA based perspective on levels of structure and emergence *Forthcoming.*

Wallis, S. E., & Valentinov, V. (2016a). The imperviance of conceptual systems: Cognitive and moral aspects. *Kybernetes, 45*(9).

Wallis, S. E., & Valentinov, V. (2016b). A limit to our thinking and some unanticipated moral consequences: A science of conceptual systems perspective with some potential solutions. *Systemic Practice and Action Research, 30*(2), 103-116.

Wallis, S. E., & Valentinov, V. (2017). What Is Sustainable Theory? A Luhmannian Perspective on the Science of Conceptual Systems. *Foundations of Science, 22*(4), 733-747. doi:10.1007/s10699-016-9496-5

Wallis, S. E., & Wright, B. (2015, March 4-6). *Strategic Knowledge Mapping: The Co-creation of Useful Knowledge.* Paper presented at the Association for Business Simulation and Experiential Learning (ABSEL) 42nd annual conference, Las Vegas, CA.

Wallis, S. E., & Wright, B. (2019). Integrative propositional analysis for understanding and reducing poverty. *Kybernetes, 48*(6), 1264-1277.

Wamba, S. F., Akter, S., Edwards, A., Chopin, G., & Gnanzou, D. (2015). How 'big data'can make big impact: Findings from a systematic review and a longitudinal case study. *International Journal of Production Economics, 165*, 234-246.

Weick, K. E. (1989). Theory construction as disciplined imagination. *Academy of Management review, 14*(4), 516-531.

Wong, E. M., Ormiston, M. E., & Tetlock, P. E. (2011). The Effects of Top Management Team Integrative Complexity and Decentralized Decision Making on Corporate Social Performance. *Academy of Management Journal, 54*(6), 1207-1228.

Wright, B., & Wallis, S. E. (2019). *Practical Mapping for Applied Research and Program Evaluation*. Thousand Oaks, CA: SAGE.

Chapter 15
Commentary on Roth—
Adding a Conceptual Systems Perspective

Wallis, S. E. (2020). Commentary on Roth: Adding a conceptual systems perspective. Systems Research and Behavioral Science, 37(1): 178-181.

Introduction

Recently, Roth (2019) provided a well-written, informative, and thought-provoking article arguing, "that the concepts of openness and closure are highly amenable to moralization with adverse implications for the development of the present-day systems theoretic thought" (Van Assche, Valentinov, & Verschraegen, 2019: 252). Plainly put, Roth explains a prejudicial tendency for scholars to write in favor of "open systems." That tendency may be understood as unnecessarily simplistic as a system may be in some ways open and closed at the same time. Roth's view generates substantial implications relating to the way we evaluate theoretical perspectives with serious implications for the development of General Systems Theory (GST), open systems theory, and other theories of the systems sciences. The present commentary will strive to explore some of those implications.

Roth's article could be enhanced by a discussion of "conceptual systems" as his argument relates to the difficulty of using one conceptual system (moral perspective/stance/code) to understand and choose another (open systems over closed). To provide some context, this commentary begins with brief definitions and reviews some common systems concepts/perspectives to clarify how they relate to conceptual systems.

Context

It is important to understand conceptual systems as systems because they are one of two general classes or categories of systems. "Systems can be either physical or conceptual, or a combination of both." (https://sebokwiki.org/wiki/System_(glossary)) Also, of equal or greater importance, we require conceptual systems to understand our other systems.

"Generally, a conceptual system is any form of theory, model, mental model, policy, etc." (Wallis, 2016: 579); while a theory is "An ordered set of assertions" (Weick, 1989: 517). That definition fits ethical/moral guidelines/stances/perspectives (Wallis, 2010). For example, "The Ten Commandments" asserts what behavior is "good;" and, a statement such as "organizations should be open systems" asserts what management behavior is "good."

Dent & Umpleby (1998) list eight key concepts as underlying assumptions of systems traditions. While this is not a comprehensive or authoritative list these concepts will serve as a starting point for some level of shared understanding.

Environment

Conceptual systems seem to exist in multiple environments. These include human minds and storage devices such as computers, books, and journals. Thus, conceptual systems may also be said to exist in one or more physical and social environments where they are created, stored, tested, applied, changed, and discarded. Theories and moral codes are developed to include concepts relevant to stakeholders and their situations (Wright & Wallis, 2019). We might also think of those conceptual systems as existing in environments of research and practice.

Self-organization

Our conceptual systems are part of an ecology of knowledge (Disessa, 2002) with many concepts (and systems or networks of concepts such as theories and morals) interacting, competing, growing, and dying. A single concept, by itself, does not seem to be a stable thing. It seems to seek connection with other concepts; similar to the way that elements tend to form bonds with other elements to form molecules. When a conceptual system achieves equilibrium, it becomes relatively stable as when a theory becomes more sustainable (Wallis & Valentinov, 2017).

Reflexivity

The reflexive process includes conceptual systems because we are always using some form of conceptual system to observe and engage the world. Where improvement of social systems is often obvious, the improvement of conceptual systems is too often opaque. That lack of clarity impedes development; a problem resolved by measuring the usefulness of conceptual systems on three dimensions of data, meaning/relevance, and logic structure (Wright & Wallis, 2019).

Observation

Each observer constructs reality by the process of observation. Indeed, "A 'system' is a set of variables selected by an observer" (Umpleby, 2009). For a conceptual system, also, the observer may choose which conceptual components seem most relevant. That choice, however, does not guarantee the creation of a highly effective theory or moral code.

Holism

Holism relates to the characteristics of a system that emerge from the interaction of its component parts. In a conceptual system, the individual concepts are the component parts. "Conceptual systems... do not directly exhibit behavior, but exhibit 'meaning'" (https://sebokwiki.org/wiki/System_(glossary)).

Multiple streams of research show that conceptual systems of more interconnected parts have more meaning (are more useful for understanding situations) (cf. Gentner & Toupin, 1986; Suedfeld, Tetlock, & Streufert, 1992; Wallis, 2016).

Relationships

In conceptual systems, a number of types or forms of relationships may be identified. Similar concepts may be clumped into categories, or connections may be identified between them. However, relationships that are merely descriptive such as "cats chase mice" (where 'chase' is the relationship) are not very useful for creating theories that are highly effective, and relationships that are vague such as, "concept A and B are related" are even less useful.

Causality

Understanding causal relationships is an important part of theory creation (Pearl, 2000) leading to more useful proposition (Dubin, 1978). Indeed, "causal explanation is the key to theoretical expla-

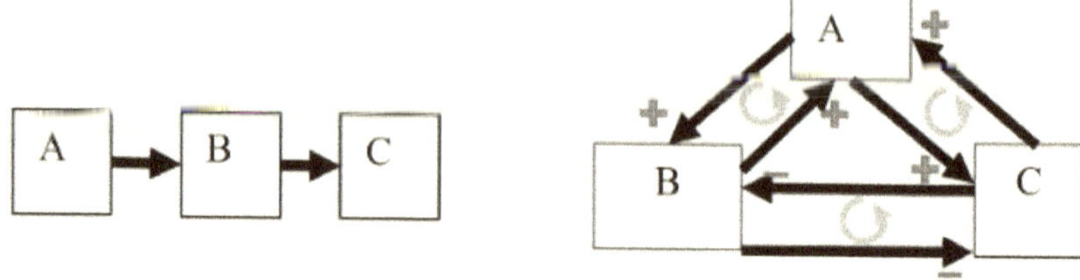

Figure 1 *Abstract conceptual systems of different levels of internal indeterminacy and systemic structure with boxes representing concepts and arrows representing causal connections*

nation" (Salmon, 1984: xi). WE should avoid theories with simple or linear causality, in favor of circular causality (Dent & Umpleby, 1998).

Indeterminism

With more intersecting loops, the paths one might take through the conceptual system becomes less deterministic and the theory becomes more useful for understanding real world situations and making effective decisions to reach desired goals (Wallis, 2019). Figure 1 shows two abstract conceptual systems with varying levels of indeterminacy in their structures. The diagram on the right shows the least determinacy and also the greatest potential usefulness based on systemic structure.

Summarizing this section (italicized words representing above context), within the *environment* of our disciplines publications and minds of our scholars, conceptual systems (such as GST, open theory, moral codes) *self-organize*. We use various conceptual systems to *reflexively* consider and strive to improve other conceptual systems using *observation* to choose relevant concepts; but progress is not always clear. If we strive more to identify multiple *relationships* between concepts that are *causal* we may accelerate the development of conceptual systems with multiple loops and so greater *indeterminacy*. The *holistic* nature of such conceptual systems will exhibit emergent behavior of meaning which reflects their usefulness in real world application.

The Openness of Theory and Its Friends

In his article, Roth (2019) applies von Bertalanffy's concept of open systems to social and natural environments to show that there are tradeoffs between the openness and closeness and that both may exist at the same time. Specifically, that systems may be open to information but closed to matter. Applying von Bertalanffy's distinction to conceptual systems, we may understand matter to represent the concepts within the system and the information to represent the data relating to measurements of real world systems.

A conceptual system with few or no casual connections between the concepts (e.g., the left hand diagram in Figure 1) tends to be less stable. That is, it will be open to concepts being added and removed easily (as we see in philosophical conversations and academic publications where new versions of old theories are put forward frequently). When a conceptual system has many casual connections between its component concepts (e.g., the right hand diagram in Figure 1), it becomes more difficult to change its component concepts. For example, no engineer would consider using Ohm's law of electricity (including volts, amps, and ohms) to design a radio, in a way that leaves volts out of consideration.

When a system has more causal connections between the concepts, as in Ohm's law of electricity, it is more open to information. That is to say, that its concepts or variables are more measurable and those causal relationships between them are supported with data. This makes them more useful in practical application. When a conceptual system has few or no connections, data seem to be secondary; as in faith-based reasoning.

The difference in stability of conceptual systems may be understood in terms of systemic structure which may be measured using Integrative Propositional Analysis (IPA), (Wallis, 2016). Those with higher levels of structure are more open to data but less open to changes in conceptual components. Those with lower levels of structure are more closed to data, but more open to changes in conceptual components.

Thus, highly structured theories are more interesting and useful for practitioners (for making decisions to reach specified goals) while less structured theories are more interesting for researchers (for investigating changes in concepts and connections between them). While research may lead to more structured conceptual systems (as it did in the scientific revolution) results are not guaranteed.

This perspective raises a question as to how well structured GST and open systems theory might be, and how well structured is the set of moral guidelines used to judge it. Using Integrative Propositional Analysis (IPA) is would certainly be possible to evaluate those although such a study is beyond the scope of the present commentary (due, in part, to the tacit nature of many moral stances and the many variations of GST). However, given my experience in measuring the structure of systems theories (Wallis, 2008, 2012), and moral/ethical systems (Wallis, 2010; Wallis & Valentinov, 2016) it seems reasonable to assume that open systems theory has a medium to low level of structure while the moral system is likely much less. Thus, moral positions are indeed poor tools for evaluating GST.

Conclusion

Roth's concern about established theories being judged by morals seems well-founded. Also, his effort to identify tradeoffs between openness and closeness and the associated dismoralization is a good step forward. That step is enhanced by the present perspective which goes beyond Roth's binary and quaternary notions to show that the structure, and therefor the usefulness, of theoretical perspectives should be measured using rigorous methods such as IPA rather than moral stances.

Another side of the present coin is that the low level of structure held by many theories makes them vulnerable to moralization, misunderstanding, attack, and faddish change. This leads to fragmentation instead of advancing the science.

In short, a theory with a low level of structure is its own enemy.

References

Dent, E. B., & Umpleby, S. A. (1998). Underlying assumptions of several traditions in systems theory and cybernetics. *Cybernetics and Systems, 29*, 513-518.

Disessa, A. A. (2002). Why "conceptual ecology" is a good idea *Reconsidering conceptual change: Issues in theory and practice* (pp. 28-60): Springer.

Dubin, R. (1978). *Theory building* (Revised ed.). New York: The Free Press.

Gentner, D., & Toupin, C. (1986). Systematicity and surface similarity in the development of analogy. *Cognitive science, 10*(3), 277-300.

Pearl, J. (2000). *Causality: Models, Reasoning, and Inference.* New York: Cambridge University Press.

Roth, S. (2019). The open theory and its enemy: Implicit moralization as epistemological obstacle for general systems theory. *Systems Research and Behavioral Science, 36*(3), 281-288.

Salmon, W. C. (1984). *Scientific explanation and the causal structure of the world.* New Jersey: Princeton University Press.

Suedfeld, P., Tetlock, P. E., & Streufert, S. (1992). Conceptual/integrative complexity. In C. P. Smith (Ed.), *Handbook of Thematic Content Analysis* (pp. 393-400). New York: Cambridge University Press.

Umpleby, S. A. (2009). Ross Ashby's general theory of adaptive systems. *International Journal of General Systems, 38*(2), 231-238.

Van Assche, K., Valentinov, V., & Verschraegen, G. (2019). Ludwig von Bertalanffy and his enduring relevance: Celebrating 50 years General System Theory. *Systems Research and Behavioral Science, 36*(3), 251-254.

Wallis, S. E. (2008). Emerging order in CAS theory: Mapping some perspectives. *Kybernetes, 38*(7).

Wallis, S. E. (2010). Towards developing effective ethics for effective behavior. *Social Responsibility Journal, 6*(4), 536-550.

Wallis, S. E. (2012, July 15-22, 2012). *Existing and Emerging Methods for Integrating Theories Within and Between Disciplines.* Paper presented at the 56th annual meeting of the International Society for Systems Sciences (ISSS), San Jose, California.

Wallis, S. E. (2016). The science of conceptual systems: A progress report. *Foundations of Science, 21*(4), 579-602.

Wallis, S. E. (2019). *Learning to Map and Act Cybernetically Without Learning Cybernetics: **How Many Loops are Needed to Enable Effective Decisions and Actions?* Paper presented at the American Society for Cybernetics, Vancouver, BC, Canada.

Wallis, S. E., & Valentinov, V. (2016). The imperviance of conceptual systems: Cognitive and moral aspects. *Kybernetes, 45*(9).

Wallis, S. E., & Valentinov, V. (2017). What Is Sustainable Theory? A Luhmannian Perspective on the Science of Conceptual Systems. *Foundations of Science, 22*(4), 733-747. doi:10.1007/s10699-016-9496-5

Weick, K. E. (1989). Theory construction as disciplined imagination. *Academy of Management review, 14*(4), 516-531.

Wright, B., & Wallis, S. E. (2019). *Practical Mapping for Applied Research and Program Evaluation.* Thousand Oaks, CA: SAGE.

www.ingramcontent.com/pod-product-compliance
Lightning Source LLC
Chambersburg PA
CBHW020458100426
42812CB00024B/2707

* 9 7 8 1 9 3 8 1 5 8 1 9 3 *